GRINDING TECHNOLOGY

Theory and Applications
of Machining with Abrasives

SECOND EDITION

Stephen Malkin

University of Massachusetts

Changsheng Guo

United Technologies Research Center

2008

INDUSTRIAL PRESS

NEW YORK

Library of Congress Cataloging-in-Publication Data

Malkin, S. (Stephen), 1941–
    Grinding technology : theory and application of machining
with abrasives/Stephen Malkin and Changsheng Guo. – [2nd ed.].
        p.   cm.
    ISBN 978-0-8311-3247-7
 1. Grinding and polishing.   I. Guo, Changsheng, 1959–   II. Title.
    TJ1280.M293 2007
    671.3'5–dc22

                                                2006053181

Industrial Press, Inc.
989 Avenue of the Americas
New York, NY 10018

Second Edition, 2008

Sponsoring Editor: Suzanne Remore
Cover Design: Janet Romano

10   9   8   7   6   5   4   3   2   1

Table of Contents

Preface

The first edition of this book was intended to provide an integrated scientific foundation for understanding of the grinding process, which can be practically utilized for enhancing and optimizing grinding operations. After 18 years in print, the first edition is still selling and is widely referenced, but many of the newer developments in grinding led us to think that the time had come for a new edition. This second edition builds upon the first edition with greatly expanded coverage of the thermal aspects of grinding, creep-feed grinding, grinding with superabrasives, fluid flow, process simulation, optimization, and intelligent control of grinding machines.

This book is written both for the researcher and the practicing engineer. As with the first edition, it is expected that the second edition will be used as a textbook or supplement for advanced courses on machining and grinding, for industrial short courses, and as a source of fundamental and practical information about the grinding process and its utilization.

Preparation of the second edition of this book was undertaken by the authors as part of their collaborative relationship which began at the University of Massachusetts in 1989. During this time, we have had the good fortune to work with many outstanding graduate students and to benefit from interactions with and support from many colleagues and friends in academia and industry who are too numerous to mention individually.

We dedicate this book to our wives, Maccabit and Ling, who encouraged us to undertake this project and cheerfully endured our excessive indulgence in grinding and abrasive processes.

Stephen Malkin
Amherst, Massachusetts

Changsheng Guo
South Windsor, Connecticut

Preface to the First Edition

Manufacturing is becoming recognized today as an important commercial activity. Competitiveness in manufacturing, utilizing the most advanced technology, is essential in order to avert serious economic chaos in most industrialized countries. It is evident that the standard of living in countries which have been slow to address this reality is suffering. Aside from mining, agriculture, and foreign tourism, the wealth of industrialized countries is generated mostly by manufacturing.

Numerous initiatives are being undertaken to develop and implement advanced manufacturing technologies. In the United States, considerable resources are being harnessed to promote manufacturing research through such bodies as the National Science Foundation, the Department of Defense, and the newly created National Center for Manufacturing Sciences. An important thrust for this effort is to foster a more scientific or analytical approach to manufacturing. The 'rules of thumb' upon which we relied in the past and continue to depend must progressively give way to analytical engineering methodologies.

The present book is intended to provide a basic analytical approach to grinding as a machining process. Grinding research during the past four decades has established a scientific foundation, thereby providing a rational understanding which can be practically utilized. It seems appropriate at this time to present a comprehensive unified treatment of this subject. As an engineering monograph, this book is written both for the researcher and for the practicing engineer. Graduates of four-year mechanical and production engineering curricula should be able to grasp the technical content. Individual chapters would be suitable reference material in a senior- or graduate-level course in materials processing or machining.

My interest in manufacturing and grinding processes dates from about 25 years ago at the time of my graduate studies. My continued fascination and involvement in grinding research have been nurtured by the challenge of integrating a broad range of diverse technical areas including plasticity, materials, mechanics, tribology, heat transfer, control, and optimization. In this endeavor it has been my good fortune to have been associated with many talented teachers, colleagues, and students in universities where I have studied and taught. My practical experience has been enriched by many engineers I have worked with in the course of technical consulting for machine-tool builders, manufacturers of abrasive products, and users of grinding processes.

I am pleased to acknowledge the support of the publisher, Mr. Ellis Horwood. For assistance with figure preparation and typing, I am grateful to Ms P. Stephan, Ms B. Craker and Ms T. Mitchell at the University of Massachusetts, and Ms M. Schreier and Ms T. Kalmar at the Technion-Israel Institute of Technology.

This book is dedicated to the memory of my father who encouraged me to undertake this project but did not live to see its completion. The book would never have been completed without the support of my wife, Maccabit, and my children.

Stephen Malkin
Amherst, Massachusetts
December, 1988

Biographical Sketches for Authors

STEPHEN MALKIN

Stephen Malkin is Distinguished Professor and former head of the Department of Mechanical & Industrial Engineering at the University of Massachusetts. He graduated from MIT with BS (1963), MS (1965), and ScD (1968) degrees in mechanical engineering. Prior to joining the University of Massachusetts in 1986, he held faculty positions at the University of Texas, State University of New York, and Technion-Israel Institute of Technology. An author of more than 200 papers, he is internationally recognized for research on grinding and abrasive processes. As an industrial consultant and lecturer, he has been a leader in the practical utilization of grinding technology for enhancing productivity and quality. Dr. Malkin is a member of the National Academy of Engineering (NAE), and a fellow of the International Institution for Production Engineering Research (CIRP), the American Society of Mechanical Engineers (ASME), and Society of Manufacturing Engineers (SME). He received the ASME Blackall Award of 1993 for best papers related to machine tools, the SME Gold Medal of 1996 for his outstanding research accomplishments and contributions to the manufacturing profession, the University of Massachusetts Outstanding Engineering Faculty Award of 1997, and the ASME William T. Ennor Manufacturing Technology Award of 2004 in recognition of his leading role in the transformation of grinding and abrasive machining from an empirical craft to an applied science.

CHANGSHENG GUO

Changsheng Guo is Principal Scientist and Project Leader at the United Technologies Research Center (UTRC) where he leads projects in modeling, simulation, and optimization of manufacturing processes. He received his Ph.D. in mechanical engineering from University of Massachusetts, a Master's degree in management from Rensselaer Polytechnic Institute, and a Master's degree in manufacturing engineering and a bachelor's degree in mechanical engineering from Northeastern University in China. Before joining UTRC, Dr. Guo was Co-Director of the grinding research program at the University of Massachusetts and Technical Director of Chand Kare Technical Ceramics. From 1985 to 1987, he was an assistant professor at Northeastern University in China. Dr. Guo's research focus has been on the fundamentals and applications of machining processes including grinding, milling, superabrasive machining, and ceramic machining. With more than 80 published papers, Dr. Guo is an associate editor for *Machining Science and Technology* and an associate member of the International Academy for Production Engineering (CIRP). He has been the recipient of numerous awards including UTRC's Outstanding Achievement Award, the Pratt & Whitney leadership award, the F. W. Taylor Medal of CIRP in 1996, the US DOE energy pioneer award in 1995, and the ASME Blackall Award in 1993.

Chapter 1

Introduction

1.1 THE GRINDING PROCESS

Grinding is the common collective name for machining processes which utilize hard abrasive particles as the cutting medium. The grinding process of shaping materials is probably the oldest in existence, dating from the time prehistoric man found that he could sharpen his tools by rubbing them against gritty rocks. Without the capability to shape and sharpen implements by grinding, we might still be living in the Stone Age.

Nowadays, grinding is a major manufacturing process which accounts for about 20–25% of the total expenditures on machining operations in industrialized countries. Society, as we know it, would be quite impossible without grinding. Almost everything that we use has either been machined by grinding at some stage of its production, or has been processed by machines which owe their precision to abrasive operations. How could we sharpen cutting tools for turning, milling, and drilling without grinding? How could we manufacture the rolling bearings for machinery and vehicles? How could we produce disk-drive components for computers?

Within the spectrum of machining processes, the uniqueness of grinding is found in its cutting tool. Grinding wheels and tools are generally composed of two materials – tiny abrasive particles called grains or grits, which do the cutting, and a softer bonding agent to hold the countless abrasive grains together in a solid mass. Prehistoric man's abrasive tool was natural sandstone, which contains grains of sand in a silicate bond matrix. Modern grinding wheels are fabricated by cementing together abrasive grains, usually from man-made materials, with a suitable bonding material. Each abrasive grain is a potential microscopic cutting tool. The grinding process uses thousands of abrasive cutting points simultaneously and millions continually.

Grinding is traditionally regarded as a final machining process in the production of components requiring smooth surfaces and fine tolerances. There is no process which can compete with grinding for most precision machining operations, but the process is far from being confined to this type of work. More abrasive is actually consumed by heavy-duty grinding operations, where the objective is to remove material as quickly and efficiently as possible with little concern for surface quality. Grinding is as essential for delicate precision slicing of silicon wafers for microelectronic circuits using paper-thin abrasive disks or saws only 20 μm thick, as it is for the heavy-duty conditioning and cleaning of billets and blooms in foundries and steel mills at removal rates of around 1600 cm³ per minute with 220 kW machines.

There are numerous types of grinding operations which vary according to the shape of the wheel and the kinematic motions of the workpiece and wheelhead. Some of the more common ones for machining flat and cylindrical surfaces are illustrated in Figure 1-1. More complex machines are used to generate other shapes. Any of these processes may be applied to fine finishing, to large-scale stock removal, or to a host of tasks between these extremes.

Another area where grinding is virtually unchallenged is for machining of materials which, because of their extreme hardness or brittleness,

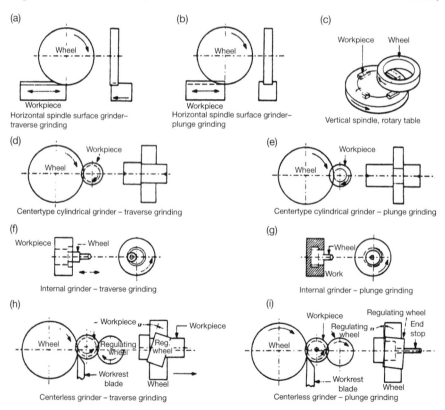

Figure 1-1 Illustration of some common grinding operations for machining flat and cylindrical surfaces [1].

cannot be efficiently shaped by other methods. In the production of hardened steel components, such as cutting tools and rolling bearing rings, grinding can be performed on either the annealed or the hardened steel, often with comparable ease, whereas other machining methods are usually restricted to the annealed material. The machining of non-metallic brittle materials, including ceramics, cemented carbides and glasses, is almost exclusively dependent on abrasive processes.

But despite its industrial importance, the grinding process is often held in low esteem. Finish grinding is usually found to be more costly than other machining processes, per unit volume of material removal, and so its use tends to be looked upon as a necessary evil. But as stock allowances for material removal have continued to decrease, owing to the development of methods for more precise casting and forging closer to the final configuration (near net shape processing), grinding has become more economical as a single process for machining directly to the final dimensions without the need for prior turning or milling.

Of all the machining processes in common use, grinding is undoubtedly the least understood and most neglected in practice. This unfortunate situation appears to have its origin in the mistaken belief that the process is too complicated to understand. Because of the multiplicity of cutting points and their irregular geometry, the high cutting speeds, and the small depths of cut which vary from grain to grain, any attempt to analyze the mechanisms of grinding might appear to be a hopeless task. Perhaps the spark stream bursting forth from the grinding wheel adds to the mystery.

The grinding process has been the subject of extensive research, especially during the past 50 years. Because of the large number of cutting events involved, it has been found that the process can be characterized by the cutting action of a typical 'average' grain, which greatly facilitates analysis and interpretation of experimental findings. Along with other machining processes, grinding has been transformed from a 'practical art' to an 'applied science'. The first edition of this book [2] was intended to bring an awareness and understanding of these developments to practicing engineers and to researchers in this field. After being in print for 18 years, we felt that it was time for an updated second edition which would incorporate many new developments since the first edition appeared.

1.2   HISTORICAL DEVELOPMENT OF THE GRINDING PROCESS

The history of grinding technology originates with the discovery of abrasive minerals, and continues with the development of abrasive products and machine tools to help satisfy man's perceived needs for manufactured products to ensure his survival and well-being. Here we present only a brief historical survey, for the purpose of placing our subject in perspective. More detailed information can be found elsewhere [3-8].

The story begins with primitive man, who probably found that he could sharpen his flint knife by rubbing it against a piece of sandstone. At a later stage, prior to the invention of pottery, abrasives were used for grinding out stones to make eating utensils. This may be the earliest use of grinding as a machining operation to obtain a desired shape, rather than just for sharpening. The huge stone blocks used in building the Pyramids of Egypt were cut to size by sawing with some crude type of grinding machine, and their surfaces were smoothed with sandstone.

Grinding of metal was begun in ancient Egypt in about 2000 BC, which corresponds to the beginnings of metallurgy. During this period, grinding skills became highly valued in the Middle East for sharpening of tools and making ornaments. According to the Bible (1 Samuel 13:19 and 13:20), one way the Philistines maintained their hegemony over the Israelites was by prohibiting them from sharpening their own tools.

Quartz, in the form of loose sand, flint, and sandstone, was probably the only abrasive known to prehistoric man. Other important natural abrasive materials known since antiquity are emery, garnet, and diamond. Emery is actually an impure form of corundum, containing aluminum and iron oxides in roughly equal proportions. The adamant (shamir in Hebrew) referred to in the Bible for engraving (Jeremiah 17:1) may have been emery. The value of emery as an abrasive was known to the ancient Greeks and Romans. Garnet abrasives, encompassing various alumina silicate minerals, were probably also known in ancient times. Diamond mining is believed to have originated in India between 800 and 600 BC, and this was the principal source of diamonds until the nineteenth century. One of the earliest recorded uses of diamond powder as an abrasive dates from fifteenth century Belgium, when it was used for cutting diamonds and for delicate finishing operations in watchmaking. Natural corundum abrasives, consisting mostly of crystalline aluminum oxide, became generally known only in the early nineteenth century. Next to diamond, corundum is the hardest naturally occurring material, which makes it especially desirable as an abrasive. The name corundum was originally applied to the ruby and sapphire gems of India. The natural diamonds and corundum for abrasives are not sufficiently transparent or perfect to be used as gemstones.

The grinding wheel originated in ancient Egypt in association with the beginnings of metallurgy, the first grinding wheels being hewn from sandstone and deriving their shape and rotary action possibly from the crude mills used to grind grain. Early grinding wheels were probably manually powered, and hand-operated wheels are still occasionally used for tool sharpening. Water-driven grinding wheels were in use at the end of the

Middle Ages, although the water wheel as a source of power was utilized much earlier by the Romans in the first century BC.

During the Middle Ages and up until the Industrial Revolution, abrasives were used for sharpening and polishing of tools, weapons, and armor. Early concepts for grinding machines appear in the drawings of Leonardo Da Vinci dating from about the year 1500. Some of his machines were designed to operate from a central power source, presumably from a water wheel, but no details were given. The grinding wheels he described contained emery abrasives held on the surface of a wooden hub with tallow. It took another 300-400 years until some of the grinding machine concepts envisioned by Leonardo were put into practice.

It was only in the beginning of the nineteenth century that the first solid-bonded abrasive wheels were known to have been manufactured in India, for hand grinding gems. The abrasive was emery or corundum, although diamond may also have been used, and the binder was a gum-resin shellac. Up until this time, the only bonded abrasive tools were from natural sandstone, like those of ancient Egypt. Shellac-bonded wheels were commercially introduced in the West only in 1880. In the middle of the nineteenth century, oxychloride-bonded wheels were invented in England, and rubber-bonded wheels in the United States and France. Later in the nineteenth century came silicate bonds, which attempted to duplicate the properties of natural-bonded abrasives. But certainly the most important development at this time was the vitrified bond, which was commercialized in the 1870s by the Norton Company. Resin-bonded wheels appeared much later in 1923, following the discovery of phenol-formaldehyde resins (Bakelite). Metal bonds for diamond grinding wheels were introduced in the early 1940s, although the idea of a metal grinding wheel may be considered to have originated with gem polishing in the late seventeenth century in Belgium using a cast-iron disk charged with diamond-powder abrasive.

Thus we see that the latter half of the nineteenth century marks the beginning of the commercial grinding wheel industry, with the introduction of bonded abrasive products of vitrified, rubber, and shellac bonds. The grinding wheel technology which became available during this period was an important factor in the development of grinding machines to utilize these tools. With the inventions of the Industrial Revolution, the demand for grinding machines and wheels grew. The first 'modern' grinding machines were introduced by the Brown & Sharpe Company in the 1860s for machining of sewing machine components. Popularization of the bicycle in the 1890s was made possible, in no small part, by the ability to produce hardened precision bearings and gears by grinding.

Near the turn of the century, we find synthetic abrasives, silicon carbide and aluminum oxide, beginning to appear. The discovery of silicon carbide abrasives is usually credited to E. G. Acheson in 1891, although this compound had been synthesized earlier. Silicon carbide is still produced by essentially the same fusion process used by Acheson. The material was referred to as carborundum, a name still occasionally used as a synonym for silicon carbide, and in 1895 the Carborundum Company was formed by Acheson and his associates to manufacture this abrasive on a large scale. Shortly thereafter in 1897, synthetic aluminum oxide (corundum) for abrasives was produced by C. B. Jacobs from bauxite by fusion in an electric furnace. The Norton Company obtained the rights to this process in 1901. Successful commercial production of fused aluminum oxide abrasives followed in 1904 with the invention of the Higgins furnace, which is very similar to the furnaces in use today. Factories to produce silicon carbide and aluminum oxide abrasives were located at Niagara Falls, owing to the availability of inexpensive electricity to power the furnaces, and the Niagara Falls area is still the center of the abrasives industry in North America.

By the early twentieth century, grinding wheels containing synthetic aluminum oxide and silicon carbide were being produced with vitrified, rubber, shellac, and oxychloride bonds. Together with the resin bond developed later in 1923, this provided a wide range of conventional abrasive tools of the types which are still in general use. Needless to say, these developments were mirrored by progress in grinding machines, and the more widespread use of the grinding process. Mass production of automobiles and other equipment with interchangeable components depended on the grinding process. More recent developments in conventional grinding wheel technologies have been directed towards providing, improved control of processing and structure of abrasive materials, the production of better and more uniform bonds, and the development of new types of aluminum oxide abrasives.

Development of the superabrasives diamond and cubic boron nitride (CBN) in the twentieth century merits special attention. Resin-bonded wheels containing natural diamond abrasive were first produced in 1930, and vitrified- and metal-bonded wheels followed about ten years later. The consumption of diamond grinding wheels grew rapidly, principally because of the need to grind tungsten carbide cutting tools, until, by the early 1940s, grinding of cemented carbides had become the single most important factor in the consumption of industrial diamond. Successful synthesis of diamond under extreme pressures was announced by the General Electric Company in 1955, although artificial diamond was actually made two years earlier in Sweden. Synthetic diamond abrasives became commercially available in the late 1950s.

Diamond abrasives, both natural and synthetic, are widely used for grinding various types of materials including cemented carbides, ceramics, metals, glasses, and fiber-reinforced composites. But in-spite of their extreme hardness, diamonds are not suitable for grinding of most ferrous metals, owing to graphitization which causes excessive wear. In searching for an alternative to diamond, cubic boron nitride (CBN) was first success-fully synthesized at General Electric in 1957 using a high-pressure process similar to that for diamond, but it became commercially available only in 1969, mainly for use on ferrous metals. CBN is the second hardest materi-al known, surpassed only by diamond. The past 30 years has seen a rapid growth in the application of CBN abrasive wheels with vitrified, resinoid, and electroplated metal bonds especially for grinding of ferrous and nickel-base alloys.

1.3 CONTENTS OF THIS BOOK

In spite of the importance of grinding as a machining process, rela-tively few books have been written on the subject. Much of the fundamen-tal information about grinding has been available only in journal articles and conference proceedings. Most previous books on grinding, as well as chapters on the subject in handbooks, tend to be somewhat descriptive, with details of grinding, operations, machines, and abrasive tools [7-23]. While providing a valuable source of practical information, they do not really help us to understand the fundamentals of the process. Notable exceptions are the book by Coes [8] on abrasive materials and tools, the book by Andrew *et al.* [16] on creep-feed grinding, and a research-oriented monograpah by Shaw[22]. Another book on grinding [23] which appeared as the present book was nearing completion, deals both with grinding theory and applications.

A number of books covering machining processes include separate discussions on what might be considered to be the rudiments of grinding theory [24-31]. It certainly seems logical to include grinding within the broader context of machining in this way. But upon closer examination, grinding seems to occupy an uneasy place within the family of machin-ing processes. Like other metal-cutting processes, material removal by grinding involves a shearing process of chip formation. Yet the theories of chip flow, cutting temperatures, and tool life, which are generally applicable to other machining operations, such as turning, milling, and drilling, are not relevant to grinding. Furthermore, the well-known opti-mization methods for selecting the best speeds and feeds are not applica-ble to grinding. With this in mind, it seems not so unreasonable that some monographs on machining make little or no mention at all of grinding [32-38].

The first edition of the present book [2] was written to satisfy the need for a comprehensive and unified treatment of grinding theory and its practical use. The intention was to present a reasonably self-consistent picture of this complex process, which reconciled differing points of view and contradictions commonly found in the literature, in order to provide practicing engineers and researchers with a logical framework to understand the process and improve grinding performance. This second edition was motivated by the extensive developments in grinding since the first edition appeared 18 years ago.

As with the first edition, the second edition's subject matter is limited to grinding as a machining process using bonded abrasive wheels, without explicitly dealing with such abrasive machining processes as loose abrasive operations, honing, lapping, polishing, superfinishing, and abrasive jet cutting. The discussion of abrasive-workpiece interactions is, for the most part, also applicable to these other abrasive processes, but the particular details of each of these processes would require separate treatment, and our fundamental understanding of these other processes is much poorer than that for abrasive processing with grinding wheels. Therefore, much of what could be added in the present work, by inclusion of these other processes, would largely be a repetition of descriptive information which is already available elsewhere.

Topics in this book are arranged in a logical sequence, starting with a description of abrasives and bonded abrasive products. We then consider the topography of the grinding wheel and its kinematic interaction with the workpiece. Abrasive-workpiece interactions are analyzed with an emphasis on specific energy (energy per unit volume of material removed), since any plausible theory of abrasive-workpiece interaction must be able to account for the magnitude of the specific energy and its dependence on the process parameters. This is followed by a much expanded coverage of thermal aspects of grinding, in three chapters instead of one, and a new chapter on fluid flow in grinding. Thermal analyses are especially important for dealing with problems of thermal damage to the workpiece and residual stresses. This is followed by chapters concerned with surface finish, wheel wear, and machine deflections and cycles. Thermal damage, residual stresses, and surface finish are the primary factors in ensuring surface integrity. Bulk wear of the grinding wheel may lead to loss of form, and attritious wear of the abrasive is often a critical factor affecting the suitability of an abrasive for a particular workpiece material. Machine deflections have important implications in the design of grinding cycles and for attaining size and shape, as well as for machine stability and the tendency for vibrations and chatter to occur. A new final chapter deals mainly with the practical utilization of grinding theory for enhancing and control of grinding performance including process simulation, optimization, and intelligent machine control.

REFERENCES

1. Kalpakjian, S., *Manufacturing Processes for Engineering Materials*, Addison Wesley, Reading, MA, 1984, p. 573.
2. Malkin, S., *Grinding Technology: Theory and Applications of Machining with Abrasives*, Ellis Horwood Ltd., Chichester, England and John Wiley and Sons, New York, 1989; reprinted by SME, 1996.
3. Jacobs, F. B., *Abrasives and Abrasive Wheels,* Henley, New York, 1919.
4. Woodbury, R. S., *History of the Grinding Machine*, Technology Press, Cambridge, MA, 1959.
5. Ueltz, H. F. G., 'Abrasive Grains—Past, Present, and Future', *Proceedings of the International Grinding Conference*, Pittsburgh, 1972, pp. 1–33.
6. Anon., *The History of Abrasive Grain*. Abrasive Grain Association, Cleveland, Booklet 1, 1967.
7. Lewis, K. B. and Schleicher, W. F., *The Grinding Wheel*, 3rd edn., The Grinding Wheel Institute, Cleveland, 1976.
8. Coes, L., Jr., *Abrasives*, Springer-Verlag, New York, 1971.
9. McKee, R. L., *Machining with Abrasives*, Van Nostrand Reinhold, New York, 1982.
10. Farago, F. T., *Abrasive Methods Engineering, Vols 1 and 2*, Industrial Press New York, 1976 (Vol. 1) and 1980 (Vol. 2).
11. Colvin, F. H. and Stanley, F. A., *Grinding Practice*, 3rd edn, McGraw-Hill, New York, 1950.
12. Drozda, T. J. and Wick, C., Eds, *Tool and Manufacturing Engineers Handbook*, 4th edn, Vol. 1, Machining, SME, Dearborn, 1983, Chapter 11.
13. *Metals Handbook*, Vol. 3, Machining, 8th edn, ASM, Metals Park, Ohio, 1967, pp. 257–298.
14. *Machinery's Handbook*, 21st edn, Industrial Press, New York, 1980.
15. *Machining Data Handbook*, 3rd eds, Machinability Data Center, Cincinnati, 1980.
16. Andrew, C., Howes, T. D. and Pearce, T. R. A., *Creep Feed Grinding*, Holt, Rinehart, and Winston, London, 1985.
17. Tawakoli, *High Efficiency Deep Grinding*, VDI-Verlag GmbH, Mechanical Engineering Publications Ltd., London, 1993.
18. Bhateja, C. P. and Lindsay, R. P., Eds, *Grinding: Theory, Techniques, and Troubleshooting*, SME, Dearborn, 1982.
19. King, R. S. and Hahn, R. S., *Handbook of Modern Grinding Technology*, Chapman and Hall, New York, 1986.
20. Borkowski, J. A., and Szymanski, A. M., *Uses of Abrasives and Abrasive Tools*, Ellis Horwood, Chichester, 1992.
21. Krar, S. F., and Ratterman, E., *Grinding and Machining with CBN and Diamond*, McGraw-Hill, New York, 1990.
22. Shaw, M. C., *Principles of Abrasive Processing*, Oxford University Press, Oxford, 1996.
23. Marinescu, I. D., Hitchiner, M., Uhlmann, E., Rowe, W. B., and Inasaki, I., *Handbook of Machining with Grinding Wheels*, CRC Press, Taylor & Francis, Boca Raton, 2007.
24. Armarego, E. J. A. and Brown, R. H., *The Machining of Metals*, Prentice-Hall, Englewood Cliffs, NJ, 1969.
25. Sen, G. C. and Bhattacharyya, A., *Principles of Metal Cutting*, 2nd edn, New Central Book Agency, Calcutta, 1969.

26. Shaw, M. C., *Metal Cutting Principles*, 3rd edn, Technology Press, Cambridge, 1956.

27. Cook, N. H., *Manufacturing Analysis*, Addison-Wesley, Reading, MA, 1966.

28. Thomas, G. G., *Production Technology*, Oxford University Press, 1970.

29. Kaczmarek, J., *Principles of Machining by Cutting, Abrasion, and Erosion*, Peter Peregrinus, Stevenage, 1976.

30. Boothroyd, G., and Knight, W., *Fundamentals of Machining and Machine Tools*, 3rd edn, CRC Press, Taylor & Francis, Boca Raton, 2006.

31. El-Hofy, H., *Machining Processes*, CRC Press, Taylor & Francis, Boca Raton, 2007.

32. Kronenberg, M., *Machining Science and Application*, Pergamon Press Oxford, 1966.

33. Zorev, N. N., *Metal Cutting Mechanics*, Pergamon Press, Oxford, 1966.

34. Shaw, M. C., *Metal Cutting Principles*, 2nd edn, Oxford University Press, Oxford, 2004.

35. Stephenson, D. A., and Agapiou, J., S., *Metal Cutting Theory and Practice*, 2nd edn, Taylor & Francis, Boca Raton, 2005.

36. Childs, T. H. C., Maekawa, K., Obikawa, T., and Yamane, Y., *Metal Machining Theory and Applications,* Arnold, London, 2000.

37. Oxley, P. L. B., *The Mechanics of Machining*, Ellis Horwood, Chichester, 1989.

38. Trent, E. M., and Wright, P. K., *Metal Cutting,* 4th edn, Butterworth Heinemann, Newton, 2000.

Chapter 2

Grinding Wheels: Composition and Properties

2.1 INTRODUCTION

Grinding wheels and abrasive segments fall under the general category of 'bonded abrasive tools'. Such tools consist of hard abrasive grains or grits, which do the cutting, held in a weaker bonding matrix. Depending on the particular type of bond, the space between the abrasive particles may only be partially filled, leaving gaps and porosity, or completely filled with binder. Aside from abrasive and bond material, fillers and grinding-aid materials may also be added. The properties and performance of bonded abrasive tools depend on the type of abrasive grain material, the size of the grit, the bond material, the properties of abrasive and bond, and the porosity.

Grinding wheels are made from many types of grit in a wide range of sizes, in conjunction with many bond materials and compositions. 'Conventional' wheels in common use contain either aluminum oxide or silicon carbide abrasive with vitrified or resinoid bonds. 'Superabrasive' wheels with diamond and cubic boron nitride (CBN) abrasives are produced with vitrified, resin, and metal bonds. Whereas conventional abrasive wheels usually comprise the entire bonded abrasive structure throughout, the abrasive-composite on superabrasive wheels is limited to a thin rim or layer on a plastic or metal hub in order to reduce the amount of costly diamond and CBN which is needed. The different types of grinding wheels, together with the requirements of a wide variety of wheel shapes and sizes to fit all the diverse grinding machines and jobs to be done, lead to an almost endless diversity of grinding wheels. A 'full line' grinding wheel company may produce tens of thousands of nominally different products to satisfy its customers' requirements.

In this chapter, the composition and properties of grinding wheels will be generally considered, as a basis for understanding the grinding process and wheel selection. The discussion will encompass wheel composition and its specification, abrasive grain materials, bond materials, and wheel testing.

2.2   GRINDING WHEEL SPECIFICATION: CONVENTIONAL ABRASIVES

As a first approach to the subject of wheel composition, it is convenient to refer to the standard marking systems for specifying conventional grinding wheels. The Wheel Specification defines the following parameters:

(1) the type of abrasive in the wheel;
(2) the abrasive grain size;
(3) the wheel's hardness;
(4) the wheel's structure;
(5) the bond type;
(6) any other maker's identification codes.

The standard marking system used in North America for conventional abrasive wheels containing aluminum oxide and silicon carbide abrasive is presented in Figure 2-1 [1]. The same or similar systems are common elsewhere in the world. The symbol A or C indicates whether the abrasive material is Aluminum oxide or silicon Carbide. In fact, there are many types of abrasive based on synthetic aluminum oxide plus two common types of silicon carbide with different chemical compositions and structural characteristics which, in turn, affect their physical and mechanical properties (section 2.4). A manufacturer's prefix (in the form of a letter or number) usually appears to the left of the abrasive letter to indicate the particular type of alumina or silicon carbide used.

The symbol to the right of the abrasive grain type indicates the abrasive grain size, by a grit number which is related to the mesh number (specified as wires per linear inch) of the screen used to sort the grains. A larger number indicates a smaller grain size. Sieving (screening) is generally used for sizing of conventional abrasive grains coarser than 240 grit size, and a sedimentation method is used with finer grits (microgrits) [2, 3]. The sieving method consists of passing abrasive grains through a stack of standard sieves from the coarser aperture sieves first through progressively finer meshes, i.e. the mesh number increases down the stack. Nominally, the aperture size decreases by a factor of $\sqrt{2}$ between adjacent sieves in a stack of standard sieves.

A standard grit number is defined in terms of grain sizes corresponding to five such sieves. For example, grit number 46 involves grains caught on sieves number 30, 40, 45, 50 and 60 using a standard sample size

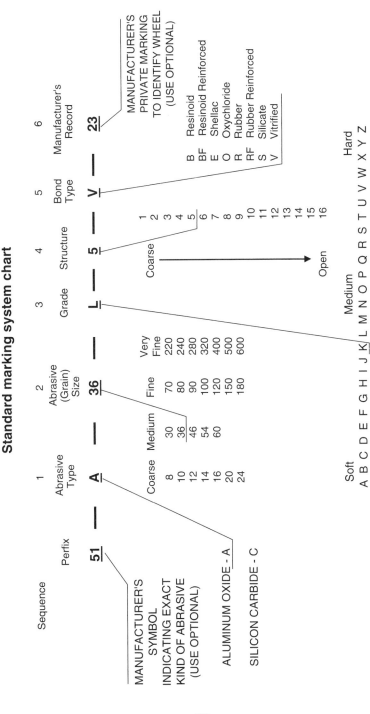

Figure 2-1 Standard marking system for aluminum oxide and silicon carbide wheels [1].

and sieve-shaking procedure. The specification requires 0% retention on the #30 sieve (i.e. no grains larger than 595 μm), not less than 70% passing the 'control' sieve #40 (not more than 30% in the size range 595–420 μm), not less than 40% retention on #45 (size range 420–354 μm), not less than 65% retention on #45 and #50 combined (not less than 65% in the size range 420–297 μm), and not more than 3% passing #60 (at least 97% in the size range 595–250 μm).

Since each nominal grit size includes a range of abrasive particle sizes, the grit dimension corresponding to a particular grit number might be characterized by an average value. However, the grit dimension d_g is often quoted in a simpler way either as equal to the aperture opening of the control sieve, or alternatively according to the relationship.

$$d_g\ (inches) = 0.6M^{-1} \tag{2-1}$$

or equivalently

$$d_g\ (mm) = 15.2M^{-1} \tag{2-2}$$

which approximates the grit dimension d_g as 60% of the average spacing between adjacent wires in a sieve whose mesh number equals the grit number M. The abrasive grain dimensions corresponding to both of these methods are plotted in Figure 2-2 as a function of grit number. When based upon the control sieve opening, the grain dimension can be approximated by

$$d_g\ (mm) = 28M^{-1.1} \tag{2-3}$$

Also included in Figure 2-2 are results obtained for the average grain dimension obtained by sieving samples of an aluminum oxide abrasive of different grit numbers with the dimension for the weight percentage retained on each sieve assigned as the average of that sieve opening and of the next coarser one [4]. This latter result can be approximated by

$$d_g\ (mm) = 28M^{-1.4} \tag{2-4}$$

Although this equation is strictly applicable only to the particular type of grit tested, it probably is a more precise indication of the actual grit dimension than either of the other two relationships. The three relationships in Figure 2-2 may appear to be rather similar, but their differences can lead to significant discrepancies in calculations related to grain packing in bulk or within a grinding wheel.

Although not indicated in Figure 2-1, some wheel manufacturers sometimes add a single digit after the grit number to indicate whether the wheel

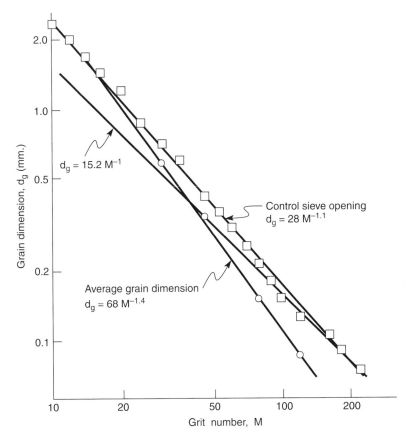

Figure 2-2 Grain dimension versus grit-number relationships based upon sieve wire spacing, control sieve opening, and average grain dimension.

contains a mixture of grit sizes. The number 1 after the grit number usually indicates that the wheel contains abrasive grains only of the indicated grit number, whereas a different number would designate a particular mixture of grit sizes. Using a wider range of grit sizes facilitates wheel manufacture, insofar as it makes it easier to pack the abrasive grains more tightly together in molding the wheel. The blending of two or three adjacent grit sizes is probably a common practice in wheel manufacture, even though it may not be explicitly indicated in the wheel specification.

Continuing on with the wheel marking system in Figure 2-1, the letter following the grit number signifies the wheel grade or hardness. The wheel grade provides a general indication of wheel strength and the degree to which abrasive grains are tightly held by the binder, as will be seen in sections 2.7 and 2.8. One method of establishing wheel grade is on the basis of porosity, regardless of the relative amounts of the abrasive and binder. According to one common scale, the hardest grade Z represents 2% porosity, Y 4%, X 6%, and so on [5]. For a given grain content, a harder wheel would thus have more binder and less porosity. Hardness scales based upon porosity are not universal, and the actual porosity or effective wheel hardness for the same letter grade will vary from one manufacturer to another.

The structure number in the wheel marking (Figure 2-1) indicates the volumetric concentration of abrasive grain in the wheel, a higher number indicating less abrasive or a more open wheel. One commonly used structure scale appears to be based upon the empirical relationship

$$V_s\ (\%) = 2(32 - S) \tag{2-5}$$

for volume percentage V_g of grain as a function of structure number S, which means that each structure number increment corresponds to a reduction by 2% in the grain content. An upper limit on the grain concentration (lower limit on structure number) is imposed by packing limitations which refers back to the grain size and its distribution. Abrasives of a given size and shape are characterized by a limiting natural packing density which can be reached by shaking and application of moderate pressures low enough so as not to cause grain crushing. Higher limiting packing densities are obtained with coarser and more equiaxed (blocky) shaped grains than with finer and less symmetrical (weaker) shapes, and the degree of natural packing (bulk density) may be used as a rather simple but effective measure of grain shape. Maximum volumetric packing densities generally range from about 40 to 60%, although somewhat higher values are obtained with broader size distributions. At the other extreme, a lower volumetric packing density limit is imposed, at least with vitrified wheels, by the need to maintain some mutual grain contact so as to minimize shrinkage and distortion during the vitrification process of wheel manufacture. In order to simplify the wheel specification and reduce the number of product variations, some manufacturers do not specify the structure number for wheels intended for general use.

The bond material is indicated by a letter, which may be followed by an additional notation to indicate a particular formulation. Most conventional abrasive wheels are made with vitreous- and resinoid-based bonds. The derivation of 'V' for Vitrified bond is obvious. Rubber-based bonds were once very common and have retained the designation 'R', while Resinoid bonds, first based on Bakelite, the original synthetic thermosetting plastic, have adopted 'B'. Shellac was formerly referred to as an Elastic bond, hence the designation 'E' for Shellac bonds. Silicate, 'S', and oxychloride, 'O', bonds are now almost extinct.

2.3   GRINDING WHEEL SPECIFICATION: SUPERABRASIVES

Wheels containing superabrasives—diamond or cubic boron nitride (CBN)–use a somewhat different wheel specification system which is illustrated in Figure 2-3 [1]. The letter indicating abrasive type for diamond (D) or cubic boron nitride (B) is usually preceded by a symbol to identify a particular abrasive material. The bond material—resin, vitrified, or metal— is also indicated by a letter, often followed by an additional manufacturer's notation to identify a particular formulation. Most superabrasive wheels have either resin or metal bonds, although vitrified bonds are also widely used for CBN wheels. As the abrasive grain is expensive, only a relatively shallow section of the active area of the wheel surface actually consists of bonded abrasive, which is attached to a metal or plastic hub. The depth of the abrasive section is indicated in the wheel marking (Figure 2-3). Less expensive metal-bonded wheels containing only a single layer of abrasive are produced with an electroplated metallic binder to hold the superabrasive grits on to a form or hub. In this case, the manufacturer usually specifies only the abrasive type and grit size.

Superabrasive grain sizes may be specified by grit numbers as with conventional abrasives, but the corresponding standard for checking the size of diamond grains is somewhat different and utilizes a two-number designation (see Table 2.1) [6]. The first number in the designation is generally considered to represent the sieve through which most of the grains would pass, and this is the number to be used when the grain size is specified by only a single number. The second number in the designation is considered to represent the sieve which would retain most of the grain, although this is only approximately true. The number in the corresponding FEPA (Federation Européenne des Fabricants de Produits Abrasifs) designation in Table 2.1, which is used by many European superabrasive wheel suppliers, indicates the approximate grain dimension d_g in microns. This dimension and the second number in the grit-size designation can be shown to very nearly follow Eq. (2-2) in the size range from 40/50 down to 325/400. Superabrasive grits finer than 325/400 (powders) are checked by other methods [7].

The letter grade in the wheel marking (Figure 2-3) provides a relative indication of the strength or hardness of the bond, as with conventional abrasives. However, resin- and metal-bonded wheels are produced with virtually no porosity, and the effective grade is obtained by changing the bond formulation (Section 2.6). With resin-bonded wheels, for example, this could involve the addition of fillers in place of porosity.

Following the grade (Figure 2-3) is the concentration number, which indicates the amount of abrasive contained in the wheel, although this is sometimes given as the related volume percentage in some systems. The

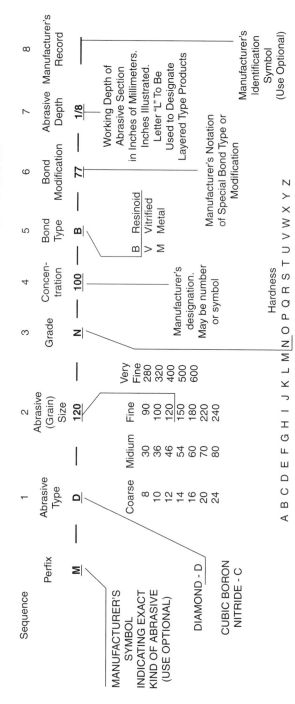

Figure 2-3 Standard marking system for diamond and cubic boron nitride grinding wheels and other bonded abrasives [1].

18

Table 2.1 - Grit-size designation for diamond and cubic boron nitride [6]

USA grit size	FEPA designation
16/18	D1181
18/20	D1001
20/30	D852
30/40	D602
40/50	D427
50/60	D301
60/80	D252
80/100	D181
100/120	D151
120/140	D126
140/170	D107
170/200	D91
200/230	D76
230/270	D64
270/325	D54
325/400	D46

concentration number is based upon a proportional scale with a value of 100 corresponding to an abrasive content of 4.4 carats/cm^3. This scale was originally developed for diamond wheels for which the concentration number divided by four equals the volumetric percentage of grit (e.g. 100 concentration is 25% by volume). The corresponding volumetric concentration for CBN is nearly the same (24%), as the density of CBN is very nearly equal to that of diamond. Typical concentrations for metal- and resin-bonded superabrasive wheels range from 50 to 150 (12.5 to 37.5 volume % for diamond). As with conventional abrasive wheels vitrified superabrasive wheels would require higher concentrations, thereby making such CBN wheels more costly. Less expensive vitrified CBN wheels are produced which contain a mixture of CBN and aluminum oxide abrasives.

2.4 CONVENTIONAL ABRASIVE MATERIALS

Abrasive grains, the cutting tools of the grinding process, are naturally occurring or synthetic materials which are generally much harder than the materials which they cut. Natural abrasives include aluminum oxide

Table 2.2 - Some properties of abrasive materials

	Material			
	Aluminum oxide (Al_2O_3)	Silicon carbide (SiC)	Cubic boron nitride (BN)	Diamond (C)
Crystal structure	Hexagonal	Hexagonal	Cubic	Cubic
Density (g/cm^3)	3.98	3.22	3.48	3.52
Melting point (°C)	2040	~2830	~3200 at 105 kbar (triple point)	~3700 at 130 kbar (triple point)
Knoop hardness[†] (GPa)	20.6	23.5	46.1	78.5

[†] *Approximate value—depends on crystal orientation, microstructure, and purity.*

(natural corundum and emery), garnet, and diamond. Technological advances in the abrasives industry have been mainly in the development of synthetic (man-made) abrasives, as discussed in Chapter 1. Some physical properties of the most important abrasive materials are summarized in Table 2.2. Conventional abrasives (aluminum oxide and silicon carbide) will be considered in this section, and superabrasives (diamond and cubic boron nitride) in the following one.

Virtually all conventional abrasives in use today for grinding wheels are synthetic materials based upon either aluminum oxide (Al_2O_3) or silicon carbide (SiC). The hard aluminum oxide phase is α-alumina having an hexagonal crystal structure like that of natural aluminum oxide abrasives (emery and corundum) which exist in various states of purity and degrees of crystallization. In addition to Al_2O_3, synthetic aluminum oxides contain various amounts of other metallic oxides either intentionally added or as impurities. Silicon carbide occurs in various polytypes, which can be generally classified as α-types having hexagonal or rhombohedral crystallographic structures and a β-type which is cubic. Silicon carbide abrasive materials consist primarily of α-SiC [8]. Several varieties of aluminum oxide abrasive, and a limited number of silicon carbide types, are in common use, each having a distinctive chemical composition and set of structural characteristics which affect the granular properties and grinding behavior and so make it useful for specific tasks.

Classically, the prime requirement of an abrasive is that it be harder than the material it is to abrade. The hardness of an abrasive is generally defined in terms of its static indentation hardness as determined by a Knoop or Vickers hardness test. Another important abrasive property is its dynamic strength or toughness. Higher toughness implies that an abrasive grain is less likely to fracture or fragment each time it engages or impacts the workpiece. On the other hand, a more friable (less tough) abrasive should regenerate sharp cutting edges (self-sharpen) as the grain dulls by attrition during use.

The comparative friability of conventional abrasives is usually evaluated by a standard comminution test wherein a sample of relatively coarse (#12 grit) material is ball milled under prescribed conditions [9]. The 'friability index' of the abrasive, indicating the degree of fragmentation caused, is defined as the percentage of milled material passing through a #16-mesh sieve, although this defines the value for only one grain size. Variants of this comminution method can be used in a more generalized way to determine the relative friability of different grain sizes of the same abrasive or of different abrasives [10]. In general, finer grits of the same material are less friable, which is to be expected since they are usually produced by crushing of coarser material. Friability of abrasive grains can also be evaluated using a single blow impact test [11].

Hardness and friability data are given in Table 2.3 for many of the common types of aluminum oxide and silicon carbide abrasives [11, 12]. Harder abrasive grains are generally more friable, which can also be seen

Table 2.3 - Hardness and friability index for aluminum oxide and silicon carbide abrasives (12 grit number) [10]

Grain type	Knoop hardness (GPa)	Friability index
Aluminum oxide		
Modified (3% Cr)	22.2	65.0
White	20.8	56.6
Monocrystalline	22.4	47.7
Regular	20.0	35.6
Microcrystalline	19.1	10.9
10% ZrO_2	19.2	10.9
40% ZrO_2	14.3	7.9
Sintered	13.4	6.5
Silicon carbide		
Green	27.9	62.5
Black	26.3	57.2

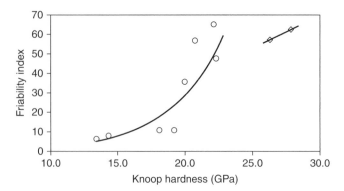

Figure 2-4 Friability index versus hardness for aluminum oxide and silicon carbide abrasives. Data are given in Table 2.3.

in Figure 2-4 where the friability index is plotted versus hardness. Silicon carbide abrasives are harder than aluminum oxide; and also tend to fall towards the upper end of the friability range. Harder and more friable abrasives are generally applied to precision-grinding operations. Tougher abrasives of larger sizes are more suitable for heavy-duty grinding.

For aluminum oxide abrasives, observed differences in properties arise from differences in chemical composition and structural characteristics associated with the manufacturing process. Starting with bauxite as the main raw-material, most aluminum oxide abrasives are made by three different methods. Bauxite dehydrated by calcination may be fused directly with coke and iron in an electric furnace, it may be first processed to form purified Bayer process alumina which is then fused, or it may be sintered after pressing.

Regular or brown aluminum oxide (Table 2.3) can be produced by the first of these methods by fusing calcined bauxite with a small amount of coke and iron. The brownish product contains about 2.7% titanium oxide retained from the bauxite, possibly as a dispersed softer β-$Al_2O_3 \cdot TiO_2$ phase, which may be responsible for the material's lower hardness and friability (higher toughness) relative to the purer white and monocrystalline varieties. The grains tend to be irregularly shaped with featureless surfaces, although second-phase inclusions can be seen with coarser-size grits [13]. This semi-friable abrasive is applied to a wide range of operations from heavy-duty grinding to roughing and semi-finishing. A tougher variation of this material referred to as microcrystalline aluminum oxide (Table 2.3), produced by more rapid cooling in smaller ingots to obtain a much finer crystal size, is used mainly for heavy-duty grinding.

Monocrystalline aluminum oxide can be produced by a similar fusion process but with the addition of iron sulfide and alkaline compounds to sweep out the titania with the other oxides. The resulting ingot consists of alumina grains in a decomposable sulfide matrix and is crushed and treated with water to release the abrasive grit. This material is much purer than the brown product, containing only minor amounts of oxide impurities, and has the great advantage that the grains are produced by a method that does not expose them to high crushing forces before they are used in a wheel. Scanning electron microscope (SEM) observations of monocrystalline aluminum oxide reveal stepwise facets which could act as sharp cutting points [13]. This abrasive material is used mainly for finish-grinding operations.

White and modified aluminum oxide abrasives (Table 2.3) are produced by the fusion of pre-purified Bayer process alumina in the Hall-Heroult electric arc furnace. The white grades of alumina are nearly 100% Al_2O_3 and modified aluminas are produced by adding small amounts of soluble metal oxides which go into solid solution and may enhance the material's hardness and normally increase its toughness appreciably. White aluminum oxide grains have sharp fracture facets, similar to those observed on monocrystalline grains. Such facets are not observed on modified (chrome and vanadium oxide) abrasives [13].

Bayer process alumina contains small amounts of sodium oxide, and both white and modified aluminum oxide abrasives can contain up to 1% residual Na_2O. Its presence can leave voids in the final abrasive from gassing due to removal of some of the original oxide present, and it produces so called β-Al_2O_3 (actually a compound approximating to the formula $Na_2O \cdot 11Al_2O_3$) in the grains, which is very soft and must be destroyed by heating to temperatures greater than 1260 °C during the firing of the final abrasive tool or wheel. These abrasives are used for finish-grinding operations.

Sintered aluminum oxide is manufactured in a completely different way by pressing or extrusion of a fine (1-5 μm) paste of calcined bauxite, granulating or chopping the compacted material, and sintering somewhat below the melting temperature. The impurities in the bauxite act as sintering agents, leading to very fine crystal size and extremely high toughness (Table 2.3). The final abrasive product generally has rounded edges without sharp corners [13]. This abrasive is applied to heavy-duty grinding operations.

The introduction of mixed alumina-zirconia abrasives has had a profound effect on heavy-duty grinding operations. Alumina and zirconia show very limited mutual solubility and form a simple eutectic at approximately 42wt% zirconia (ZrO_2). Although ZrO_2 has intrinsically a hardness value about half that of Al_2O_3, this eutectic structure is very tough because of the ability of the softer dispersed zirconia phase to stop cracks, preventing the grains from fracturing under quite high loads and so both allowing them to be used under more arduous conditions and to last longer at a given load. Another factor contributing to the superior performance of alumina-zirconia abrasives for heavy-duty grinding, relative to fused or sintered aluminas, is the higher melting point of ZrO_2 (2720 °C) as compared with that of Al_2O_3 (2040 °C), which suggests a greater degree of chemical stability [14]. Mixed alumina-rich abrasives, consisting of distinct hard Al_2O_3 and tough Al_2O_3-ZrO_2 eutectic, are produced with zirconia contents up to the eutectic composition by fusing calcined bauxite, zircon sand, coke and iron in an electric arc furnace. The melt is rapidly quenched and solidified to a fine dendritic structure by pouring it in a thin layer over a water-cooled steel plate and crushing. Different ZrO_2 contents give abrasives with different mechanical properties, materials richer in Al_2O_3 tending to be harder and more friable and those nearer the eutectic being tougher and less friable. The Al_2O_3-ZrO_2 eutectic is a very interesting material since, on quenching, the zirconia crystals may be locked into their high-pressure form and revert to the low-pressure form with an increase in volume if a crack approaches, thus enhancing toughness by filling the crack and making it more difficult for it to move and split the material [15, 16].

'Sol-gel' alumina abrasives represent a more recent development in the technology of abrasive synthesis [17]. Such materials are made neither by fusing nor by sintering, but instead by converting a colloidal dispersion of hydrosol ('sol') containing goethite ($Al_2O_3 \cdot H_2O$) to a semi-solid 'gel', drying this gel to a glassy state, crushing to the required grain size, and firing at 1200 °C to 1600 °C. The final product consists of abrasive grains with a randomly oriented structure comprising alumina microcrystals that are less than 5 μm and, in some instances, less than 1 μm in size. This very fine polycrystalline structure is considered to enable microfracturing of the abrasive grains, thereby promoting self sharpening and continual generation of new cutting edges. One variation of sol gel abrasives is produced by alloying the alumina with magnesia, yttria, and rare earth metal oxides, which react during firing to form a magnetoplumbite phase in the form of platelets to reinforce the alumina matrix [18]. A somewhat different sol gel process uses dispersed alpha alumina 'seed' particles in the gel which results in the formation of submicron alpha alumina crystals during the firing process [19].

Silicon carbide is made by reducing sand (SiO_2) with excess coke (C) in an electric furnace at temperatures above 2000 °C according to

$$SiO_2 + 3C \rightarrow SiC + 2CO$$

The desired product is the hexagonal form of silicon carbide (α-SiC), which is green-to-black in color, although the final reaction mass contains a mixture of unreacted coke, partly reduced 'firesand', and silicon carbide. This is carefully separated and the silicon carbide fraction collected for further processing, the firesand and coke being returned to the furnace for further treatment as part of a subsequent charge. Green silicon carbide is purer than black and, being a semiconductor, is a premium product also used to make heating elements for furnaces. The black material is preferred for grinding on a cost basis, although it is slightly less hard.

Both varieties of silicon carbide are intrinsically harder than alumina and comparable in friability to the hardest alumina abrasives (Table 2.3 and Figure 2-4), and this combination of physical properties would suggest that silicon carbide might be better than alumina oxide for fine grinding processes. This is generally so for nonferrous metals and for most ceramics, but silicon carbide is inferior for most ferrous applications, because of its chemical reactivity with iron and steel alloys, leading to poor attrition resistance and low grinding ratios (Chapter 11). Silicon carbide is, however, better for some hard cast irons, where the high carbon content in the metal minimizes chemical interaction with the wheel.

2.5   SUPERABRASIVE MATERIALS

Superabrasive materials include diamond and cubic boron nitride. Diamond is the hardest known material, and cubic boron nitride is the second hardest. As an abrasive, diamond is used in both its natural and its synthetic forms, although the trend is generally towards the synthetic material. Boron nitride, in both its cubic and its soft hexagonal forms, is a synthetic material.

In the case of natural diamonds, their shape and size are determined by nature during their geological formation, although they can be reshaped by man using mechanical and thermal methods. For synthetic diamond, the same reshaping techniques can be used, but their intrinsic strength and structure can be altered by varying the processing conditions. Synthetic diamond is produced by subjecting graphite to high temperatures at extremely high pressures in the presence of a catalyst solvent such as nickel or other metals from group VII of the periodic table [20-23]. With nickel as a catalyst, operating conditions might be about 2000 °C at 75-95 kbar. Depending on the particular temperature, pressure, and processing time, diamonds are made with varying crystal sizes and structures. Synthetic diamond abrasives range from weak, friable, irregularly shaped polycrystalline grains with a skeletal structure to tough blocky-shaped cubo-octahedral single crystals. The weaker shapes are applied mainly to grinding of cemented carbides with resin-bonded wheels. For this application, the diamonds are usually coated with nickel comprising about 55% by weight of the grain and coating. The purpose of the coating is to more strongly hold the diamond grit in the resin binder, in addition to providing some protection from the atmosphere. Stronger blockier monocrystalline diamond grits are used mainly with metal bonds for cutting of ceramics, stone, glass, and other hard brittle materials.

In spite of its extreme hardness, diamond had been found not to be economical for grinding of most ferrous materials, except for some hard cast irons, owing to graphitization and carbon diffusion into the iron causing excessive diamond wear. Cubic boron nitride has emerged as an important alternative superabrasive for grinding of steels and some non-ferrous high-strength alloys. Boron nitride was first made as the hexagonal polymorph, isostructural with and soft and slippery as graphite. This structural analogy with graphite, combined with the fact that boron is chemically very similar to carbon, inspired the thought that it might be possible to synthesize a cubic form of boron nitride analogous to diamond, and this was achieved at temperatures of 1500 – 2000 °C at pressures in the range 50-90 kbar (5-9 GPa) using alkali metals as catalytic solvents [24].

Almost all CBN grits produced today are monocrystalline, although polycrystalline (microcrystalline) abrasives with submicron crystal size have also been introduced. In its polycrystalline form, CBN is claimed to be significantly tougher. Monocrystalline CBN grits tend to be blocky shaped with sharp edges and smooth faces, which make bonding difficult, As with diamond, a nickel coating is added to retain the grits more strongly in resin-bonded wheels.

In comparison with diamond, one important advantage of CBN is its thermal stability. Both diamond and CBN are stable in vacuum up to temperatures in excess of 1400 °C. In normal atmosphere, a B_2O_3 protective layer on CBN is credited with preventing oxidation up to 1300 °C, and no conversion from the cubic to hexagonal form occurs up to 1400 °C. By contrast, diamond is thermally stable only to a much lower temperature of about 800 °C in normal atmosphere. An important consequence of this is related to the possibilities for vitrified superabrasive wheels. CBN wheels with vitrified bonds can be fired to a much higher temperature than diamond, and so a much wider range of vitreous bonds can be considered for their manufacture. Some CBN grits are specially coated to protect their surface from chemical reaction above 800 °C with the alkali and water present in most glass frits used in vitrified wheel manufacture. While vitrified bonds are only occasionally used with diamond, they have become commonly used with CBN for precision grinding of metallic materials.

2.6 BOND MATERIALS

Abrasive grains are held together with various kinds of bond materials. In general, the bond must be strong enough to withstand grinding forces, temperatures, and centrifugal forces without disintegrating, while resisting chemical attack by the cutting fluid. Additional bond requirements may include wheel rigidity, and the ability to retain abrasive grains during cutting yet release dulled grains.

According to the wheel marking system in Figure 2-1, there are six general types of bond materials for conventional abrasive wheels: resinoid (including reinforced), shellac, oxychloride, rubber (including reinforced), silicate, and vitrified. Most conventional abrasive wheels have either vitrified or resinoid bonds. Superabrasive wheels (Figure 2-3) are produced with three bond types: resinoid, vitrified, and metal.

Vitrified wheels probably account for about half of all conventional abrasive wheels, although the trend towards higher wheel speeds has led to some replacement by resinoid wheels, especially for heavy-duty grinding. Historically, the use of vitrified wheels was restricted to peripheral speeds of about 30 m/s owing to strength limitations, although methods for wheel reinforcement now make it possible to use these wheels at much higher peripheral speeds. The highest speed used in production with vitrified wheels is at present 120 m/s.

Vitreous bonds are formed from mixtures of a clay, a feldspar, and a frit, normally using locally available materials, in amounts mainly determined by the nature of the wheel to be built but also affected by the mineralogy and detailed chemistry of the clays and feldspars used, in particular minor phases and trace elements present. The frit is man-made and its composition is under better control. Such mixtures soften and melt in the temperature range 950 to 1400 °C with mixtures richer in clay melting at higher temperatures, those with more frit melting at lower temperatures. It is, thus, possible to prepare bond mixtures with different viscosities, and hence different surface tensions at a given temperature, and so tailor the bond to the required structure of the final wheel. In particular, it becomes possible to help control the porosity in, and provide the strength to, the wheel by careful choice of bonding mix. The mixtures are prepared by milling the raw materials together with about 1–5% water containing an organic binder, such as dextrin, until a plastic mass ensues.

Vitreous bonds are almost always used with alumina grits, and this plastic mixture is next added to between two and six times its weight of the requisite abrasive. Again, the exact weight percentage depends on the nature of the wheel and the use to which it will be put and the clays and feldspars used. A combustible filler, such as sawdust, may be added if a very porous wheel is required. The mix is pressed to the required shape, dried, and fired in a traveling kiln to a temperature in excess of 1260 °C in a regime extending over some days. The heating stage is relatively rapid, taking 1 to 2 days to reach the maximum temperature, which is held for about 12 hours, but cooling must be slow and carefully controlled to avoid building thermal stresses into, or even cracking, the wheels. Very large wheels can take weeks to cool.

At the firing temperature, the bonding mix melts, partly wets the abrasive grits, and surface tension pulls them together [25]. At the same time, chemical reactions occur at the grit-melt interface with interpenetration of grit and melt, the β-Al_2O_3 phase in the white grits is dispersed into the surrounding matrix (removing a potential source of weakness in the grit particles), and a titania-rich phase is exsolved from the brown Al_2O_3 grains, causing them to turn blue. On subsequent cooling, glassy 'necks' of solidified bonding material, called bond posts, develop between the grains holding them firmly together, and these are anchored both by a mechanical bond, where the molten material has flowed into irregularities in the grit surfaces, and by chemical bonds due to the new phases formed at the grit-bond interface [26]. An example of a vitrified wheel structure is shown in Figure 2-5.

Resinoid-bonded wheels are produced by mixing abrasive grains with phenolic thermosetting resins and plasticizers, molding to shape, and baking (curing) at 150–200 °C. The bond hardness is varied by controlling the amount of plasticizer and by addition of fillers. Conventional abrasive resinoid wheels are widely used for heavy-duty grinding (snagging) operations because of their high strength and ability to withstand shock load. Another important application is for cut-off wheels, which are usually reinforced with fiberglass for added strength and high-speed operation up to about 100 m/s. For superabrasive wheels, resinoid bonds are the most popular, the most important applications being with diamond abrasives for grinding of cemented carbides and CBN for grinding of steels.

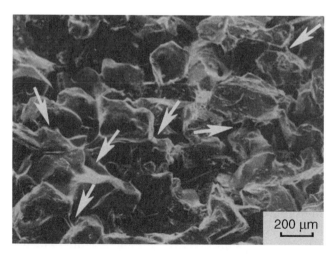

200 μm

Figure 2-5 SEM micrograph of vitrified wheel 32A54I8VX. Abrasive is monocrystalline alumina. Arrows point to bond posts.

Resinoid wheels are susceptible to chemical attack by alkaline cutting fluids which adversely affect their strength, especially with prolonged exposure at elevated temperature [27]. The fluid not only lowers the strength of the resin itself, but can weaken its bonding to the abrasive, which is one reason why aluminum oxide grains for resinoid wheels are specially treated with a thin coating. Grinding fluid attack may not be a problem with heavy-duty and cut-off wheels, insofar as they are often used dry. The strength of superabrasive resinoid wheels generally does not depend on the resinoid bond, since it is only in a thin outer layer. However, alkaline grinding fluids will likely degrade the wheel performance over a period of time.

Rubber bonds consist of vulcanized natural or synthetic rubber. The main applications are thin wheels for wet cut-off operations to produce nearly burn-free cuts, and regulating wheels for centerless grinders. Rubber wheels were once popular for finishing operations on bearings and cutting tools, but their use for these purposes has declined. The manufacture of thin rubber-bond wheels involves mixing together the rubber and abrasive with sulfur added as a vulcanizing agent, rolling out in sheets to the required thickness, cutting out the required shape, and vulcanizing under pressure at 150–275 °C. Thick wheels can be manufactured in a similar way, but by stacking of thin sheets after cutting.

Silicate-bonded wheels are manufactured by mixing sodium silicate with abrasive, tamping in a mold, drying and baking. The historical advantage of silicate in comparison with vitrified wheels is the much lower processing temperature and shorter heating cycles. At one time, the silicate process was popular with small grinding wheel manufacturers lacking facilities for producing vitrified products. The process might still be occasionally used for producing extra-large slow-speed wheels for some sharpening and finishing operations.

Shellac is a natural organic material which is only rarely used today as a bond material. The wheels can be manufactured by mixing abrasive grain with shellac, shaping under pressure in heated molds, and baking. At one time this bond was used for flexible cut-off wheels, which is probably why they were referred to as elastic wheels. The use of shellac-bond wheels is mainly for fine finishing of mill rolls, camshafts, and cutlery.

Another less common type of bond is oxychloride, which is a cold-setting cement from a mixture of magnesium oxide and an aqueous solution of magnesium chloride. Apparently it was very popular about a hundred years ago, but its only use today might be for disk grinding. It is susceptible to chemical attack by grinding fluids, so it is used dry.

Metal bonds are extensively used with superabrasive wheels. The most common are from sintered bronze, which are produced by powder metallurgy methods. Variation of the wheel grade is controlled by adding modifiers and altering the bronze composition. Other powder metal bonds, which are generally stronger, include iron and nickel. Segmented diamond saws for cutting stone and granite typically have sintered nickel bonds. Tungsten powder infiltrated with a low melting point alloy is used in diamond wheels for grinding diamond tools. Still stronger bonds consisting of WC-Co cemented carbide are used in diamond abrasive tools for geological drilling.

A different type of metal bond for superabrasive wheels is manufactured by electroplating. These grinding wheels consist of a single layer of diamond or CBN held in place on a form or hub by an electroplated nickel binder. Figure 2-6 shows the surface of an electroplated CBN wheel with abrasive grain tips protruding above the electroplated nickel. Typically the nickel layer is equal to about 30% of the grain dimension, although its thickness may be varied. Because these wheels contain only a single layer of supereabrasive, electroplated superabrasive wheels are generally less expensive than bonded types. Much of the expense is associated with precise machining of the hub, which can be reused. Electroplated CBN wheels have become widely used especially in automotive and aerospace industries for grinding of metallic materials. Extremely thin metal-bonded diamond wheels for slicing and dicing of electronic materials are produced by electroplating a single layer of diamond onto a substrate which is subsequently discarded.

Figure 2-6 SEM micrograph of electroplated CBN wheel surface (120 grit).

2.7   VITRIFIED WHEEL COMPOSITION
AND PHASE DIAGRAMS

In sections 2.2 and 2.3, the wheel composition was described in terms of its marking system. The grinding wheel can, however, be described more objectively and with more references to its structure and composition in terms of a 'grinding wheel phase diagram'. If we neglect possible additions of fillers and grinding aids, a grinding wheel can be considered as a three-phase system consisting of abrasive grains, bonding medium, and porosity. We can then write

$$V_g + V_b + V_p = 100 \qquad (2\text{-}6)$$

where V_g, V_b and V_p are the volume percentages of the grit, bond and pore or soft phases, respectively. This composition relationship can conveniently be represented in the form of a standard, equilateral-triangle-shaped three-phase diagram, as shown in Figure 2-7 [28]. In such a diagram, each

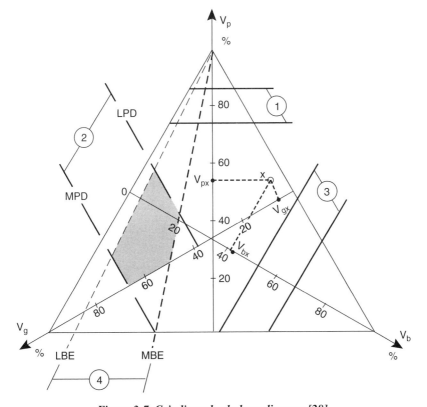

Figure 2-7 Grinding wheel phase diagram [28].

apex represents 100% of one component with the opposite side corresponding to 0% and intermediate percentages represented by the fractional distance from the side to the apex. Three axes can be drawn from the apexes to the opposite sides to illustrate this relationship, and these are shown in Figure 2-7.

Lines in the phase diagram drawn perpendicular to these axes, and so parallel to the sides, represent compositions all with the same percentage of the component represented by the opposite apex. Thus, lines '1' represent compositions all with $V_p = 75\%$ (lower line) or 86% (upper line), lines '2' with $V_g = 68\%$ (MPD) or 38% (LPD), and lines '3' with $V_b = 54\%$ or 68%. Such lines are known as 'iso-lines', either iso-porosity (1) or iso-grain (2) or iso-bond (3). Note that the sum of the percentages of the other two components is constant across an iso-line but the individual percentages vary. Any point 'x' within the triangle in Figure 2-7 represents a specific wheel composition, i.e. V_{gx}, V_{bx}, and V_{px}.

Also of some significance are the lines shown as '4' in Figure 2-7. These are lines that connect an apex to points on the opposite side and represent sets of compositions with a constant ratio of the components at the ends of the side and varying amounts of the apex component. Thus, lines '4' represent compositions with constant grit-to-bond ratios, of about 12:1 (LBE) and 3:1 (MBE), but with different porosity contents.

In practical terms, iso-grain lines, corresponding to particular grain volume percentages and varying amounts of bond and pore phases, generally define the 'structure' or 'packing number' in the wheel markings for conventional wheels (Figure 2-1) or the concentration number for superabrasive wheels (Figure 2-3). Iso-porosity lines are sometimes considered to indicate the 'wheel hardness' or 'grade' of conventional wheels, but this will be discussed further below. Note that a filler may take the place of the pore phase in a resinoid-bonded wheel.

Actual wheel compositions do not cover the whole composition range, represented by the phase diagram but are restricted to a limited range by technological and practical factors. For example, a typical useful composition range for conventional vitreous-bonded wheels is represented by the shaded region in Figure 2-7, and shown expanded in Figure 2-8. This is bounded by two iso-grain lines (MPD and LPD) and three constant ratio lines (LBE, MBE, and the V_g axis where $V_p = V_b$). The limiting upper iso-grain line, the 'maximum packing density' or MPD line, represents the natural maximum packing density imposed by the shapes and sizes of the abrasive grains, and its exact value (position on the diagram) will depend on the shape and size distribution of the abrasive grains used. The 'lower packing density' or LPD line is linked to the condition at which there are just enough grit-grit contacts, with their attendant bond posts, to give the required strength to the wheel. Again, the exact position of this line depends on details of the abrasive grains used. The 'maximum bond equivalent' or MBE line corresponds to the bond-grit ratio above which the bond simply coats the grit

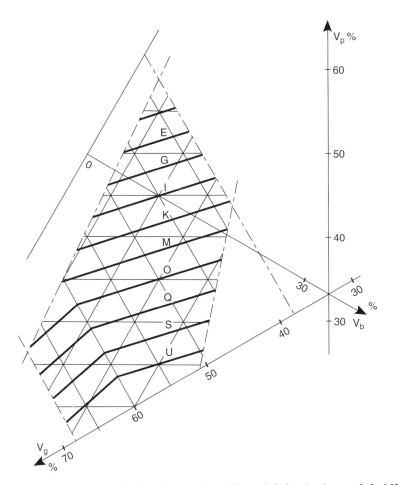

Figure 2-8 Shaded region of phase diagram from Figure 2.7 showing iso-grade loci [28].

particles without forming additional bond posts, while the 'lower-bond equivalent' or LBE line is again associated with the provision of a minimal amount of bond phase to give the required strength. The shaded area in Figure 2-7 is bounded by $V_g = 68\%$ for the MPD line, $V_g = 38\%$ for LPD, $V_b/V_g = 0.31$ for MBE and $V_b/V_g = 0.08$ for LBE, but the actual boundaries for other types of wheel will depend on the particular abrasive-bond system used.

Most wheel manufacturers use different bond mixes and proportions of abrasive and bond to achieve a given letter grade (hardness) of wheel, and so iso-grade lines do not usually coincide with iso-porosity lines. One manufacturer's gradings for wheels of different compositions are plotted in Figure 2-8. Wheels with the same grade plot as straight lines against overall composition which are uniformly inclined to the iso-porosity lines, rather than coinciding with them, and they show a marked discontinuity in relative slope at $V_g = 60\%$. Below this grit content, i.e. in the majority of useful wheels, the percentage porosity can be expressed as a function of grade number in the form

$$V_p(\%) = \frac{2(99.5 - 2n) - V_g}{3} \qquad (2\text{-}7)$$

where n is an integer (n = 1, 2, 3, 4, . . .) corresponding to the letter grade (E, F, G, H, ...), respectively. If the structure number relationship of Eq. (2-5) applies, the wheel porosity can be expressed in terms of the structure number and letter grade by the relationship

$$V_p(\%) = 45 + \frac{s - 2n}{1.5} \qquad (2\text{-}8)$$

Accordingly, wheels with the same grade number but containing less grain (higher S value) should be more porous, and this is in accordance with the predictions of the phase diagram.

Again, each manufacturer uses its own characterization standards to specify particular wheel grades and structure numbers. Moreover, bond formulations and processing methods, as indicated earlier, vary from one manufacturer to another, and so wheels from different manufacturers, although with apparently identical abrasive grit, bond and pore contents and having the same indicated letter grade and structure number, can be expected to perform differently. The grinding wheel marking system has been standardized, as also are the abrasive grits to some extent, but the wheels produced are not.

2.8   GRINDING WHEEL TESTING

Various testing procedures have been developed for evaluating grinding wheel performance, checking quality in wheel production, and ensuring wheel safety. Wheel performance is generally evaluated by actual grinding tests, as will be seen in Chapter 8. Here we will be concerned with non-grinding tests for identifying inherent properties of conventional vitrified wheels.

One of the more elusive characteristics of a grinding wheel is its hardness or grade. As seen in the previous section, a harder-grade wheel having a given abrasive content contains more binder and less porosity. Therefore, harder wheels should be stronger, and the abrasive grits should be more firmly held by the binder.

Hardness testing of vitrified grinding wheels was originally introduced not for the purpose of checking the wheel grade, but rather for actually grading the wheel. Owing to lack of technical sophistication for controlling the vitrified wheel manufacturing process, wheels were assigned grades only after their manufacture. The test method consisted of scratching the grinding wheel with a screwdriver-like hand tool. The operator would determine the wheel grade by the resistance he felt and the sound emitted.

Similar modern methods for testing and grading of grinding wheels involve measurement of forces while removing wheel material with a tool. One such test involves the use of a triangular-pointed tool to scratch a groove in a wheel with the depth of penetration set equal to the grit dimension [29]. Force pulses are measured which are each assumed to be associated with dislodgement of a single grit, and the average force is taken as an indication of wheel grade. However, grit fracture is also likely to occur in addition to dislodgement, especially with harder wheels. Another test uses a conical metal tool instead of a grooving tool [30]. The conical tool is free to rotate, so it crushes the wheel instead of being ground away as it is fed across the rotating wheel face. These two test methods, groove scratching and crushing, were found to agree quite well in their ability to distinguish between wheels of different hardness [31]. However, the measured forces with both methods are dependent not only on the bond strength but also on grit toughness, so the measured grade indication cannot be generally adopted as an intrinsic wheel grade property.

Numerous other wheel grade tests have been proposed, but only a few have been used to a significant extent. Two relatively simple ones, which were adopted many years ago in industry, measure penetration depths due to a rotating chisel driven vertically into the wheel or by sandblasting under standard conditions. While these tests provide a relative indication of bond strength and its variation over the wheel surface, the results are difficult to physically interpret.

A more fundamental parameter which has been proposed for characterizing wheel grade is the elastic modulus [28]. Relatively simple methods have been developed for determining the elastic modulus based upon measuring the natural frequency of a grinding wheel excited by impact [28, 32]. For a disk (wheel) of outer diameter d_s, with a central hole of bore diameter d_o, the relationship between the elastic modulus E and the frequency f for the two nodal diameter vibration mode is given to a good approximation for $d_o/d_s < 0.25$ by [28]:

$$E = \frac{1.07(1 - v^2)\rho \; d_s^4 \; f^2}{b^2\left[1 - \left(\dfrac{d_o}{d_s}\right)^2\right]} \tag{2-9}$$

where ρ is the mass density, v is Poisson's ratio, and b is the disk thickness.

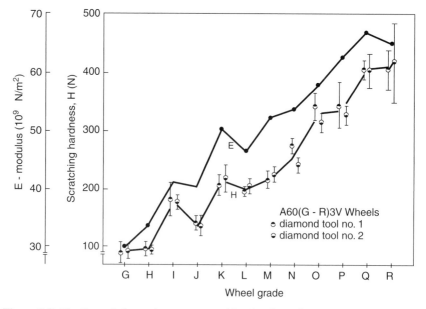

Figure 2-9 Elastic modulus and groove-scratching hardness for a series of vitrified
wheels of differing grades [28].

Experimental results are presented in Figure 2-9 for both the elastic modulus and the groove-scratching hardness for a series of vitrified wheels of differing grades [28]. Both tests show the same relative trend of increasing 'hardness' with letter grade. Good correlation was also found with the popular sandblast hardness test. Advantages of the sonic test are that it does not consume any wheel, is much simpler to perform, and the measured modulus is insensitive to grit toughness. The discontinuities observed in the results in Figure 2-9 indicate deviations in wheel hardness.

On the basis of these results, it would appear that a rational wheel grade scale might be based upon elastic modulus. From tests on aluminum oxide wheels over a range of compositions, it has been found that iso-modulus lines on a ternary phase diagram are as shown in Figure 2-10, which are somewhat different than the iso-grade lines in Figure 2-8. Although this proposed grading system has not been adopted, the sonic testing of grinding wheels has become popular mainly as a tool for quality control. Wheel manufacturers use this method for monitoring their production process, and some wheel users have adopted it for acceptance testing and matching of wheels. The test method is supposedly applicable to both vitrified- and resinoid-bond wheels [28], although it is reported not to function well on resinoid wheels because of their lower elastic module and higher damping [32]. Vitrified wheels tend to display more variability in grade and grinding performance,

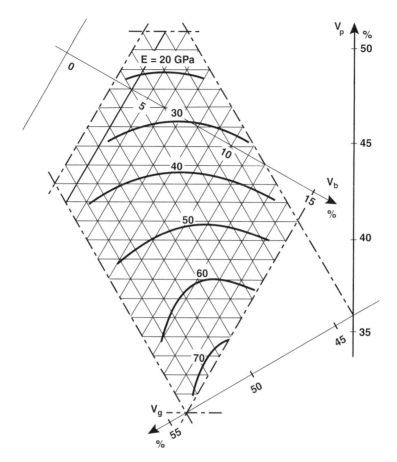

Figure 2-10 Iso-modulus loci for vitrified wheels shown on the ternary phase diagram [28].

and the use of this test method has assisted wheel manufacturers in their efforts to provide more consistent products.

Aside from the wheel grade, another important wheel property is strength. Grinding wheels are operated at speeds which generate high stresses due to centrifugal loading. In order to ensure the safety of the operator and the machine, it is essential that wheel speeds be maintained within safe limits. For this reason, maximum operating speeds are clearly indicated on all grinding wheels, and the wheels are proof tested at faster speeds (minimum 1.2-1.5 times their rated speed depending on bond type) to ensure safety [33]. Operating speeds on grinding machines have been historically limited by wheel strength, rather than by the process itself. The search for more efficient grinding methods by employing higher wheel speeds prompted the development of stronger wheels with higher bursting speeds. Some approaches which have been explored for raising allowable wheel speeds include wheel reinforcement, segmental wheel designs, solid center wheels, higher strength bonds, and higher core strengths.

Speed and safety considerations are an especially important factor with vitrified wheels. Since these wheels are brittle bodies, they can be expected to burst when the rotationally induced tensile stress reaches a critical value, releasing loose fragments with only a negligible portion of their kinetic energy absorbed by fracture. Considering a grinding wheel in the form of a disk of diameter d, with bore diameter d_0 rotating at a peripheral velocity v_s, the maximum tensile stress developed, which is a tangential (hoop) stress at the bore diameter, can be written as [34]

$$\sigma_{max} = \rho \left(\frac{3 + v}{4} \right) \left[1 + \left(\frac{1 - v}{3 + v} \right) \left(\frac{d_o}{d_s} \right)^2 \right] v_s^2 \qquad (2\text{-}10)$$

where ρ is the mass density and v is Poisson's ratio. For typical ratios of bore to outer diameter, the second term within the brackets is generally much less than unity, which means that the maximum tensile stress developed is relatively insensitive to the bore and hole diameters and proportional to the velocity squared.

Aside from centrifugal forces, additional stresses are induced by wheel clamping and grinding forces [35, 36]. In general, these effects appear to be much less significant than the rotational stresses, although bursting speeds of improperly clamped wheels can be significantly lowered [37]. With thin resinoid (reinforced) wheels, sideways loading tending to bend the wheel may be a more significant factor than centrifugal loading. The bending strength depends on the orientation of the fiber-reinforcing layers [38]. The grinding fluid may also lower resinoid wheel strength [27].

Some results for bursting speeds of non-reinforced vitrified grinding wheels, compiled by one manufacturer over a number of years, are summarized in Figure 2-11. According to Eq. (2-10), the observed bursting speeds of 90 m/s to 130 m/s would correspond to maximum tensile stresses of about 13 MPa to 26 MPa, which are comparable to measured short-time tensile strengths [39]. These bursting speeds far exceed the normal operating limit of 30 m/s with these wheels, and the relative margin of safety is even bigger when compared in terms of strength. It should be noted, however, that these results are for average bursting speeds. As is typical of many brittle materials, the tensile strength of vitrified wheels is found to exhibit significant scatter which can be descried by a Weibull distribution [39]. Furthermore, the grinding fluid and moisture in the air may also degrade wheel strength over a period of time. The higher strengths observed with harder-grade wheels (Figure 2-11) are a consequence of the higher bond content which provides bigger bond bridges with larger cross-sections. Finer-grit wheels may be

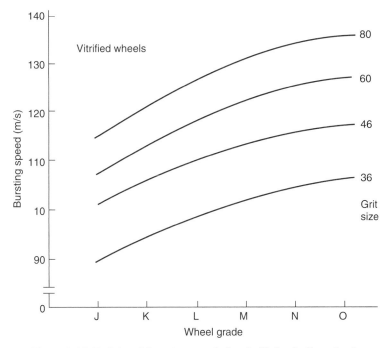

Figure 2-11 Peripheral bursting speeds for vitrified grinding wheels.

stronger owing to proportionally smaller defect sizes for initiation of brittle fracture. More recent bending tests show somewhat higher strengths [40], which enables wheel use at higher speeds. Vitrified bonded abrasive wheels are readily available nowadays for operation at speeds up to 120 m/s.

A more fundamental approach to ensuring reliability of grinding wheels against failure combines fracture mechanics, finite element stress analysis, and Weibull statistics [41]. With this method, bonded abrasive specimens are tested in four-point bending to measure the fracture strength, its statistical variability, and its time-dependent degradation (static fatigue). The stress distribution in the grinding wheel during operation is obtained using a finite element analysis. The overall time-dependent probability of wheel failure can be statistically predicted by coupling the stress distribution over the wheel with the fracture-strength results, while taking into account the increased probability of encountering strength-impairing flaws in larger stressed volumes. The results of this analysis provide a rational basis for overspeed proof testing to ensure that a grinding wheel will not fail during its useful life.

Bursting of a wheel operating within its rated speed limit is an extremely unlikely event. The causes of wheel failure are generally associated with overspeeding, shock loading, improper mounting, and mishandling [37]. With proper guarding, a wheel failure should not pose a serious danger to the operator.

REFERENCES

1. 'Markings for Identifying Grinding Wheels and Other Bonded Abrasives', American National Standard ANSI B74.13-1982.
2. 'Specifications for the Size of Abrasive Grain-grinding Wheels, Polishing and General Industrial Uses', American National Standard ANSI B74.12-1976 (R1982).
3. 'Specifications for Grading of Abrasive Microgrits', American National Standard ENSI B74.10 (R1983).
4. Malkin, S. and Anderson, R. B., 'Active Grains and Dressing Particles in Grinding', *Proceedings of the International Grinding Conference*, Pittsburgh, 1972, p. 161.
5. Coes, L., Jr., Abrasives, Springer-Verlag, New York, 1971, Chapter 3.
6. 'Checking the Size of Diamond Abrasive Grain', American National Standard ANSI B74.10 - 1982.
7. 'Specifications for Grading of Diamond Powder in Sub-sieve Sizes', American National Standard ANSI B74.20–1981.
8. Ueltz, H. G. G., 'Abrasive Grains—Past, Present, and Future', *Proceedings of the International Grinding Conference,* Pittsburgh, 1972, p. 1.
9. 'Ball Mill Test for Friability of Abrasive Grain', American National Standard B7418 1965.
10. Cadwell, D. E. and Duwell, E. J., 'Evaluating Resistance of Abrasive Grits to Comminution', *Amer. Ceram. Soc. Bull.*, 39, 1960, p. 663.
11. Brecker, J. N., Komanduri, R. and Shaw, M. C., 'Evaluation of Unbonded Abrasive Grains', *Annals of the CIRP*, 22/2, 1973, p. 219.
12. Brecker, J. N., 'The Fracture Strength of Abrasive Grains,' *Trans. ASME, J. Eng. Ind.,* 96, 1974, p. 1253.
13. Komanduri, R. and Shaw, M. C., 'Scanning Electron Microscope Study of Surface Characteristics of Abrasive Materials', *Trans. ASME, J. of Eng. Matls. Tech.,* 96, 1974, p. 145.
14. Coes, L., Jr., *Abrasives*, Springer-Verlag, New York, 1971, Chapter 8.
15. Lange, F. F., Transformation Toughing - 4. Fabrication, Fracture Toughness, and Strength of Al_2O_3-ZrO_2 Composites, *J. Mater. Sci.*, 17, 1982, p. 247.
16. Hever, A. H., 'Transformation Toughening in ZrO_2-containing Ceramics', *J. Am. Ceram. Soc.*, 70, 1987, p. 689.
17. Leitheiser, M. A. and Sowman, H. G., 'Non-fused Aluminum Oxide-based Abrasive Mineral', U.S. Patent 4,314,827, February 9, 1982.
18. Yarbrough, W. A., Bauer, R, van de Merwe, R. H., and Cottringer, T. E., Abrasive Material and Method, U.S. Patent 5,383,945, January 24, 1995.
19. Bange, D. W. and Orf, N., Sol Gel Abrasive Makes Headway, *Tools and Production*, March 1998.
20. Bundy, F. P., Hall, H. T., Strong, H. M. and Wentorf, R. F., Jr., 'Man-made Diamonds', Nature, 176, 1955, p. 55.
21. Coes, L., Jr., Abrasives, Springer-Verlag, New York, 1971, Chapter 10.

22. Komanduri, R. and Shaw, M. C., 'Surface Morphology of Synthetic Diamonds and Cubic Boron Nitride', *Int. J. Mach. Tool Des. Res.*, 14, 1974, p. 63.

23. O'Donovan, K. H., 'Synthetic Diamond', *Annals of the CIRP*, 24/1, 1975, p. 265.

24. Wentorf, R. H., 'Synthesis of Cubic Form of Boron Nitride', *J. Chem. Phys.*, 34, 1961, p. 809.

25. Brecker, J. N., 'Analysis of Bond Formation in Vitrified Abrasive Wheels', *J. Amer. Ceram. Soc.*, 57, 1974, p. 486.

26. Kingery, W. D., Sidhwa, A. P. and Waugh, A., 'Structure and Properties of Vitrified Bonded Abrasives', *Amer. Ceram. Soc. Bull.*, 42, 1963, p. 297.

27. Ellendman, M., 'How Coolants Affect the Performance of Resin-bonded Abrasive Wheels', *Machine and Production Engineering*, 115, No. 2975, 1969, p. 812.

28. Peters, J., Snoeys, R. and Decneut, A., 'Sonic Testing of Grinding Wheels', *Proceedings of the Ninth International Machine Tool Design and Research Conference*, 1968, p. 1113.

29. Peklenik, J. and Opitz, H., 'Testing of Grinding Wheels', *Proceedings of the Third International Machine Tool Design and Research Conference*, 1962, p. 163.

30. Colwell, L. V., Lane, R. O. and Soderlund, K. N., 'On Determining the Hardness of Grinding Wheels', *J. of Eng. for Ind., Trans. ASME*, 84, 1962, p. 113.

31. Peklenik, J., Lane, R. and Shaw, M. C., 'Comparison of Static and Dynamic Hardness of Grinding Wheels', *J. of Eng. for led., Trans. ASME*, 86, 1964, p. 294.

32. Brecker, J. N., 'Grading Grinding Wheels by Elastic Modulus', *Proceedings, First North American Metalworking Research Conference*, Vol. 3, 1973, p. 149.

33. 'Safety Requirements for the Use, Care and Protection of Abrasive Wheels', American National Standard ANSI B 7.1-1978.

34. Timoshenko, S. and Goodier, J. N., *Theory of Elasticity*, 2nd edn, McGraw-Hill, New York, 1951, p. 71.

35. Miyamoto, H., Shibahara, M., Oda, J and Kazama, E., 'Three-dimensional Stress Analysis of Grinding Wheel', *Bull. Japan Soc. of Prec. Engg*, 4, 1970, p. 79.

36. Oda, J., Shibahara, M. and Miyamoto, H., 'Calculation of Stresses in Grinding Wheel Caused by Grinding Force', *Bull. Japan Soc. of Prec. Engg*, 6, 1972, p. 25.

37. Sabberwal, A. J. P., 'Review of Codes of Practice on Safety in Grinding', Annals of the CIRP, 21/2, 1972, p. 187.

38. Rajagopal, S. and Kalpakjian, S., 'Properties of Reinforced Abrasive Disks in Flexure', *J. Eng. Ind., Trans. ASME*, 99, 1977, p. 318.

39. Yamamoto, A., 'Strength and Static Fatigue of Vitrified Grinding Wheels under Various Environments. II', *Bull. Japan Soc. of Prec. Engg*, 10, 1976, p. 45.

40. Stabenow, R., 'Hardness Influencing Effects in Grinding Wheels', *1st European Conference on Grinding*, Aachen, Germany, 2003, p. 5-1.

41. Ritter, J. E., Jr., 'Assuring Mechanical Reliability of Ceramic Components', *J. Ceramic Society Japan*, 93, 1985, p. 341.

BIBLIOGRAPHY

Armarego, E. J. A. and Brown, R. H., *The Machining of Metals*, Prentice-Hall, Englewood Cliffs, NJ, 1969, Chapter 11.

Borkowski, J. and Szyma_ski, A, *Uses of Abrasives and Abrasive Tools*, Ellis Horwood, Chichester, 1992, Chapter 1–4.

Coes, L., Jr., *Abrasives*, Springer-Verlag, New York, 1971.

DeVries, R. C., 'Cubic Boron Nitride: Handbook of Properties', GE Report No. 72CRD178, 1972.

Drozda, T. J. and Wick, C., Eds, *Tool and Manufacturing Engineers Handbook*, Vol. 1, Machining, 4th edition, SME, Dearborn, 1983, Chapter 11.

Jacobs, F. B., *Abrasives and Abrasive Wheels*, Henley, New York, 1919.

Komanduri, R. and Shaw, M. C., 'Scanning Electron Microscope Study of Surface Characteristics of Abrasive Materials', *Trans. ASME, J. of Eng. Matls. Tech.*, 96, 1974, p. 145.

Komanduri, R. and Shaw, M. C., 'Surface Morphology of Synthetic Diamonds and Cubic Boron Nitride', *Int. J. Mach. Tool Des. Res.*, 14, 1974, p. 63.

Lewis, K. B. and Schleicher, W. F., *The Grinding Wheel*, 3rd edn, The Grinding Wheel Institute, Cleveland, 1976.

Metzger, J. L., *Superabrasive Grinding*, Butterworths, London, 1986, Chapters 3 and 4.

Moser, M, *Microstructure of Ceramics: Structure and Properties of Grinding Tools*, Akadémiai Kiadó, Budapest, 1980.

O'Donovan, K. H., 'Synthetic Diamond', *Annals of the CIRP*, 24/1, 1975, p. 265.

Peters, J., Snoeys, R. and Decneut, A., 'Sonic Testing of Grinding Wheels', *Proceedings of the Ninth International Machine Tool Design and Research Conference*, 1968, p. 1113.

Spur, H. C. and Stark, C., 'Methods for Testing Grinding Wheel Quality', *Proceedings, Twelfth North American Manufacturing Research Conference*, 1984, p. 339.

Stewart, I. J., 'On the-Safety of High Speed Grinding', *Proceedings of the International Grinding Conference*, Pittsburgh, 1973, p. 649.

Ueltz, H. F. G., 'Abrasive Grains—Past, Present, and Future', *Proceedings of the International Grinding Conference*, Pittsburgh, 1973, p. 1.

Whitney, E. D. and Shepler, R. E., 'Ceramics in Abrasive Processes', *Materials Science Research,* 7, 1974, p. 167.

Chapter 3

Grinding Geometry and Kinematics

3.1 INTRODUCTION

Material removal by grinding occurs mainly by a chip formation process, similar to that of other machining methods such as turning or milling, but on a much finer scale. While the cutting-tool geometry and its interaction with the workpiece is well defined for most machining processes, the situation for grinding is not readily discernible. A grinding wheel has a multitude of geometrically undefined cutting points (tools) which are irregularly distributed on its working surface and which are presented to the workpiece at random orientations and positions. Consequently, there is significant variation in the cutting geometry from point to point.

In spite of these complexities, there have been many attempts to analyze chip geometry, beginning almost 90 years ago [1, 2], usually in terms of what occurs at a 'typical' or 'average' cutting point rather than at each separate cutting point. Some of these analyses also attempt to describe the point-to-point variability in cutting geometry by using either statistical models or computer simulations to describe how the non-uniform wheel surface interacts with the workpiece.

This chapter presents mathematical analyses of the cutting geometry during grinding, arising from consideration of the kinematic interactions between the topography of the grains in the wheel surface and the workpiece. Aside from leading to fundamental parameters, such as the depth of cut taken by a cutting point (undeformed chip thickness) and the size of the grinding zone (contact length), it also provides a basis for analyzing the mechanisms of abrasive interactions with the workpiece (Chapter 5), the grinding temperatures (Chapters 6–8), and the geometry of the ground surface generated (Chapter 10).

3.2   GEOMETRICAL WHEEL-WORKPIECE CONTACT LENGTH

The grinding geometry is illustrated in Figure 3-1 for straight, external, and internal cylindrical grinding. For straight surface grinding (Figure 3-1(a)), a wheel of diameter d_s rotating with a peripheral velocity v_s takes a wheel depth of cut a from the workpiece as it translates past at velocity v_w. A similar situation applies to cylindrical grinding (Figures 3-1(b) and 3-1(c)) with the workpiece velocity v_w obtained by rotation instead of translation. For straight surface grinding, the depth of cut a corresponds to the machine downfeed, whereas in cylindrical grinding it is equal to the radial infeed at velocity v_f during one revolution of the workpiece ($a = \pi d_w v_f/v_w$). For cylindrical grinding, typical depths of cut are $a \approx 2–20 \ \mu m$, and for straight surface grinding $a \approx 10–50 \ \mu m$. Obviously, the relative size of the wheel depth of cut is highly exaggerated in Figure 3-1. Typical wheel velocities are $v \approx 30 \ m/s$, although faster velocities even up to 120 m/s are used in some extreme cases, and somewhat slower velocities may be used for some difficult-to-grind materials. The workpiece velocity v_w is always much slower than the wheel velocity, the ratio v_s/v_w being typically in the range 100–200 for straight surface grinding and 5–100 for cylindrical grinding.

Penetration of the grinding wheel into the workpiece results in an apparent area of contact where the grinding action occurs. The arc length of the contact area is indicated by l_c in Figures 3-1(a)–(c). Neglecting motions and deformations of the wheel and workpiece, the arc length of contact for each type of grinding can be generally expressed as

$$l_c = AB = \frac{d_s \theta}{2} \tag{3-1}$$

For straight surface grinding (Figure 3-1(a)) it can be readily shown that

$$\theta = \cos^{-1}\left(1 - \frac{2a}{d_s}\right) \tag{3-2}$$

Since $2a \ll d_s$, the small-angle approximation would apply

$$\cos \theta = 1 - \frac{\theta^2}{2} \tag{3-3}$$

which combined with Eqs. (3-1) and (3-2) leads to the result

$$l_c = (ad_s)^{1/2} \tag{3-4}$$

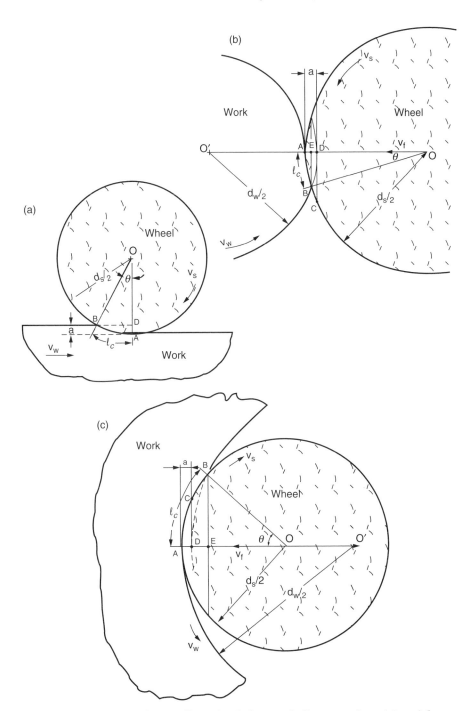

Figure 3-1 *Illustration of two—dimensional plunge grinding operations: (a) straight surface, (b) external cylindrical, and (c) internal cylindrical.*

This expression for the arc length can be shown to be identical to the chord length AB. Since motions as well as deformations are neglected, the parameter l_c is often referred to as the static contact length.

The same analysis can also be applied to external and internal cylindrical grinding (Figures 3-1(b) and 3-1(c)). As with straight grinding, the wheel depth of cut a in each case is equal to AD and the arc length of contact $l_c \approx$ AB. By analogy with straight grinding it is apparent that Eq. (3-4) applied to external or internal grinding would result in the chord length AC. The actual contact length AB is less than AC in the case of external grinding, and greater than AC in the case of internal grinding. The opposing curvatures of the wheel and workpiece in external grinding reduce the contact length, whereas the conforming curvatures with internal grinding elongate the contact length.

As with straight surface grinding, the contact length AB for external and internal grinding (Figures 3-1(b) and (c)) can be approximated by the chord length AB. In both cases, it can be seen by analogy with Eq. (3-4) that

$$l_c \approx AB = [(AE)d_s]^{1/2} \tag{3-5}$$

From the geometry of external grinding (Figure 3-1(b)):

$$AE = \frac{a}{1 + d_s/d_w} \tag{3-6}$$

and internal grinding (Figure 3-1(c)):

$$AE = \frac{a}{1 - d_s/d_w} \tag{3-7}$$

Therefore, the results for straight, external and internal grinding can all be combined into a single equation:

$$l_c = (ad_e)^{1/2} \tag{3-8}$$

where d_e is called the 'equivalent wheel diameter' and is defined by

$$d_e \equiv \frac{d_s}{1 \pm d_s/d_w} \tag{3-9}$$

The plus sign in the denominator is for external grinding, the minus sign for internal grinding, and $d_w = \infty$ for straight grinding ($d_e = d_s$). The equivalent diameter for external grinding it is always less than both d_s and d_w. For internal grinding it is always bigger than d_s, and also bigger than d_w provided that $d_s > d_w/2$. For example, with internal grinding with $d_s = 2/3\ d_w$, the equivalent wheel diameter is three times the wheel diameter and twice the workpiece diameter. Typically, the contact length l_c ranges from about 0.1 mm up to 10 mm, although it can be somewhat bigger for operations which utilize very large depths of cut such as creep feed grinding.

3.3   CUTTING PATH

For the purpose of analyzing the cutting geometry, it is convenient to liken the grinding wheel action to that of a milling cutter, with the cutting points corresponding to cutter teeth. For the idealized wheel, the cutting points around the wheel periphery are equally spaced apart by a distance L. The situation is illustrated in Figure 3-2(a) for straight grinding (plain horizontal milling), and for cylindrical external grinding in Figure 3-2 (b). In each case, the wheel velocity v_s is shown in the opposite direction to the workpiece velocity v_w at the grinding zone. This is known as 'up-grinding', when the tangential directions of motion of wheel and workpiece are opposed. The case in which these motions are in the same direction is known as 'down-grinding'. As previously mentioned, the actual wheel depth of cut is very small and the wheel velocity is much faster than the workpiece velocity, so the dimensions shown in the grinding zone are highly distorted.

A cutting point in up-grinding begins its contact with the workpiece at point F′, and follows the curved path to point A′. For down-grinding, it begins at A′ and ends at F′. The cutting path F′B′CA′ relative to the workpiece is a trochoid consisting of the superposition of the circular motion at velocity v_s, and tangential motion along the workpiece at velocity v_w. The previous cutting point followed the same geometrical path shape but displaced along the workpiece surface by the distance AA′ which is the feed per cutting point s. The feed per cutting point is equal to the product of the workpiece velocity v_w and the time between successive cuts (L/v_s):

$$s = \frac{Lv_w}{v_s} \qquad (3\text{-}10)$$

The undeformed chip for up-grinding is shown by the cross-hatched area AF′A′ for each case in Figure 3-2.

Relative to an *x-y* coordinate system with its origin at B′ fixed to the workpiece in Figure 3-2 (a), it can be shown [3] that the trochoidal path of a cutting point initially at the origin moves horizontally:

$$x = \frac{d_s}{2} \sin\theta' \pm \frac{d_s}{2} \frac{v_w}{v_s} \theta' \qquad (3\text{-}11)$$

(a)

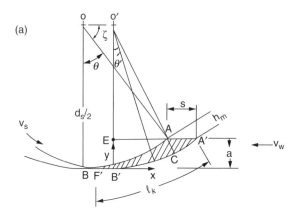

(b)

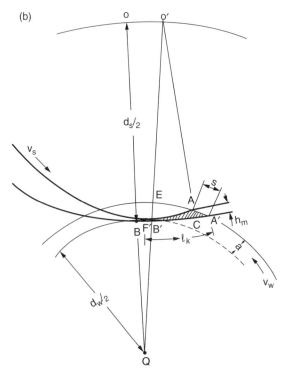

Figure 3-2 Undeformed chip geometry for (a) straight surface and (b) external cylindri-cal grinding. The cutting path of length l_k begins at F' and ends at A' as the wheel center moves from O to O'.

and vertically

$$y = \frac{d_s}{2}(1 - \cos \theta') \tag{3-12}$$

as the wheel rotates through the angle θ'. The plus sign in Eq. (3-11) refers to up-grinding as shown, and the negative to down-grinding with the workpiece velocity in the opposite direction. Since θ' is a very small angle ($\theta' < \theta$), Eqs. (3-11) and (3-12) can be simplified to

$$x = \left(1 \pm \frac{v_w}{v_s}\right)\frac{d_s}{2}\theta' \tag{3-13}$$

and

$$y = \frac{d_s\theta'^2}{4} \tag{3-14}$$

Eliminating θ' leads to the result for the cutting path

$$y = \frac{x^2}{\left[d_s\left(1 \pm \dfrac{v_w}{v_s}\right)^2\right]} \tag{3-15}$$

which is a parabola instead of a trochoid.

An equation for the trochoidal cutting path has also been derived for external and internal cylindrical grinding [4]. As with straight grinding, the trochoidal cutting path can be approximated by a parabola:

$$y = \frac{x^2}{D} \tag{3-16}$$

where the parameter D is given for external grinding by

$$D = \frac{d_s\left(1 \pm \dfrac{v_w}{v_s}\right)^2}{\left[1 \mp \dfrac{v_w d_s}{v_s d_w}\left(2 \pm \dfrac{v_w}{v_s}\right)\right]} \tag{3-17}$$

and for internal grinding by

$$D = \frac{d_s\left(1 \pm \dfrac{v_w}{v_s}\right)^2}{\left[1 \pm \dfrac{v_w}{v_s}\dfrac{d_s}{d_w}\left(2 \pm \dfrac{v_w}{v_s}\right)\right]} \tag{3-18}$$

When double signs appear, the upper is for up-grinding and the lower for down-grinding. For straight grinding ($d_w = \infty$), both of the above expressions for D reduce to the denominator of Eq. (3-15).

The radius of curvature of the parabolic cutting path at its origin ($\theta' = 0$) is equal to $D/2$, and it grows by a negligible amount up to $\theta' = \theta$. Furthermore, the radius of curvature is bigger than the wheel radius for up-grinding and smaller than the wheel radius for down-grinding. It has been proposed [4] that the degree of non-conformity between the wheel radius and the cutting path radius R can be defined by their curvature difference:

$$\Delta = \frac{2}{d_s} - \frac{1}{R} \tag{3-19}$$

which for the parabolic approximation to the cutting path is

$$\Delta = \frac{2}{d_s} - \frac{2}{D} \tag{3-20}$$

Combining Eqs. (3-17) and (3-18) with Eq. (3-20):

$$\Delta = \frac{2\left(\dfrac{v_w}{v_s}\right)^2 \pm 4\left(\dfrac{v_w}{v_s}\right)}{d_e\left(1 \pm \dfrac{v_w}{v_s}\right)^2} \tag{3-21}$$

which for $v_w \ll v_s$ can be simplified to

$$\Delta = \pm \frac{4v_w}{d_e v_s} \tag{3-22}$$

where the plus sign is for up-grinding and the minus sign for down-grinding. Eqs. (3-21) and (3-22) apply to straight, external, and internal grinding with the appropriate definition of equivalent diameter d_e according to Eq. (3-9).

The length of the cutting path F′B′A′ for straight grinding (Figure 3-2(a)) can be obtained from the equation of the cutting-path motion. The length F′B′ in each case can be taken as half the feed per cutting point. The total cutting-path length l_k can be expressed as

$$l_k = \int_0^\theta dl_k + \frac{s}{2} \tag{3-23}$$

where

$$dl_k \equiv \left[\left(\frac{dx}{d\theta'} \right)^2 + \left(\frac{dy}{d\theta'} \right)^2 \right]^{1/2} d\theta'$$

Substituting for x and y from Eqs. (3-13) and (3-14) and integrating lead to the result

$$l_k = \left(1 \pm \frac{v_w}{v_s} \right) \frac{d_s \theta}{2} + \frac{\theta^3}{6 \left(1 \pm \dfrac{v_w}{v_s} \right)} + \frac{s}{2} \tag{3-24}$$

Since θ is a small angle, the second term is negligible compared with the first one, and the quantity $d_s\theta/2$ corresponding to the arc length AB can be approximated by its chord length. Therefore

$$l_k = \left(1 \pm \frac{v_w}{v_s} \right)(ad_s)^{1/2} + \frac{s}{2} \tag{3-25}$$

or

$$l_k = \left(1 \pm \frac{v_w}{v_s} \right) l_c + \frac{s}{2} \tag{3-26}$$

where l_c is given by Eq. (3-4). Repeating the analysis for external and internal grinding leads to the same result as Eq. (3-26) with l_c given by Eq. (3-8) with d_s replaced by d_e. The cutting path is longer for up-grinding (plus sign) than for down-grinding (minus sign) although the difference is extremely small for most practical speed ratios v_w/v_s. Also, the contribution of $s/2$ to the total path length may be negligible, in which case

$$l_k \approx l_c = (ad_e)^{1/2} \tag{3-27}$$

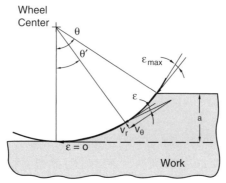

Figure 3-3   Illustration of infeed angle ε of material flow relative to a cutting point on the wheel periphery.

For this reason, there is often no distinction made between the contact length l_c and the cutting-path length l_k.

The cutting-path length l_k as given by Eq. (3-26) may be considered as a kinematic correction to the static contact length l_c, in which case l_k is called the kinematic contact length. Conversely, l_c may be considered to be a static approximation to the cutting-path length or undeformed chip length.

As a cutting point on the periphery of the rotating grinding wheel passes through the grinding zone, its motion will interfere with that of the moving workpiece, which is of course a necessary condition for cutting. The degree of interference at any location along the cutting path can be defined by the infeed angle ε between the peripheral velocity vector and the workpiece velocity vector, as illustrated in Figure 3-3 for straight grinding. The infeed angle increases from $\varepsilon = 0$ at the bottom of the cutting path ($\theta' = 0$) to a maximum value $\varepsilon = \varepsilon_{max}$ at the top of the cutting path ($\theta' = \theta$).

For the purpose of deriving an expression for ε, it is convenient to imagine the wheel as being stationary, in which case the material velocity vector relative to a cutting point on the wheel surface can be obtained as the workpiece velocity vector minus the wheel velocity vector. With reference to Figure 3-3, the radial and tangential components of this relative velocity vector, v_r, and v_θ, are given at an arbitrary intermediate position θ' by

$$v_r = v_w \sin \theta' \qquad (3\text{-}28)$$

and

$$v_\theta = v_s \pm v_w \cos \theta' \tag{3-29}$$

The infeed angle ε is

$$\tan \varepsilon = \frac{v_r}{v_\theta} \tag{3-30}$$

which combined with Eqs. (3-28) and (3-29) leads to

$$\tan \varepsilon = \frac{v_w \sin \theta'}{v_s \pm v_w \cos \theta'} \tag{3-31}$$

The plus sign in Eqs. (3-29) and (3-31) is for up-grinding, and the minus sign for down-grinding. Using the small-angle approximation ($sin \theta' \approx \theta'$) and neglecting $v_w \cos\theta'$ compared to v_s ($v_w \ll v_s$), the expression for tan ε is simplified to

$$\tan \varepsilon = \frac{v_w \theta'}{v_s} \tag{3-32}$$

The maximum infeed angle ε_{max} at $\theta' = 0$, the end of the cutting path for up-grinding and the beginning for down-grinding, can be obtained in terms of the grinding parameters by combining Eq. (3-32) with Eqs. (3-2) and (3-3):

$$\tan \varepsilon_{max} = 2\frac{v_w}{v_s}\left(\frac{a}{d_s}\right)^{1/2} \tag{3-33}$$

This relationship was derived for straight surface grinding, but it can be generalized to also include external and internal cylindrical grinding by replacing the wheel diameter d_s by the equivalent diameter d_e:

$$\tan \varepsilon_{max} = 2\frac{v_w}{v_s}\left(\frac{a}{d_e}\right)^{1/2} \tag{3-34}$$

Since ε is an extremely smell angle, even much smaller than θ', its average value $\bar\varepsilon$ half way along the contact length ($\theta' = \theta/2$) can be written as

$$\tan \bar\varepsilon = \frac{v_w}{v_s}\left(\frac{a}{d_e}\right)^{1/2} \tag{3-35}$$

3.4   MAXIMUM CUTTING DEPTH (UNDEFORMED CHIP THICKNESS)

The maximum cutting depth (undeformed chip thickness) taken by a cutting point is indicated by h_m in Figure 3-2. This parameter is often referred to as the 'grain depth of cut', but since one grain may have multiple cutting points (Chapter 4), this name may be misleading. For the idealized wheel with cutting points equally spaced around the wheel periphery, an expression for h_m can be obtained for the case of a trochoidal cutting path, and a similar result can be obtained for a parabolic path. However, such analyses are extremely cumbersome, and the physical interpretation becomes muddled.

It is sufficient for virtually all grinding conditions to approximate the cutting path by a circular arc. This assumption implies an intermittent motion in which the workpiece remains stationary during an individual cut, and then moves suddenly by the distance OO′ before the next cutting point engages. For straight grinding (Figure 3-2 (a)) the maximum undeformed chip thickness h_m corresponds to the length AC, so that

$$h_m = O'C - O'A = \frac{d_s}{2} - O'A \qquad (3\text{-}36)$$

From the triangle OO′ A, where OA is equal to the wheel radius and OO′ is equal to s, the length O′A can be written as

$$O'A = \left[\left(\frac{d_s}{2} \right)^2 + s^2 - sd_s \cos \xi \right]^{1/2} \qquad (3\text{-}37)$$

or since ξ and θ comprise a right angle,

$$O'A = \left[\left(\frac{d_s}{2} \right)^2 + s^2 - sd_s \left(1 - \cos^2\theta \right)^{1/2} \right]^{1/2} \qquad (3\text{-}38)$$

From triangle OAB or Eq. (3-2)

$$\cos \theta = 1 - \frac{2a}{d_s} \qquad (3\text{-}39)$$

which combined with Eq. (3-38) leads to

$$O'A = \frac{d_s}{2} \left[1 - \left(\frac{8s}{d_s} \left(\frac{a}{d_s} \right)^{1/2} \left(1 - \frac{a}{d_s} \right)^{1/2} - \frac{4s^2}{d_s^2} \right) \right]^{1/2} \qquad (3\text{-}40)$$

The second of the two terms within the brackets is much less than unity, so this expression can be further simplified and combined with Eq. (3-36). The resulting expression for the undeformed chip thickness is

$$h_m = 2s\left(\frac{a}{d_s}\right)^{1/2}\left(1 - \frac{a}{d_s}\right)^{1/2} - \frac{s^2}{d_s} \tag{3-41}$$

and with $a/d_s \ll 1$,

$$h_m = 2s\left(\frac{a}{d_s}\right)^{1/2} - \frac{s^2}{d_s} \tag{3-42}$$

The analysis can be repeated for external and internal cylindrical grinding. For external cylindrical grinding (Figure 3-2 (b)) the geometry is similar to that for straight grinding, except that the wheel axis moves relative to the workpiece axis along a circular path of radius $(d_s+d_w)/2$ about the workpiece center Q. For a feed per cutting point s at the workpiece surface, the arc length OO′ traveled by the wheel center between successive cutting-point engagements is

$$OO' = s\left(\frac{d_w + d_s}{d_w}\right) \tag{3-43}$$

Similarly for internal grinding

$$OO' = s\left(\frac{d_w - d_s}{d_w}\right) \tag{3-44}$$

Expressions for h_m can be derived from Eq. (3-38) for external and internal cylindrical grinding by substituting the appropriate value of OO′ (Eq. (3-34) or Eq. (3-44)) in place of s and replacing 'a' in Eq. (3-39) by the appropriate value of AE (Eq. (3-6) or Eq. (3-7)). This can be readily shown to be identical to replacing d_s by the equivalent diameter d_e, as given by Eq. (3-9). Therefore Eq. (3-2) for the undeformed chip thickness can be rewritten more generally as

$$h_m = 2s\left(\frac{a}{d_s}\right)^{1/2} - \frac{s^2}{d_e} \tag{3-45}$$

or substituting for s from Eq. (3-10):

$$h_m = 2L\frac{v_w}{v_s}\left(\frac{a}{d_s}\right)^{1/2} - \frac{L^2}{d_e}\left(\frac{v_w}{v_s}\right)^2 \tag{3-46}$$

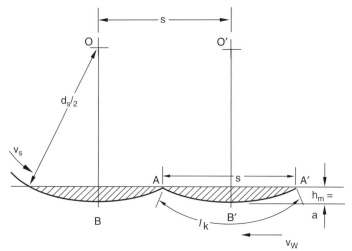

Figure 3-4 Undeformed chip shapes $h_m = a$.

Again, the equivalent diameter allows for the use of one equation to describe all three cases.

For this analysis, no distinction is made between up-grinding and down-grinding, and Eqs. (3-45) and (3-46) would apply in both cases. With a trochoidal or parabolic cutting path, the chip thickness would be slightly smaller for up-grinding than for down-grinding, although these differences are generally insignificant.

The maximum possible value of h_m is equal to the wheel depth of cut a. The limiting condition where $h_m = a$ is shown in Figure 3-4 (a) for straight grinding. From the geometry with $a/d_s \ll 1$, or from Eq. (3-45), it can be shown that $h_m = a$ when

$$s = (ad_e)^{1/2} \tag{3-47}$$

or substituting for s and $(ad_e)^{1/2}$ from Eqs. (3-10) and (3-8):

$$L = \left(\frac{v_s}{v_w}\right)l_c \tag{3-48}$$

In other words, such a condition ($h_m = a$) would arise when the distance between successive points is at least v_s/v_w times the geometrical arc length of contact, at which point the cutting-path length is

$$l_k \approx \frac{3}{2}l_c = \frac{3}{2}(ad_e)^{1/2} \tag{3-49}$$

With still-longer values of L, h_m would remain the same but l_k would increase up to the point where the successive cuts no longer interact. This latter condition is shown in Figure 3-4(b) when the path length becomes

$$l_k \approx 2l_c = 2(ad_e)^{1/2} \tag{3-50}$$

The magnitude of h_m in Eq. (3-46) is typically an order of magnitude smaller than the wheel depth of cut a. In this case, $h_m << l_c$, so that the undeformed chip shape (AF'A' in Figure 3-2) is very nearly a triangle. For $h_m < a/3$ the second term in Eqs. (3-45) and (3-46) for h_m can be neglected with less than 10% error, in which case

$$h_m = 2s\left(\frac{a}{d_e}\right)^{1/2} \tag{3-51}$$

or

$$h_m = 2L\left(\frac{v_w}{v_s}\right)\left(\frac{a}{d_e}\right)^{1/2} \tag{3-52}$$

In order to calculate the undeformed chip thickness, h_m, we need to estimate the spacing, L, between successive cutting points. Measurements of the wheel topography often provide information on the numbers of cutting points per unit area, C, as will be seen in Chapter 4, and we must derive a conversion formula. If the average effective cutting width for each cutting point is denoted by $\bar{b}_c$, the number of cutting points K around any line on the wheel periphery is equal to C times the area given by the wheel circumference multiplied by the effective cutting width:

$$K = C(\pi d_s \bar{b}_c) \tag{3-53}$$

But since

$$K = \frac{\pi d_s}{L} \tag{3-54}$$

then

$$L = \frac{1}{C\bar{b}_c} \tag{3-55}$$

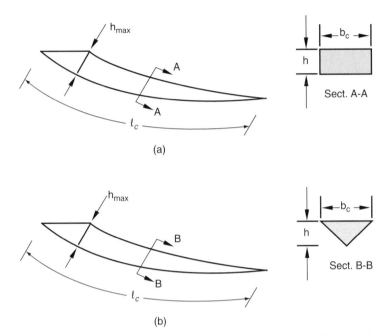

(a)

(b)

Figure 3-5 Undeformed chip with (a) rectangular cross-section and (b) triangular cross-section.

The effective average width $\bar{b}_c$ depends on the maximum undeformed chip thickness and its cross-sectional shape normal to the cutting path. For simplicity, the undeformed chip can be modeled as having a rectangular cross-section as seen in Figure 3-5 (a), whose width b_c is assumed to be proportional to the average undeformed chip thickness h_a:

$$\bar{b}_c = b_c = rh_a \tag{3-56}$$

For $h_m \ll l_c$ the average undeformed chip thickness in this case is half the maximum value so that

$$\bar{b}_c = \frac{rh_m}{2} \tag{3-57}$$

and

$$L = \frac{2}{Crh_m} \tag{3-58}$$

Substituting this result for L into Eq. (3-52) and solving for h_m:

$$h_m = \left[\frac{4}{Cr}\left(\frac{v_w}{v_s}\right)\left(\frac{a}{d_e}\right)^{1/2} \right]^{1/2} \tag{3-59}$$

3.5 UNDEFORMED CHIP THICKNESS-CONTINUITY ANALYSIS

A different approach to calculating the undeformed chip thickness is based upon making a material balance between the volume of the chips produced at the cutting points and the overall removal rate. For continuity, the product of the number of chips generated per unit time and the volume per chip is equal to the volumetric removal rate, or

$$(Cbv_s)V_c = av_w b \tag{3-60}$$

where Cbv_s is the number of chips per unit time for an overall grinding width b, V_c is the volume per chip, and $av_w b$ is the volumetric removal rate. For the chip geometry as in Figure 3-5 (a) with $h_m \ll l_c$, the chip volume is obtained as the product of the average cross-sectional area $(h_a b_c)$ and the length l_c:

$$V_c = h_a b_c l_c \tag{3-61}$$

Combining Eqs. (3-60) and (3-61), substituting for b_c and l_c from Eqs. (3-56), (3-57) and (3-8), and recalling that h_a is half of h_m in this case:

$$h_m = \left[\frac{4}{Cr}\left(\frac{v_w}{v_s}\right)\left(\frac{a}{d_e}\right)^{1/2} \right]^{1/2} \tag{3-62}$$

which is identical to Eq. (3-59).

The continuity analysis can also be applied to other idealized undeformed chip shapes. For example, consider a chip with a triangular cross-section (Figure 3-5 (b)) instead of a rectangular one. In this case, the parameter r is the ratio of chip width to thickness at any point along the cutting path. For $h_m \ll l_c$, the volume of the undeformed chip in Figure 3-5(b) can be approximated, analogous to that of a triangular pyramid, as one-third times the product of the maximum cross-sectional area $(rh_m^2/2)$ and the length l_c:

$$V_c = \frac{rh_m^2 l_c}{6} \tag{3-63}$$

Using this expression for the chip volume in Eq. (3-60) and substituting for l_c from Eq. (3-8) leads to

$$h_m = \left[\frac{6}{Cr} \left(\frac{v_w}{v_s} \right) \left(\frac{a}{d_e} \right)^{1/2} \right]^{1/2} \tag{3-64}$$

which is of the same form but slightly larger than Eq. (3-62). Since Eq. (3-61) would also apply to an undeformed chip with a triangular cross-section, it can be readily shown by equating Eqs. (3-61) and (3-63) that the maximum undeformed chip thickness is $\sqrt{3}$ times the average value ($h_m = \sqrt{3}h_a$), instead of twice the average value with the rectangular undeformed chip cross-section. It will be seen in Chapter 4 that the actual undeformed chip shape appears to be nearly trapezoidal, which is intermediate between rectangular and triangular.

3.6 NON-UNIFORM WHEEL TOPOGRAPHY

For analyzing the cutting path and the undeformed chip geometry in the previous sections, an idealized grinding wheel was assumed with cutting points uniformly distributed over the wheel surface. The actual situation is much more complex, as the cutting points are not equally spaced apart and do not protrude uniformly (Chapter 4). Therefore, the relationships obtained above for the cutting-path length and undeformed chip thickness would represent an 'average' condition.

The influence of a non-uniform wheel topography is illustrated in Figure 3-6 for straight grinding. Successive cutting points take depths of cut a_0, a_1, and a_2, where $a_2 > a_1 > a_0$, with the wheel center at O_0, O_1, and O_2, respectively. The corresponding feeds per cutting point are s_0 (not shown), s_1, and s_2, denoted by the distances $O_0 O_1$, and $O_1 O_2$, respectively. Since $a_0 > a_1$, the cutting point corresponding to wheel center O_1 protrudes less than the preceding one by the depth δ_1 such that $a_1 = a_0 - \delta_1$. Had both points been at the same elevation, the undeformed chip thickness could be approximated by Eq. (3-51) but now it is reduced by δ_1:

$$h_{m1} = 2s_1 \left(\frac{a_0}{d_e} \right)^{1/2} - \delta_1 \tag{3-65}$$

Likewise, h_{m1} would be bigger by the same amount if the cutting point protruded more than the preceding one. For any cutting point n, the maximum undeformed chip thickness is

$$h_{mn} = 2s_n \left(\frac{a_{n-1}}{d_e} \right)^{1/2} - \delta_n \tag{3-66}$$

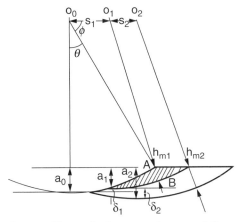

*Figure 3-6 Influence of non-uniform wheel topography on undeformed chip geometry
for straight surface grinding.*

where δ_n is positive if the cutting point protrudes less and negative if it protrudes more than the preceding one. If h_{mn} is negative, the prospective cutting point does not cut and can be ignored. This condition prevails when

$$\frac{\delta_n}{s_n} > 2\left(\frac{a_{n-1}}{d_e}\right)^{1/2} \tag{3-67}$$

or using Eqs. (3-10) and (3-34), and noting that $a_{n-1} \approx a$:

$$\frac{\delta_n}{L_n} > 2\left(\frac{v_w}{v_s}\right)\left(\frac{a}{d_e}\right)^{1/2} = \tan \varepsilon_{max} \tag{3-68}$$

where L_n is the spacing along the wheel surface from the prospective cutting point to the proceeding active one. Therefore, the number of active cutting points depends not only on the wheel topography, but also on the grinding conditions. In this way, a distinction can be made between the 'static' cutting-point density or spacing, as determined from direct measurements of the wheel topography, and the 'dynamic' cutting-point density or spacing, which takes into account the kinematics of successive cutting points.

The non-uniform distribution of cutting points on the wheel surface leads to variation in the size of the undeformed chips. One approach to studying this effect has been to kinematically simulate the grinding action on a computer [5-13]. For this purpose, the static grinding wheel topography is input to the computer program either as a statistical model of the cutting-point spacing and protrusion or by the wheel profile itself. From

Eq. (3-68) it can be appreciated that an increase in the dimensionless quantity $(v_w/v_s)(a/d_e)^{1/2}$ will raise the limiting value of δ_n, thereby allowing points deeper below the outermost grains to cut. The increased density of active cutting points reduces their average spacing L so that the mean value of h_m will be influenced to a lesser degree than indicated by Eq. (3-52) for the idealized uniform wheel surface. In addition to the size distribution of h_m, other results obtained from computer simulations include distributions for the active grain spacing, cutting-path length, and cross-sectional cutting area, as well as the ground surface topography and roughness (see Chapter 10).

The effect of a non-uniform radial distribution of active cutting points can also be seen using a continuity analysis, similar to that in the previous section for calculating the undeformed chip thickness [14-16]. For this purpose, it is convenient to assume a power function relationship for the cumulative radial distribution of active (dynamic) cutting points per unit area C_{dyn}

$$C_{dyn} = C_0 z^m \qquad (3\text{-}69)$$

where C_0 and m are constants, and z is the radial distance into the wheel from the outermost protruding point. The exponent m is a measure of the steepness of the distribution, a larger value indicating a relatively greater accumulation of cutting points with radial distance into the wheel. A uniform cutting-point distribution would correspond to $m = 0$, whereas the condition $m > 1$ would apply in many actual cases. (Experimental results for the cumulative distribution of cutting points with radial depth into the wheel are presented in Chapter 4.) The radial working depth z into the wheel may be assumed to correspond to the maximum value of h_m. Using this condition and assuming a triangular undeformed chip cross-section (Figure 3-5 (b)), the mean value of the maximum undeformed chip thickness can be written as [16]

$$\bar{h}_m = H\left[\frac{6}{C_0 r}\left(\frac{v_w}{v_s}\right)\left(\frac{a}{d_e}\right)^{1/2}\right]^{1/(m+2)} \qquad (3\text{-}70)$$

where

$$H \equiv (m + 1)^{1/2}\left[\frac{(m + 1)(m + 2)(m + 3)}{6}\right]^{1/m} \qquad (3\text{-}71)$$

and r is the ratio of the undeformed chip width to its thickness as before. The mean value is defined such that half of the material is removed by cutting points for which h_m exceeds $\bar{h}_m$ and the other half by cutting points for which h_m is less than $\bar{h}_m$. For the uniform distribution ($m = 0$ and $C_{dyn} = C_0 = C$), $H = 1$ and the relationship for h_m reduces to Eq. (3-64). From this analysis, it is again apparent that a steeper cutting-point distribution reduces the sensitivity of $\bar{h}_m$ to the quantity $(v_w/v_s)(a/d_e)^{1/2}$ owing to an increasing number of cutting points, but the size distribution for h_m becomes broader. According to this model, the relationship between $\bar{h}_m$ and its maximum value h_{mm} is [16]

$$h_{mn} = \left[\frac{(m+2)(m+3)}{6} \right]^{1/2} \bar{h}_m \qquad (3\text{-}72)$$

which is also equal to the radial working depth into the wheel surface.

This analysis of the undeformed chip thickness takes into account the effect of the radial distribution of active cutting points. However, it is apparent from the kinematic analysis (Eq. (3-70)) that the dynamic cutting-point density for a given wheel topography should uniquely depend on the quantity $(v_w/v_s)(a/d_e)^{1/2}$, which in turn is directly related to the infeed angle ε (section 3.3) rather than the radial distance z. With this in mind, a special technique was developed to measure C_{dyn} over a range of infeed angles ε, as will be seen in Chapter 4 [17]. For numerous wheels and dressing conditions it was found that

$$C_{dyn} = C_0 (\tan \varepsilon)^p \qquad (3\text{-}73)$$

where C_0 and p are constants for a particular wheel and dressing condition. Values of the exponent p range from about 0.4 to 0.8. From Eqs. (3-73) and (3-32), the distribution of active cutting points along the grinding zone is obtained as

$$C_{dyn} = C_0 \left(\frac{v_w}{v_s} \right)^p (\theta')^p \qquad (3\text{-}74)$$

An expression can now be obtained for the undeformed chip thickness with this cutting-point distribution. The continuity condition can be written as

$$N \bar{A}_n v_s = v_w a \qquad (3\text{-}75)$$

On the left side of this equation, N is total number of cutting points per unit width and $\bar{A}_n$ is the average cross-sectional area swept out by a cutting point, so the product $N\bar{A}_n v_s$ is the volumetric removal rate per unit width. N is obtained by integrating C_{dyn} along the grinding zone:

$$N = \int_0^\theta C_{dyn} \frac{d_s}{2} d\theta' = \frac{C_0 d_s}{2(p+1)} \left(\frac{v_w}{v_s}\right)^p \theta^{p+1} \qquad (3\text{-}76)$$

Assuming a triangular undeformed chip cross-section (Figure 3-5 (b)):

$$\bar{A}_n = \frac{r\bar{h}_a^2}{2} \qquad (3\text{-}77)$$

where $\bar{h}_a$ is the mean value of the average undeformed chip thickness. In this case, the maximum undeformed chip thickness is $\sqrt{3}$ bigger than the average (section 3.4) so that

$$\bar{A}_n = \frac{r\bar{h}_m^2}{6} \qquad (3\text{-}78)$$

Combining Eqs. (3-75), (3-76) and (3-78) leads to the final result:

$$\bar{h}_m = \left[\frac{3(p+1)}{rC_0}\right]^{1/2} \left[2\left(\frac{v_w}{v_s}\right)\left(\frac{a}{d_e}\right)^{1/2}\right]^{(1-p)/2} \qquad (3\text{-}79)$$

With a larger exponent p, the sensitivity of $\bar{h}_m$ to $(v_w/v_s)(a/d_e)^{1/2}$ is reduced. For the uniform distribution ($p = 0$ and $C_{dyn} = C_0 = C$), this relationship for $\bar{h}_m$ becomes identical to Eq. (3-64).

Returning to Figure 3-6, it can be seen that all the active cutting points do not contribute directly to the final surface generated by grinding. For example, the cutting point corresponding to O_2 completely masks the cutting path of the preceding cutting point corresponding to O_1, so that it leaves no trace on the envelope of the cutting paths defining the ground surface which is generated. It can be shown that the geometric condition for the nth active cutting to be masked by the successive one is

$$\frac{\delta_n}{s_n} - \frac{\delta_{n+1}}{s_{n+1}} > \frac{s_n + s_{n+1}}{d_e} \qquad (3\text{-}80)$$

or using Eq. (3-10):

$$\frac{\delta_n}{L_n} - \frac{\delta_{n+1}}{L_{n+1}} > \left(\frac{L_n + L_{n+1}}{d_e}\right)\left(\frac{v_w}{v_s}\right)^2 \qquad (3\text{-}81)$$

with the parameters as defined above. For a given wheel topography, it can be appreciated from Eq. (3-81) that a bigger speed ratio v_w/v_s will reduce the percentage of the active cutting points contributing to the ground surface geometry. Even though an increase in v_w/v_s, increases the number of active cutting points, the number of active cutting points actually contributing to the ground surface geometry will decrease, owing to the stronger influence of v_w/v_s in Eq. (3-81) as compared with Eq. (3-68).

Of all the active cutting points, only the outermost ones directly affect the geometry of the surface generated, and it is apparent that their cutting paths may be significantly longer than the theoretical path length for an idealized uniform wheel (Eq. (3-26) or (3-27)). The least protruding of the active grains further into the wheel surface will have path lengths which are much shorter than the corresponding theoretical value. With the uniform wheel surface, all active points cut identical undeformed chips with a path length essentially identical to the wheel-workpiece contact length. The differences in path length from point to point due to a more realistic non-uniform wheel topography suggest a dynamic variability in the wheel workpiece contact length depending on which active cutting points are instantaneously engaging the workpiece. Measured values of the 'contact length', using a fine thermocouple embedded in the workpiece to observe the heat pulse duration of individual cutting points [18] and an explosive device to suddenly separate the wheel from the workpiece during grinding [19], are found to exceed the theoretical value by about 50%, although the discrepancy may be somewhat bigger with extremely fine depths of cut. These experimental techniques actually measure only the longest path lengths generated by the outermost cutting points, so it would appear that the apparent elongation of the contact length may be attributable to the non-uniform wheel topography. Another factor contributing to elongation of the contact length is elastic deformation of the grinding wheel under the grinding forces, and most attempts to analyze the apparent elongation of the contact length have considered elastic deformation as the only cause.

3.7   TRAVERSE GRINDING

The grinding operations considered so far were all two dimensional with wheel and workpiece motions in a common plane. Such operations are often referred to as 'plunge grinding'. 'Traverse grinding' involves the addition of crossfeed (traverse) motion of the workpiece relative to the grinding wheel in a direction perpendicular to the plane of wheel rotation. This type of grinding would apply to straight and cylindrical operations analogous to those for plunge grinding (Figure 3-1).

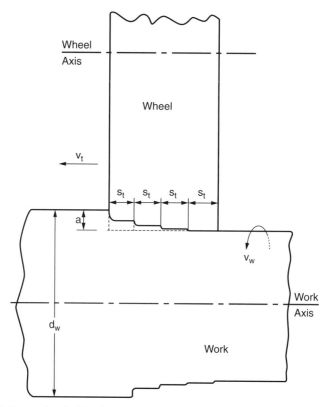

Figure 3-7 External cylindrical traverse grinding showing stepped wheel surface due to progressive wear across wheel width.

Traverse grinding is illustrated in Figure 3-7 for external cylindrical grinding. In this case, a continuous traverse velocity component v_t gives a crossfeed s_t per revolution of the workpiece:

$$s_t = \frac{\pi d_w v_t}{v_w} \tag{3-82}$$

Essentially the same situation would apply to internal cylindrical grinding. With straight surface grinding, the crossfeed s_t is usually incremented at the end of the stroke while the wheel is out of contact with the workpiece.

In the absence of any wheel wear, the total wheel depth of cut a (Figure 3-7) would be taken by the leading edge of the wheel over a width s_t.

Provided that $s_t \ll d_w$, the process would be very similar to plunge grinding a width s_t. Owing to wheel wear, however, part of the depth remains behind to be removed by a second wheel area of width s_t adjacent to the first one during the next workpiece rotation. Wheel wear on this second area leaves behind a third width s_t, and so on [20]. Those portions of the wheel closer to the leading edge cut more material and, therefore, wear more rapidly. This leads to steps across the wheel width and corner rounding at the leading edge of each width s_t, as shown in Figure 3-7. Neglecting corner rounding, the grinding operation can be likened to plunge grinding operations with decreasing wheel depths of cut. If the traverse direction is reversed at the end of the crossfeed path, the same process repeats itself at the opposite edge of the wheel. This leads to crowning of the wheel, which can cause a form error especially when grinding to a shoulder.

Plunge grinding is usually preferred to traverse grinding in production, as it provides for simultaneous grinding over a wider area and is easier to control. Traverse grinding operations are most often found in workshops and tool rooms. One notable exception where traverse grinding becomes very efficient is centerless grinding (Figure 1-1).

3.8　PROFILE (FORM), ANGLE, AND HELICAL-GROOVE GRINDING

A common feature of all the preceding grinding operations has been that the wheel profile across the grinding width has been a straight line parallel to the wheel axis, with the wheel axis parallel to the workpiece axis in cylindrical grinding, or perpendicular to the direction of motion of the workpiece in surface grinding. In this section, the analysis is extended to include arbitrary wheel profiles (form grinding) and grinding operations with the wheel axis rotated relative to that of the workpiece (angle grinding). Although these operation are more complex, it will be seen that the same relationships derived for two-dimensional plunge grinding may also be applied with appropriate geometrical corrections [21].

Consider the case of straight grinding of a flat surface inclined at angle β with machine downfeed a, as shown by the end view in Figure 3-8. The geometrical situation in the grinding zone at any point across the grinding width can be appreciated from an inclined projected side view at angle β of a circular section through the wheel. This is illustrated for a section where the wheel diameter is d_s. In this view, the circular wheel shape is projected onto a plane which is inclined at angle β to the plane of wheel rotation such that it is perpendicular to the workpiece surface where the wheel diameter is equal to d_s. Within the plane of the projected view, the wheel has the shape of an ellipse of major axis d_s and minor axis $d_s \cos\beta$.

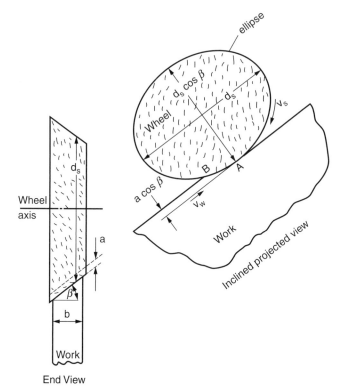

Figure 3-8 Straight surface grinding of an inclined surface. In the inclined projected view, the process is equivalent to straight surface grinding.

For $a \ll d_s$, a cutting point on the wheel periphery can be considered to move through the arc length of contact AB at the peripheral wheel velocity v_s, as with straight plunge grinding. At point A, the radius of curvature of the ellipse is $d_s/2\cos\beta$, so the wheel at this point has an 'effective' diameter

$$d_{se} = \frac{d_s}{\cos\beta} \tag{3-83}$$

Within this projected view, the wheel takes an 'effective' depth of cut

$$a_e = a\cos\beta \tag{3-84}$$

and the workpiece velocity is v_w.

The contact length l_c, curvature difference Δ, and maximum undeformed chip thickness h_m can now be calculated as before with the wheel diameter and the wheel depth of cut replaced by their effective values. For the contact length (Eq. (3-8)), this leads to

$$l_c = \left[(a \cos \beta) \left(\frac{d_s}{\cos \beta} \right) \right]^{1/2} = (a d_s)^{1/2} \qquad (3\text{-}85)$$

This result does not depend on the profile angle β and is the same as before for the two-dimensional case of straight grinding, although l_c will vary across the grinding width insofar as d_s is not constant. For the curvature difference (Eq. (3-22)):

$$\Delta = \frac{4v_w}{(d_s/\cos \beta)v_s} = \left(\frac{4v_w}{d_s v_s} \right) \cos \beta \qquad (3\text{-}86)$$

which indicates a reduced value (increased conformity) with bigger profile angles. For the maximum undeformed chip thickness (Eq. (3-52)):

$$h_m = 2L \left(\frac{v_w}{v_s} \right) \left(\frac{a \cos \beta}{(d_s/\cos \beta)} \right)^{1/2} = 2L \left(\frac{v_w}{v_s} \right) \left(\frac{a}{d_s} \right) \cos \beta \qquad (3\text{-}87)$$

which is also smaller than in the two-dimensional case. The same influence of $\cos\beta$ on h_m can also be seen from Eq. (3-59) if, in addition to the substitution of effective values for a and d_s, a correction is made for the cutting-point density C. Because of the inclination, the grinding width normal to the projected view is increased by the factor $1/\cos\beta$, and the number of cutting points is also increased by this same factor.

While the preceding analysis has been developed for grinding of a flat inclined surface, the same result would also apply to any arbitrary profile shape as illustrated in Figure 3-9. At any point P on the profile, the angle β would now correspond to the angle between the tangent to the profile shape and the wheel axis. The wheel diameter variation across the profile may be sufficiently small so that an average value can be used.

The analysis can be readily extended to include cylindrical grinding. For a projected view at angle β, as in Figure 3-8, both the wheel and the workpiece would appear as ellipses with their radii of curvature at the grinding zone both bigger than their circular radii by the factor $1/\cos\beta$. The effective wheel and workpiece diameters are

$$d_{se} = \frac{d_s}{\cos \beta} \qquad (3\text{-}88)$$

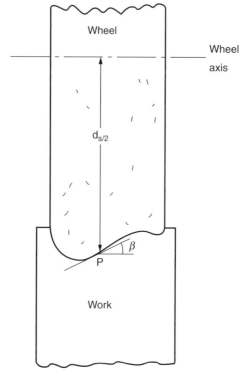

Figure 3-9 Illustration of profile grinding and profile angle β at point P.

and

$$d_{we} = \frac{d_w}{\cos \beta} \tag{3-89}$$

The effective equivalent diameter would also be bigger by the same factor:

$$d_{ee} = \frac{d_e}{\cos \beta} \tag{3-90}$$

but the effective wheel depth of cut would be proportionally smaller:

$$a_e = a \cos \beta \tag{3-91}$$

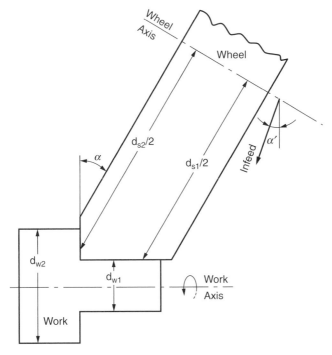

Figure 3-10 Illustration of angle grinding.

Eqs. (3-85), (3-86) and (3-87) would apply to cylindrical external as well as internal grinding with d_s replaced by d_e.

With angle grinding, the situation becomes slightly more complicated. For this type of operation, the wheel axis and the work axis are not parallel, but are turned relative to each other through the angle α. Such operations are commonly used to simultaneously grind an external cylindrical surface and a flat shoulder (annulus) as illustrated in Figure 3-10. The infeed direction is defined by the angle α'. For a wheel depth of cut per workpiece revolution a in the infeed direction, the effective wheel depth of cut is

$$a_{e1} = a \cos \alpha' \qquad (3\text{-}92)$$

at the external cylindrical surface and

$$a_{e2} = a \sin \alpha' \qquad (3\text{-}93)$$

at the shoulder.

Grinding on the external cylindrical surface can be analyzed, by analogy with Figure 3-8, in a projected view from the side looking parallel to the workpiece axis at a projected plane where the wheel diameter is d_{s1}. The wheel would appear as an ellipse of major axis d_{s1}, and minor axis $d_{s1}\cos\alpha$, and the workpiece would appear as a circle of diameter d_{w1}. The radius of the wheel curvature at its point of contact is $d_{s1}/2\cos\alpha$, so the effective equivalent diameter can be approximated from Eq. (3-9) as:

$$d_{e1} = \frac{d_{s1}/\cos\alpha}{1 + \dfrac{d_{s1}/\cos\alpha}{d_{w1}}} = \frac{d_{s1}}{\cos\alpha + \dfrac{d_{s1}}{d_{w1}}} \qquad (3\text{-}94)$$

Therefore, the previous equations for l_c, Δ and h_m would still apply using effective values of a_{e1} and d_{e1} from Eqs. (3-92) and (3-94). Furthermore, the variation in wheel diameter d_{s1}, across the grinding width is usually relatively small, such that average values of d_{s1} and v_s can be used to characterize this area.

Grinding on the shoulder resembles straight surface grinding. A projected view of the wheel section, directly from above (or below) perpendicular to the workpiece axis at the plane where the wheel diameter is d_{s2}, would appear as an ellipse of major axis d_{s2} and minor axis $d_{s2}\sin\alpha$. The effective equivalent wheel diameter at the 'point' of contact with the straight shoulder would be

$$d_{e2} = \frac{d_{s2}}{\sin\alpha} \qquad (3\text{-}95)$$

Using this value for the equivalent diameter and the effective wheel depth of cut from Eq. (3-93), the grinding parameters l_c, Δ and h_m can be calculated. As before, it is often sufficient to use average values of d_{s2} and v_s at the shoulder for these calculations. However, the variation in the workpiece diameter $(d_{w2} - d_{w1})$ may be so big as to necessitate taking into account the variation in workpiece velocity along the shoulder.

In many angle-grinding operations, the wheel infeed direction is normal to the wheel axis and 60° to the workpiece axis, such that $\alpha = \alpha' = 30°$. Both the undeformed chip thickness and the curvature difference tend to be much bigger on the external cylindrical surface than on the shoulder, whereas the cutting-path length on the shoulder is much longer than on the cylindrical surface.

The same approach used here for profile and angle grinding can be generally applied to most grinding operations with arbitrary grinding profiles and orientations between the wheel and workpiece motions. The grinding action is analyzed in a projected plane which is normal to the surface generated and includes the peripheral wheel velocity at the contact area, which enables estimation of the effective equivalent diameter. The effective wheel depth of cut in each case is its component normal to the generated surface, and the effective workpiece velocity is its component collinear with the wheel velocity in the same (down) or opposite (up) direction. Another approach would be to analyze the grinding action in a projected plane normal to the wheel surface instead of the workpiece surface, which leads to essentially the same result. Aside from the assumptions in the two-dimensional analysis, it has also been tacitly assumed that the working surface of the grinding wheel does not conform completely to the generated cutting path such that $\Delta \neq 0$ (see section 3.3).

This generalized method is especially helpful for visualizing complex grinding operations as equivalent plunge grinding processes. In many practical cases, however, quantitative analysis becomes much more tedious than for profile or angle grinding. As an example, consider the case of grinding a helical groove in a cylindrical body, as illustrated in Figure 3-11. In the plan view as shown, the grinding wheel is located above the workpiece. Practical grinding operations which can be categorized as helical grinding include thread grinding and flute grinding. In general the grinding wheel axis is oriented at angle ψ to the workpiece axis, and the helix angle

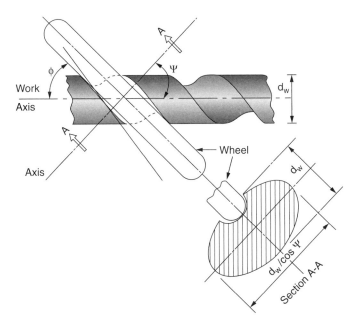

Figure 3-11 Grinding of a helical groove in a cylindrical workpiece shown in a plan from above the workpiece.

ϕ of the groove is defined at the outer diameter d_w of the workpiece. Relative to the grinding wheel, the workpiece has a circumferential peripheral velocity v_w and an axial (traverse) velocity v_t such that

$$\tan \phi = \frac{v_w}{v_t} \tag{3-96}$$

The pitch corresponds to the traverse cross feed s_t as given by Eq. (3-82). The angle ψ may be set such that the grinding wheel axis is perpendicular to the helix direction at the workpiece periphery ($\psi = 90° - \phi$), although a slightly larger angle is often used.

Now to a first approximation, helical-groove grinding would appear similar to cylindrical external grinding with a profile. Using the concept of elliptical sections, it can be appreciated that the effective radius of curvature of the workpiece would become enlarged owing to the influence of both the rotation of the wheel axis by the angle ψ and the workpiece profile angle. However, further complications arise owing to the curvature of the helical path as seen in the plan view of Figure 3-11, as well as differences in orientation between the plane of the grinding wheel and the groove. This results in wheel interference and undercutting of the groove profile relative to the wheel profile, as seen in Section A-A of Figure 3-11. In this sectional view, the wheel and groove axes coincide only at the point of intersection between the wheel and workpiece axes in the plan view. Even if the wheel plan is oriented along the helix angle ($\psi = 90° - \phi$), undercutting will still occur. The same problem arises in the milling of helical flutes, for which geometrical methods have been developed to calculate the actual profile generated or the cutter profile required to obtain a desired flute geometry [22–25]. A user-friendly interactive computer program has been developed for design and analysis of helical profile grinding operations [24, 25].

3.9 GRINDING OPERATIONS WITH TOTAL CONFORMITY

For grinding operations with total conformity, the grinding wheel shape corresponds to the cutting-path radius of curvature ($\Delta = 0$). The grinding geometry for such operations cannot be analyzed using the results in the previous section. Examples of such operations include face grinding, vertical-spindle grinding, and cut-off. The total conformity condition applies when the grinding wheel maintains constant contact with the surface it is grinding either because there is no lateral motion between the workpiece and the wheel axis (face grinding) or because the two approach each other along the line that is the direction of the cut being made (vertical spindle, cut-off).

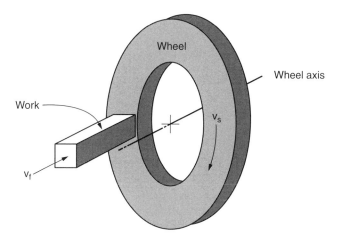

Figure 3-12 Illustration of face grinding of a rectangular workpiece.

For face grinding, a flat wheel surface generates a flat workpiece surface, as seen in Figure 3-12. The cutting-path length corresponds to the circular arc across the workpiece along which the undeformed chip thickness h can be considered to be constant. Assuming a triangular undeformed chip cross-section with its geometry defined by the ratio of its width to thickness, as before, continuity requires that

$$v_f A_w = C A_w \left(\frac{rh^2}{2} \right) v_s \qquad (3\text{-}97)$$

where v_f is the infeed velocity, A_w is the cross-sectional area of the workpiece, and $rh^2/2$ is the cross-sectional area of a cutting point. Therefore, the undeformed chip thickness is

$$h = \left[\frac{2}{Cr} \left(\frac{v_f}{v_s} \right) \right]^{1/2} \qquad (3\text{-}98)$$

For vertical-spindle grinding with linear workpiece motion (Figure 3-13) and cut-off grinding (Figure 3-14), conformity occurs at the step being cut laterally into, or the groove being cut through, the workpiece. In vertical-spindle grinding, the surface generated on the workpiece corresponds to the circular section on the wheel periphery whose width equals the depth of cut a. The maximum undeformed chip thickness is obtained at the middle of the vertical step being cut and can be approximated by the feed per cutting point s:

$$h_m = s = \frac{v_w}{v_s} L \qquad (3\text{-}99)$$

Grinding geometry and kinematics

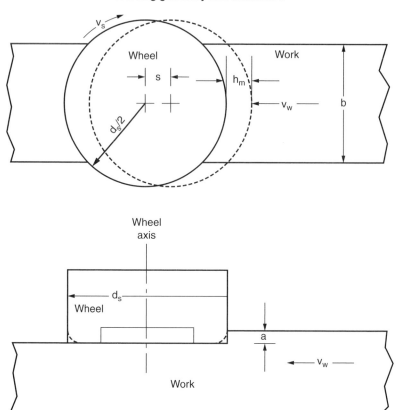

Figure 3-13 Illustration of vertical-spindle surface grinding.

Another expression for h_m may be obtained in terms of C and r for a triangular undeformed chip cross-section by substituting for L from Eq. (3-55) and noting that $\bar{b} = rh_m/\sqrt{3}$:

$$h_m = \left[\frac{\sqrt{3}}{Cr} \left(\frac{v_w}{v_s} \right) \right]^{1/2} \tag{3-100}$$

Because the cutting load is concentrated near the edge of the wheel, the wheel corner will become rounded and the actual cutting profile will

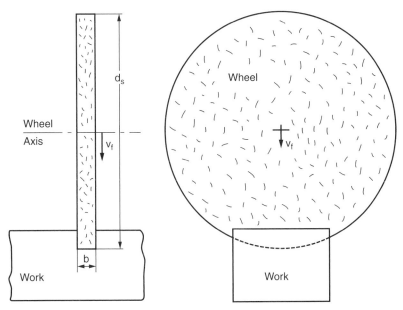

Figure 3-14 Illustration of an abrasive cut-off operation.

change. In this case, the flat underside of the wheel becomes more important. The geometrical situation for cut-off grinding would be very similar to that shown for vertical-spindle grinding. With this operation (Figure 3-14), the infeed velocity v_f would be equivalent to the workpiece velocity for vertical-spindle grinding, and the grinding width on the wheel periphery would correspond to the wheel depth of cut. The maximum undeformed chip thickness would be given by Eq. (3-99) or (3-100), with v_w replaced by v_f.

REFERENCES

1. Alden, G. I., 'Operation of Grinding Wheels in Machine Grinding', *Trans. ASME*, 35, 1919, p. 45.
2. Guest, J. J., *Grinding Machinery*, Edward Arnold, London, 1914.
3. Martelloti, M. E., 'An Analysis of the Milling Process', *Trans. ASME*, 63, 1941, p. 677.
4. Hahn, R. S., 'The Effect of Wheel-Work Conformity in Precision Grinding' *Trans. ASME*, 77, 1955, p. 1325.
5. Baul, R. M. and Shilton, R., 'Mechanics of Metal Grinding with Particular Reference to Monte Carlo Simulation', *Proceedings of the Eighth International Machine Tool Design and Research Conference*, Part 2, 1967, p. 923.

6. Yoshikawa, H. and Peklenik, J., 'Simulated Grinding Process by Monte Carlo Method', *Annals of the CIRP*, 26, 1968, p. 297.

7. Yoshikawa, H. and Peklenik, J., 'Three Dimensional Simulation Techniques of the Grinding Process–II. Effects of Grinding Conditions and Wear on the Statistical Distribution of Geometrical Chip Parameters', *Annals of the CIRP*, 18, 1970, p. 361.

8. Law, S. S. and Wu, S. M., 'Simulation Study of the Grinding Process', *Trans. ASME, J. of Eng. for Ind.,* 95, 1973, p. 972.

9. Law, S. S., Wu, S. M. and Joglekar, A. M., 'On Building Models for the Grinding Process', *Trans. ASME, J. of Eng. for Ind.,* 95, 1973, p. 983.

10. Peklenik, J., 'The Statistical Mechanism of Chip Formation in Grinding Process', *Proceedings of the International Conference on Production Engineering*, Part 2, Tokyo, 1974, p. 51.

11. König, W. and Lortz, W., 'Properties of Cutting Edges Related to Chip Formation in Grinding', *Annals of the CIRP*, 24/1, 1975, p. 231.

12. König, W. and Steffens, K., 'A Numerical Method to Describe the Kinematics of Grinding, *Annals of the CIRP*, 31/1, 1982, p. 201.

13. Brough, D., Bell, W. F. and Rowe, W. B., 'A Re-examination of the Uncut Models of Grinding and its Practical Implication', *Proceedings of the Twenty Fourth International Machine Tool Design and Research Conference*, 1983, p. 261.

14. Kassen, G., 'Beschreibung der elementaren Kinematik des Schleifvorganges', Dissertation, TH Aachen, 1969.

15. Werner, G., 'Kinematic and Mechanik des Schleifprozesses', Doctoral Dissertation, TH Aachen, 1971.

16. Pecherer, L., Grinding of Steels with Cubic Boron Nitride (CBN) Abrasive', MS Thesis, Technion–Israel Institute of Technology, 1983.

17. Tigerström, L., 'A Model for Determination of Number of Active Edges in Various Grinding Processes', *Annals of the CIRP*, 24/1, 1975, p. 271.

18. Verkerk, J., 'The Real Contact Length of Cylindrical Plunge Grinding', *Annals of the CIRP*, 24/1, 1975, p. 259.

19. Brown, R. H., Wager, J. G. and Watson, J. D., 'An Examination of the Wheel-Workpiece Interface Using an Explosive Device to Suddenly Interrupt the Surface Grinding Process', *Annals of the CIRP*, 26/1, 1977, p. 143.

20. Pekelharing, A. J., Verkerk, J. and van Beukering, F. C., 'A Model to Describe Wheel Wear in Grinding', *Proceedings of the International Grinding Conference*, Pittsburgh, 1972, p. 412.

21. Graham, W. and Falconer, D., 'Wheel-Workpiece Conformity in Form Grinding', *Proceedings of the Nineteenth International Machine Tool Design and Research Conference*, 1979, p. 615.

22. Friedman, M. Y. and Meister, I., 'The Profile of a Helical Slot Machined by a Form-Milling Cutter', *Annals of the CIRP*, 22/1, 1973, p. 29.

23. Agullo-Batlle, J., Cardona-Foix, S. and Vinas-Sanz, C., 'On the Design of Milling Cutters or Grinding Wheels for Twist Drill Manufacture: A CAD Approach', *Proceedings of the Twenty Fifth International Machine Tool Design and Research Conference*, 1985, p. 315.

24. Sheth, D. S. and Malkin, S., 'CAD/CAM for Geometry and Process Analysis of Helical Groove Machining', *Annals of the CIRP*, 39/1, 1990, p. 129.

25. Shi, Z. and Malkin, S. Malkin, 'Valid Machine Tool Setup for Helical Groove Machining', *Trans. of NAMRI/SME*, 31, 2003, p. 193.

BIBLIOGRAPHY

Armarego, E. J. A. and Brown, R. H., *The Machining of Metals*, Prentice-Hall, Englewood Cliffs, NJ, 1969, Chapter 11.

Nakayama, K., 'Grinding Wheel Geometry', *Proceedings of the International Grinding Conference*, Pittsburgh, 1972, p. 197.

Okamura, K., 'Fundamental Analysis of Grinding Process', *Ingenieursblad*, 40/4, 1971, p. 82.

Reichenbach, G. S., Shaw, M. C., Mayer, J. E., Jr. and Kalpakcioglu, S., 'The Role of Chip Thickness in Grinding', *Trans. ASME*, 78, 1956, p. 519.

Saljé, E., Matsuo, T. and Lindsay, R. P., 'Transfer of Grinding Research Data for Different Operations in Grinding', *Annals of the CIRP*, 31/2, 1982, p. 519.

Sen, G. C. and Bhattacharyya, A., *Principles of Metal Cutting*, 2nd edn, New Central Book Agency, Calcutta, 1969, Chapter 8.

Shaw, M. C., *Principles of Abrasive Processing*, Clarendon Press, Oxford, 1996.

Shaw, M. C., *Metal Cutting Principles*, 3rd edn, The Technology Press, Cambridge, Massachusetts, 1960, Chapter 18.

Chapter 4

Wheel Truing, Dressing, and Topography

4.1   INTRODUCTION

Grinding is a machining process which utilizes a grinding wheel consisting of abrasive grains held together with a binder material. The actual cutting points on abrasive grains at the wheel surface are micro-cutting tools which interact with the workpiece material. The spatial distribution of abrasive grains over the wheel surface and their morphology comprise the grinding wheel topography.

With the exception of some heavy-duty and cut-off operations, the grinding wheel topography and the macroscopic wheel shape are generated by preparing the wheel prior to grinding and periodically during the course of grinding. Wheel preparation generally includes truing and dressing. Truing usually refers to removal of material from the cutting surface of a grinding wheel so that the spinning wheel runs true with minimum run-out from its macroscopic shape, although truing may also include profiling of the wheel to a particular shape. Dressing is the process of conditioning the wheel surface so as to achieve a certain grinding behavior. With conventional abrasive wheels, both truing and dressing are usually done by the same process, and the combination is commonly called dressing. With superabrasive wheels, separate truing and dressing processes may be used.

It was seen in the previous chapter how the wheel topography and the grinding parameters affect the kinematic interaction between the abrasive grains and the workpiece. In subsequent chapters, it will be shown that the wheel topography and consequently the wheel preparation conditions have a profound effect on the grinding performance as characterized by forces, power consumption, temperatures, and surface finish. We should therefore attempt to understand what happens during wheel preparation, as this can provide a logical basis for controlling what is perhaps the most significant as well as the most often neglected factor in grinding.

4.2   DRESSING OF CONVENTIONAL WHEELS

Grinding wheels containing conventional ceramic abrasives are usually prepared by feeding a dressing tool across the rotating wheel surface, as illustrated in Figure 4-1 for a simple grinding wheel with a flat peripheral surface. During each pass of the dressing tool across the wheel face, a layer of depth a_d is removed from the radius. This type of dressing motion is analogous to turning on a lathe. The axial feed (pitch) of the dressing tool per wheel revolution is called the dressing lead, s_d, and is given by

$$s_d = \frac{\pi d_s v_d}{v_s} \qquad (4\text{-}1)$$

where v_d is the crossfeed (traverse) velocity of the dresser across the wheel, v_s is the wheel velocity, and d_s is the wheel diameter. Between two and five dressing passes are usually taken beyond what is required to true the wheel, although it may be necessary to dress off more wheel material to eliminate artifacts from previous grinding. As a final step, 'spark-out' passes may also be taken, in which the dressing tool is traversed across the wheel without further incrementing the dressing depth setting. Each successive spark-out pass removes less wheel material and imparts an increasingly finer wheel topography.

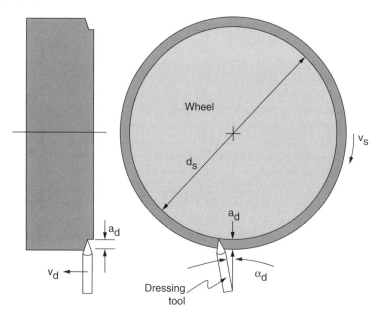

Figure 4-1 Single-point dressing of a grinding wheel.

The dressing tool is usually a mounted natural diamond (single point) or a multipoint tool (cluster, nib, chisel, blade) consisting of one or more layers of diamonds set or impregnated in a metal binder. Single-point diamonds are usually tilted relative to the wheel with a drag angle $\alpha_d = 10–15°$ (Figure 4-1). With continued use, the exposed point on a single-point diamond tool becomes duller, thereby changing the dressed wheel topography and the subsequent grinding behavior. Eventually it becomes necessary to discard or reset the diamond to expose a new point. With multipoint diamond tools the grinding performance during the life of the dressing tool tends to remain more consistent, which is a definite advantage in automated production. Commonly applied wheel dressing leads s_d and radial dressing depths a_d are [1]:

single-point diamond	$s_d < 0.2$ mm	$10 < a_d < 30$ μm
multipoint diamond	$s_d < 0.5$ mm	$10 < a_d < 50$ μm
multipoint diamond (cluster)	$s_d < 2$ mm	$10 < a_d < 50$ μm

Another dressing method which is especially used for generating profiles is rotary diamond dressing. A rotary diamond dressing tool (roll) consists of an axisymmetric body with diamond particles impregnated in a metal matrix or held by an electroplated metal coating on its outer surface. The dressing roll has the same profile as that required on the workpiece, so the wheel is dressed with the reverse profile. For most applications, the dressing roll is driven at a peripheral velocity v_r while being fed radially into the rotating wheel at an infeed velocity v_i corresponding to a depth per wheel revolution a_r (Figure 4-2). With this arrangement, the rotary dresser appears to be 'grinding' the cylindrical grinding wheel surface. The peripheral

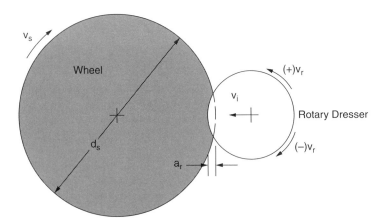

Figure 4-2 Rotary diamond dressing of a grinding wheel

dresser velocity v_r may be in the same ($+$) or opposite ($-$) direction as the grinding wheel velocity at the dresser-wheel contact area. Typically, the rotary dresser velocity at its maximum diameter might be 20–50% of the wheel velocity in the opposite (up) direction, although the dresser is often rotated in the same direction (down) especially with creep feed grinding at speeds up to 80% of the grinding wheel velocity. The dresser infeed per wheel revolution is typically $a_r = 10$–30 μm and the total wheel depth removed is 50–200 μm. Prior to retracting the dressing tool, its infeed may be deliberately or unintentionally stopped (dwell), analogous to 'spark-out' with single-point dressing. Rotary dressing may also be applied continually during creep feed grinding processes, rather than intermittently, in order to maintain the wheel sharpness and profile shape. Continuous-dress-creep-feed-grinding (see Chapter 7) is commonly applied to grinding of complex profiles on jet engine components.

Other less-used profile dressing methods include diamond block dressing and crush dressing. With diamond block dressing, the rotating wheel is traversed along a fixed diamond impregnated block having the required workpiece profile. With crush dressing, the wheel is usually forced under pressure into a hardened steel or cemented carbide axisymmetric crush roll mounted on bearings. In order not to grind away the roll, the rotational speed of the wheel during crush dressing is slowed down to only a few per cent of the normal wheel velocity.

4.3   TRUING AND DRESSING OF SUPERABRASIVE WHEELS

Virtually all types of superabrasive wheels undergo pre-grinding preparation by truing, which may be followed by dressing. An exception arises in the case of electroplated wheels which may only require occasional 'cleaning' or 'touching up' with an abrasive stick. While the primary aim in truing is to produce the required macroscopic wheel shape, flat or profiled, the truing process also influences the microscopic wheel topography. Therefore accurate mounting of electroplated wheels with a minimum run-out of only a few microns is critical.

One popular truing method for diamond wheels utilizes a vitrified green (friable) silicon carbide 'grinding' wheel mounted on a brake-controlled truing device. The rotating motion of the truing wheel obtained by direct contact with the diamond grinding wheel is resisted by a centrifugal brake on the truing wheel spindle, thereby imparting a 'slip' velocity between the truing wheel and grinding wheel. With peripheral grinding wheels, the truing wheel is operated as if it cylindrically traverse grinds the grinding wheel (Figure 4-3) with the axes of the dressing wheel and grinding wheel parallel to each other. The same device is also used in a similar way to true other types of wheels (e.g., cup wheels). A new truing wheel is typically about 120 mm in diameter, but it rapidly wears down virtually by the amount of its total infeed into the grinding wheel. The silicon carbide truing wheel is expendable, since its cost is negligible relative to that of a diamond wheel. The rotational speed of the braked spindle might be 1400–1500 revolutions per minute, corresponding to a peripheral velocity of about $v_b = 9$ m/s with a new truing wheel, although this will vary according to the particular dressing device and the force developed between the grinding and truing wheels. An additional slip velocity between the grinding and truing wheel surfaces can be obtained by orienting the axis of the truing wheel such that its peripheral velocity is not collinear with that of the grinding wheel at their point of contact. This latter concept is sometimes applied to truing of diamond wheels using a silicon carbide wheel on a freely rotating shaft without any brake.

With reference to Figure 4-3, typical brake-controlled truing conditions might be $a_b = 10$–20 μm depth increment after each traverse across the wheel face with a crossfeed velocity v_c corresponding to a lead of $s_b = 0.1$–0.2 mm per grinding wheel revolution. This truing process is continued until it is apparent that the whole surface of the diamond wheel has been in contact with the truing wheel, at which point the wheel run-out should be reduced to about 3–5 μm. The radial depth of diamond wheel removed by

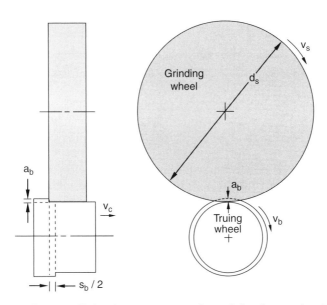

Figure 4-3 Brake-controlled truing arrangement for peripheral superabrasive wheels.

truing in this way is usually comparable to the initial run-out, which normally should not exceed 10 μm.

Aside from the use of a brake-controlled truing device, silicon carbide dressing wheels may be used with motor-driven spindles to true diamond wheels. Essentially the diamond wheel is ground by the silicon carbide wheel. Some suppliers of superabrasive wheels also true them, prior to shipment, by grinding. If such a pre-trued wheel is carefully mounted in the machine, no further truing may be required.

Numerous other truing methods utilizing diamond tools, similar to those applied to conventional abrasive wheels, have been tried with diamond wheels, but with limited success. Diamond truing of diamond wheels removes wheel material much faster than silicon carbide truing, but there is a danger of damaging both the truing tool and wheel. Truing with single-point and multipoint diamond tools may also necessitate excessive wheel loss, and the truing tools wear out rapidly. The use of diamond-impregnated blocks has shown promise, especially for large peripheral wheels which would require very long truing times with silicon carbide abrasives [2]. Diamond rolls and diamond cup wheels have also been tried, but their high cost may preclude general application.

Diamond tools are widely used for truing of resin and metal bonded CBN wheels, although brake-controlled truing with silicon carbide wheels is also common. CBN is much softer than diamond (Chapter 2), so diamond truing tools wear much less than with diamond wheels, although the truing process is still much more difficult than with conventional wheels. Single-point diamonds are usually not used, except with some small vitrified CBN wheels for generating profiles. Multipoint diamond clusters (nibs) are more commonly used, which may be specially manufactured with a steel binder for added wear resistance. With resin-bonded CBN wheels, the volumetric wear of the multipoint diamond dresser is typically less than one per cent of the volumetric wheel removal. However, the cross-sectional area of these truing tools is only a small fraction of the wheel surface, so a significant portion of the truing depth is taken up by wear of the dressing tool. This can make it extremely difficult to accurately true a large grinding wheel. This problem may be overcome by the use of diamond rolls and diamond cup wheels, which have much larger tool areas and also wear much less.

After truing, dressing of diamond wheels and resin and metal bonded CBN wheels is usually accomplished by infeeding a fine-grained vitrified abrasive stick into the wheel surface either manually or with a holding device. Silicon carbide sticks are usually used with diamond wheels and aluminum oxide sticks with CBN wheels. This process is generally considered to 'open up' the wheel surface and 'expose' the abrasive grains by removing binder, without significantly affecting the abrasive grains themselves. With

CBN wheels, more aggressive dressing methods are often needed to sharpen the abrasive grains which may be flattened down by truing (see section 4.5). The simplest sharpening method consists of grinding a block of mild steel. Some other methods utilize fine-grained conventional abrasives applied either in a slurry between a steel dressing roll and the wheel surface, within a pressurized jet, or impregnated in a wax stick applied to the wheel surface during grinding [3]. A steel-wire brush mounted on a brake-controlled truing device in place of the truing wheel has also been found to sharpen CBN wheels effectively [4].

Vitrified CBN wheels are much easier to true and profile than resin bonded CBN wheels. This has been an important factor leading to their widespread adoption in place of conventional aluminum oxide wheels in production. The preferred method for truing of vitrified CBN wheels utilizes a thin rotating diamond disk as illustrated in Figure 4-4 [5, 6]. The disc is typically 100–150 mm diameter and 2–3 mm thick with a diamond composite layer on its periphery which is produced by reverse plating, chemical vapor deposition (CVD), electroplating, or brazing. Analogous to rotary dressing illustrated in Figure 4-3, the disk dresser is rotated with its peripheral velocity v_r either in the same direction (+) or opposite direction (−) as the grinding wheel velocity. The dresser is also given an axial traverse velocity v_d corresponding to a feed (lead) s_d per wheel revolution given by Eq. 4-1, in order to dress off a depth a_r from the wheel surface. Typical dressing conditions use a speed ratio v_r/v_s in the range of +0.4 to +0.8

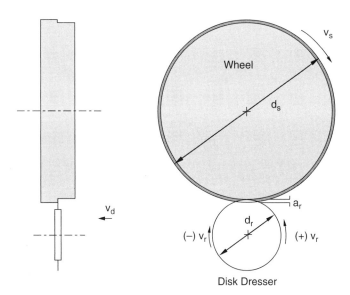

Figure 4-4 Diamond disk dressing of a grinding wheel.

(same direction), a dressing lead s_r which is 60–80% of the CBN grain dimension in the wheel, and a radial dressing depth of 0.5–3 μm. An additional radial motion with CNC control can be used to generate a profile as the wheel is traversed. For this purpose, the dressing disk would be manufactured with a radius rather than a flat surface on its periphery

4.4 GENERATION OF WHEEL TOPOGRAPHY-CONVENTIONAL WHEELS

During dressing of conventional wheels with a single-point diamond tool, the dresser moves across the wheel surface with a lead s_d per wheel revolution while removing a depth a_d (Figure 4-5). Therefore, the dressing tool follows a path which would appear to be cutting a thread on the abrasive grains with a pitch (lead) s_d, as illustrated in Figure 4-2. For a dressing diamond with a radius of curvature r_d at its tip, the theoretical peak-to-valley height (roughness) of the thread profile generated on the grain tips can be written [1]:

$$R_{ts} = \frac{s_s^2}{8r_d} \qquad (4\text{-}2)$$

provided that $r_d \gg s_d/2$, and $R_{st} \ll a_d$. According to this equation, a bigger dressing lead s_d and, to a lesser extent, a sharper or more pointed

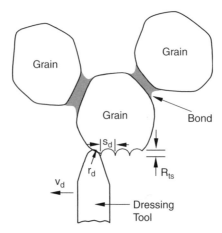

Figure 4-5 Cutting path of a single-point diamond dressing tool through an abrasive grain

dressing tool (smaller r_d) should lead to a rougher (coarser) wheel surface, but the dressing depth a_d should have no influence.

Replication of the dressing lead on a grinding wheel has been clearly identified from profiles across the surfaces of ground workpieces [7-12]. However, measurements of the size distribution of particles removed from the wheel by dressing would seem to contradict this concept of thread cutting by dressing, at least for vitreous bonded wheels containing friable abrasives [13-18]. Generation of the 'theoretical' profile in Figure 4-5 would require a ductile flow-type cutting mechanism between the dressing tool and the wheel, but the abrasive particles removed by dressing appear to be produced mainly by brittle fracture. Examples of size distributions for particles removed by single-point dressing are shown in Figure 4-6 in terms of their cumulative weight distributions obtained by sieving for vitreous-bonded G, I, and K grade wheels [15]. Also included for comparison is the particle-size distribution of the original 46 grit monocrystalline aluminum oxide used in manufacturing these wheels. A higher elevation of the size distribution curve in this type of graph indicates a coarser particle-size distribution. It is apparent that the particles removed by dressing are somewhat finer than the original grain material, and that a harder wheel grade also gives somewhat finer dressing particles. On the other hand, virtually the entire weight of material dressed off the wheel consists of particles which

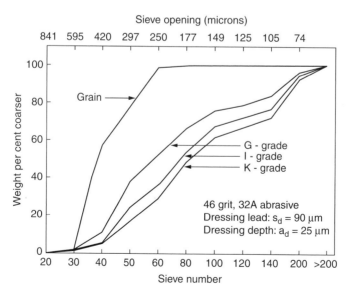

Figure 4-6 Cumulative size distribution of particles removed by single-point dressing for three wheel grades [15]. The size distribution of the original grain is included for comparison.

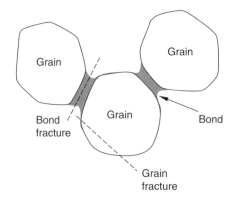

Figure 4-7 Illustration of grain fracture and bond fracture

are much bigger than the dressing depth ($a_d = 25$ μm). Therefore, wheel material is mostly removed by brittle fracture to a depth below that of the nominal path of the dressing tool (Figure 4-5).

Because the dressing particles are generally much bigger than the depth of dressing but smaller than the original grains, it has been postulated that their removal from the wheel involves a two-step process of grain fracture followed by bond fracture, as schematically illustrated in Figure 4-7 [15]. A harder wheel having more bond material holds the grains more strongly, thereby resulting in a greater degree of grain fracture (fragmentation) prior to final dislodgement by bond fracture. This explains why a finer particle-size distribution is obtained with harder wheels.

The weight fraction of the dressing particles associated with bond fracture can be estimated by comparing their particle size distribution with that of the original grain. This calculation is based upon the following assumptions [14, 15]:

(1) the largest particles are removed by bond fracture;
(2) there is only one bond fracture per grain;
(3) the weight of a particle or grain is proportional to the cube of its dimension as measured by sieving;
(4) the bond material is a negligible fraction of the total weight removed.

For the dressing particle distributions in Figure 4-6, the percentage bond fracture is 35 by weight for the K wheel, 39 for the I wheel, and 49 for the G wheel, again indicating that less grain fracture occurs with softer wheels. This same trend has also been observed with other grain sizes. Finer-grained wheels are also found to undergo less fragmentation (larger percentage bond fracture) during dressing, which suggests that finer grains are tougher [15].

On the basis of the third assumption, the average dimension d_b of the particles removed by bond fracture is

$$d_b = B^{1/3} d_g \qquad (4\text{-}3)$$

where B is the weight fraction of dressing particles removed by bond fracture and d_g is the original grain dimension. The dimension d_b would also represent the radial depth into the wheel affected by dressing.

We can now proceed to consider how the dressing particle-size distribution relates to the number of active grains on the wheel surface. Let us begin with an imaginary limiting case of infinitely small particles removed by dressing, as if the wheel surface were very finely polished down. The wheel surface generated in this case would correspond to that of a random plane or section through the bulk of the wheel. For simplicity, the grains are approximated as uniform spheres of diameter d_g randomly distributed throughout the wheel volume. On the average, the expected value of an intersected area $\overline{A}_g$ through a sphere is two-thirds the area of a circle though its center, or

$$\overline{A}_g = \frac{\pi d_g^2}{6} \qquad (4\text{-}4)$$

The overall area fraction at any random section through the bulk of the wheel is identical to the volume fraction V_g of grain in the wheel, so that

$$V_g = G_0 \overline{A}_g \qquad (4\text{-}5)$$

where G_0 is the number of grains intersected per unit area. By combining Eqs. (4-4) and (4-5), this theoretical packing density (grains/area) G_0 can be expressed in terms of V_g and d_g [19]:

$$G_0 = \frac{6V_g}{\pi d_g^2} \qquad (4\text{-}6)$$

As might be expected, G_0 is directly proportional to V_g and varies inversely with d_g^2.

The parameter G_0 would represent the number of active grains on the hypothetical finely polished wheel surface. It would also correspond to the number of grains per unit area for the theoretical uniformly turned wheel surface (Figure 4-5). However, because many dressing particles are actually rather large, the number of active grains per unit area of wheel surface is usually much less than G_0. An estimate of the number of active grains can be made based upon the concept of 'unavailability' due to bond fracture.

During successive dressing passes with $a_d \ll d_g$, a typical grain is reduced in size by the accumulated dressing depth from its initial dimension d_g by grain fracture and deformation down to the dimension d_b, at which point it is dislodged by bond fracture. Since the radial working depth into the wheel (Chapter 3) is generally much smaller than d_b, a potentially active grain will become unavailable if bond fracture occurs. For steady-state dressing conditions and assuming that grain unavailability is due only to bond fracture, the number of grains per unit area remaining available G_a as a fraction of the theoretical area packing density G_0 is equal to the ratio of the accumulated dressing depth per grain before bond fracture to the grain dimension:

$$\frac{G_a}{G_0} = \frac{d_g - d_b}{d_g} \tag{4-7}$$

or substituting for d_b from Eq. (4-3):

$$\frac{G_a}{G_0} = 1 - B^{1/3} \tag{4-8}$$

Experimental verification of Eq. (4-8) can be seen in Figure 4-8 which covers four grit sizes (30, 46, 80, and 120 mesh) and three wheel grades (G, I, and K). The number of active grains after dressing in each case was determined from optical microscopy of the wheel surfaces after taking only two plunge grinding passes across a steel workpiece [15]. Active grains were identified using optical microscopy by the presence of adhering metal on their tips (see section 4.6.5). Considering the complexity of the dressing process and the measuring technique, as well as the simplifying assumptions which were made, the agreement is quite remarkable. For values of B ranging from 30 to 90%, the corresponding number of active grains decreases from about 25% of the theoretical packing number down to only about 5%. With larger values of B, corresponding to the finer grain sizes, the measured values of B tend to exceed the prediction of Eq. (4-8). This may be attributed to the finite radial working depth into the wheel (Chapter 3), which has been neglected in this analysis. With smaller values of B corresponding to coarser-grained wheels, the measured values of G_a tend to fall below Eq. (4-8), which might suggest that some grains also become unavailable by grain fracture.

While bond fracture is mainly responsible for determining how many potentially active grains remain at the wheel surface, the morphology of these grains is largely controlled by grain fracture on a much finer scale,

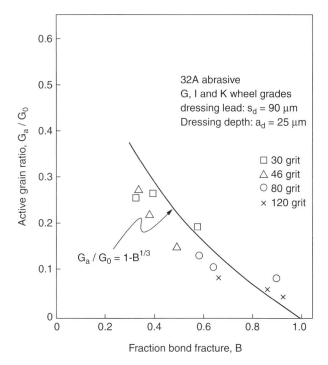

Figure 4-8 Active grain ratio versus fraction bond fracture for dressing. Includes results for three wheel grades and four grit sizes.

and even by plastic deformation. Although conventional ceramic grain materials are brittle, they can still flow plastically during dressing, as can be seen in the SEM micrograph of Figure 4-9(a). In this case, the dressing lead and dressing depth were rather small (fine dressing). For coarser dressing with a much bigger dressing lead and dressing depth (Figure 4-9(b)), there is much less deformation and the grains appear more fractured. In general, both grain fracture and plastic deformation play important roles. With finer dressing, localized plastic flow results in flattening and smoothing of some grain tips not fractured away. A similar effect is obtained by the addition of spark-out passes without incrementing the dressing depth. The pitch of the dressing lead (Eq. (4-1)) is probably contained within the morphology of the small flattened areas (Figure 4-9(a)), although it is not apparent in these micrographs. Coarser dressing causes more grain fracture and a sharper wheel.

It will be seen (Chapters 5 and 10) that coarser wheel dressing generally results in reduced grinding forces and rougher workpiece finishes, whereas finer dressing leads to bigger forces and smoother finishes. Of the

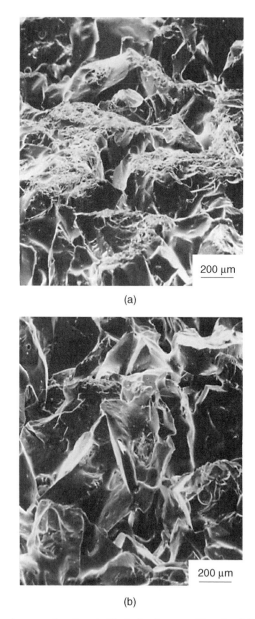

(a)

(b)

Figure 4-9 SEM micrographs of a vitrified aluminum oxide wheel (32A46I8VBE) after (a) fine dressing and (b) coarse dressing with a single-point diamond [20].

two single-point dressing parameters, the dressing lead is found to have a much bigger relative influence than the dressing depth, as judged by grinding performance [20].

This same concept of coarser versus finer dressing is also applicable to other dressing tools and processes. With multipoint diamond dressing tools, the dressing lead and depth have a similar influence as with single-point dressing. With rotary diamond dressing (Figure 4-2), it has been suggested that the relative dressing severity depends on the angle δ at which the diamonds on the dresser surface initially cut into the wheel, which can be written, following the definitions given in section 4.2, as [21]

$$\delta = \tan^{-1}\frac{v_i}{|v_s - v_r|} \qquad (4\text{-}9)$$

A larger angle δ tends to favor more fracture and consequently less plastic deformation of the grain material. Coarser dressing (larger δ) is obtained with a bigger radial infeed velocity and smaller difference between the wheel velocity v_s and dresser velocity v_r. At the extreme condition of equal wheel and dresser velocities in the same direction ($v_r/v_s = +1$), there is only normal relative motion between the wheel and dresser surfaces ($\delta = 90°$), thereby producing a wheel surface which is very rough and very sharp. A more coarsely dressed wheel using crush dressers or impregnated diamond dressing blocks is obtained by feeding the wheel more rapidly into the dressing tool.

Numerous other factors are also likely to affect the dressed wheel topography or dressing severity. With continued use, a single-point diamond tends to dull at its tip and its average radius becomes bigger, thereby reducing the localized stresses as it cuts into the abrasive grain. This should increase the dressing force and the likelihood of bond fracture instead of grain fracture, thereby leaving fewer active grains at the wheel surface, but the prevailing effect appears to be a greater degree of wheel dulling due to plastic flow instead of fracture at the grain tips. This uncontrolled variation in the grinding behavior may cause difficulties, especially in automated production.

4.5   GENERATION OF WHEEL TOPOGRAPHY - SUPERABRASIVES

While the discussion in the preceding section deals with conventional abrasive wheels, the same general concepts might also apply to superabrasive wheels. Much effort has been directed toward the development of truing and dressing methods for superabrasive wheels, but little has been reported concerning the mechanisms whereby the wheel topography is generated.

Brake-controlled truing of resin and metal bonded diamond wheels with silicon carbide wheels is generally considered to occur by erosion of bond material to the point that diamond grains fall out. But more severe dressing conditions, obtained by slowing down the diamond wheel velocity so as to reduce the 'slip' velocity between the grinding and dressing wheels, result in fracture of the diamond grains and a less open wheel structure [22]. As grinding wheels are normally trued at their full operating speed, whole grain dislodgement is likely to prevail, at least with the common types of diamond wheels used for grinding hard and brittle materials. This would imply a large percentage of bond fracture B, perhaps approaching 100%, and consequently a very small active grain ratio ($G_a/G_0 \ll 1$) according to Eq. (4-8).

The situation with CBN wheels is very different from diamond wheels. Brake-controlled truing of resin-bonded CBN wheels with silicon carbide wheels causes flattening of the grain tips, as if the CBN grains are being polished down. This can be seen in Figure 4-10 which shows SEM micrographs of the wheel surface after brake-controlled truing [4]. This behavior is rather surprising insofar as CBN is almost twice as hard as silicon carbide. Many flattened CBN grains are also fractured (Figure 4-10(b)). Grain dislodgement appears to be a rare event, although it becomes an important factor during the early stages of subsequent grinding (Chapter 5). This explains why the number of exposed CBN grains per unit area of wheel surface after brake-controlled truing has been found to be comparable to the theoretical packing density G_0 (Eq. (4-6)). Multipoint diamond truing of resin-bonded CBN wheels also leads to flattening down of the

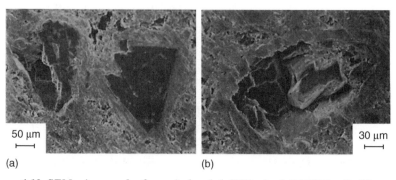

(a)   (b)

Figure 4-10 SEM micrograph of a resin-bonded CBN wheel (100/120 grit, 75 concentration) after brake-controlled truing [4]. Note grit flattening in (a) and flattening and fracture in (b).

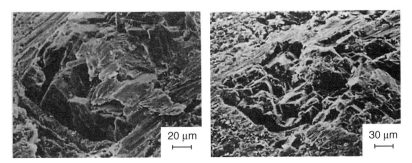

Figure 4-11 SEM micrograph of a grain in a CBN wheel as in Figure 4.10 after sharpening with a wire brush [4]. Note multiple sharp cutting points.

grain tips and grain fracture. In this case, however, the 'polished' grain tips are somewhat rougher than with brake-controlled truing.

After truing, dressing of resin and metal bonded diamond and CBN wheels with abrasive sticks exposes the superabrasive grains by removing binder, together with metal coating on the grains in the case of resin-bonded wheels. This dressing process seems to have little or no effect on the diamond or CBN itself. It is often claimed that stick dressing provides clearance for chip removal, as well as improved access of the grinding fluid to the cutting points. It also reduces or eliminates frictional contact between the binder and the workpiece.

While truing and dressing as described above are usually adequate for diamond wheels, resin-bonded CBN wheels usually require an additional or alternative dressing process to sharpen the wheel (section 4.3). An example of what can be obtained using one of these methods (a wire brush mounted on a brake-controlled truing device) is seen in Figure 4-11 [4]. A sharpened wheel cuts much more efficiently with lower forces and power, but the surface finish is somewhat poorer (see Chapter 10).

With vitreous-bonded CBN wheels, flattening of the grains would also be likely to occur. However, this is much less of a problem than with resin-bonded CBN wheels especially with rotary disk dressing using positive ratios of dresser velocity to wheel velocity. This can be seen in Figure 4-12 which show the fragmented surface of a CBN grain which was trued with a diamond disk dresser using a speed ratio of $v_r/v_s = +0.8$ [6]. The use of a bigger (more positive) speed ratio tends to cause more grain fragmentation and a sharper wheel, analogous to what is found with rotary diamond dressing of conventional abrasive wheels, and this is also reflected in the grinding behavior (see Chapter 5). A bigger dressing lead and dressing depth also tend to result in a somewhat sharper CBN wheel, although the speed ratio has a much greater effect.

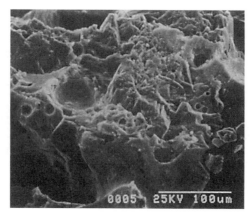

Figure 4-12 SEM micrograph showing fragmentation of a 80/100 mesh CBN grain in a vitrified wheel for disk truing with $v_r/v_s = +0.8$ [6].

4.6   MEASUREMENT OF WHEEL TOPOGRAPHY

Numerous techniques have been developed for the purpose of measuring and characterizing grinding wheel topography. The various approaches to the problem include: profilometry, imprint methods, scratch methods, dynamometry methods, thermocouple measurements, and microscopy. Each measuring technique has its advantages and limitations, according to such factors as resolution, measuring depth, ease of application, and data analysis and interpretation. With each new technique, additional insight is gained into the nature of the grinding wheel surface. No single approach to the problem provides a complete description of the grinding wheel topography in a three-dimensional space.

In the remainder of this chapter, the various techniques for measuring wheel topography will be described. For the purpose of illustrating and comparing some of the methods, results will be cited from a CIRP cooperative study involving measurements at different research laboratories with the same wheels dressed in the same way using the same dressing tools [23].

4.6.1 Profilometry methods

Profilometry of wheel surfaces is similar, in principle, to surface roughness measurement. A stylus coupled to a displacement transducer is dragged over the wheel surface to obtain a profile trace [12, 22-33]. An example of a wheel profile is shown in Figure 4-13. For computer analysis, the profile is usually digitized, so some information is invariably lost. Abrasive grains and cutting points (edges) are identified within the profile according to arbitrarily defined criteria. For example, a peak in the profile might be counted as a cutting point only if it protrudes by at least 5 μm above adjacent valleys ($\Delta h = 5$ μm), and cutting points may be defined as belonging to the same grain if

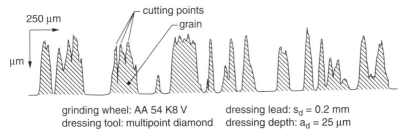

grinding wheel: AA 54 K8 V   dressing lead: s_d = 0.2 mm
dressing tool: multipoint diamond   dressing depth: a_d = 25 μm

Figure 4-13 Profile trace of dressed grinding wheel [23].

they are spaced closer than one grain dimension d_g. The profile signal can be stochastically decomposed into a primary component whose wavelength corresponds to the average grain spacing, and a secondary component of much shorter wavelength which might represent the cutting-point spacing [34].

The results obtained from wheel profilometry can be used directly as the input for simulation of the grinding process, as discussed in Chapter 3. A radial distribution of cutting points and grains can also be presented as a plot of the cumulative number of cutting points per unit length along the wheel versus the radial distance into the wheel. (The inverse of the number of cutting points per unit length corresponds to the cutting-point spacing L.) An example is shown in Figure 4-14. In this case C'_{stat} and G'_{stat}, refer to the static number of cutting points and grains per unit length, respectively. The *dynamic* number of cutting points and grains per unit length, C'_{dyn}, and

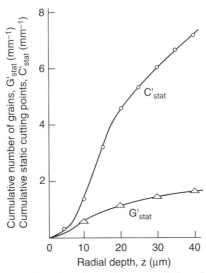

Figure 4-14 Cumulative number of cutting points per unit length (C'_{stat}) and grains per unit length (G'_{stat}) versus radial depth into wheel. Corresponds to conditions in Figure 4-13 [23].

G'_{dyn}, would represent the actual number in each case, taking into account the kinematic trajectories of successive cutting points for the actual grinding conditions (see section 3.6.).

There are numerous practical drawbacks and limitations associated with wheel profilometry. It is apparent that a wheel profile does not present a three-dimensional picture of the wheel-surface, but is rather intended to provide a cross-section of the wheel topography. It must also be remembered that the finite thickness of the stylus tip (of the order of 10-20 μm) introduces lateral thickening of the peaks, since a peak is detected by the right side of the tip on one side and by the left on the other [23, 25]. The finite stylus radius, as well as digitizing the data for subsequent analysis limits the resolution such that the detailed morphology of the sharp edges and jagged peaks where actual cutting occurs is not fully recorded. Also, the radial penetration of the stylus into the wheel is limited to about 70 μm, or less, which is greater than the typical radial working depth into the wheel but smaller than the dressing-affected depth. Special profilometers have been developed to provide deeper profile depths, while also avoiding stylus hang-up or sticking in the wheel crevices [36, 37].

4.6.2 Imprint methods

Imprint methods generally consist of replicating or imprinting the wheel surface on a second body. One of the earliest reported measurements of wheel topography of this type was the soot-track method, which was developed to provide a value for the number of cutting points per unit area (C) to calculate undeformed chip thickness (Eqs. (3-59), (3-62), and (3-64)) [38]. The soot-track method consists of rolling a grinding wheel over a soot-coated glass slide. Each cutting point is assumed to remove a particle of soot, so counting the number of points where soot was pulled away provides an estimate for the number of cutting points per unit area. For the particular wheel tested (32A46H8VBE), there were found to be approximately three cutting points per square millimeter. Similar techniques have also been described using glass coated with dye instead of soot [39] and carbon paper between the wheel surface and the glass [40].

Another type of imprint method was subsequently developed which also measures the radial distribution by rolling the wheel surface against a plastic tapered roller mounted on a freely rotating axis parallel to the wheel axis [41]. Owing to the taper (1°), the penetration depth of the wheel surface into the softer plastic increases across the wheel width towards the larger end of the roller. The imprints left behind by contact with the abrasive grain tips are recorded by carbon paper against smooth white paper inserted between the plastic roller and the grinding wheel. The carbon spots on the white paper are counted, and the results are reported either as a number per unit area or per unit length.

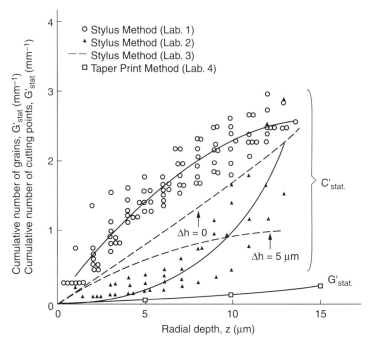

Figure 4-15 **Cumulative number of cutting points per unit length (C'_{stat}) and grains per unit length (G'_{stat}) from various laboratories using profilometry (stylus method) and taper print method [23]. Curves without data points are shown for Lab. 3, which include all peaks detected for $\Delta h = 0$ and only those protruding 5 μm or more from adjacent valleys for $\Delta h = 5$ μm.**

It is not clear whether these imprint methods actually measure grain density, cutting-point density, or perhaps something between the two. In the cooperative CIRP study involving several laboratories, a comparison was made of various profilometer (stylus) and taper print measurements on identical wheels dressed in the same way [23]. The results in Figure 4-15 show a much lower density with the taper print method, from which it was concluded that it actually measures the number of grains (G'_{stat}), not cutting points (C'_{stat}). It is also apparent that the profilometer (stylus) results from the various laboratories show a rather large scatter, which may be mostly attributed to the criteria adopted in analyzing the profile. For example, it is seen in the results from Lab. 3 (data points not shown) that the cumulative cutting-point density is significantly higher when counting 'all' the peaks detected in the profile ($\Delta h = 0$) rather than only those which protrude 5 μm or more above adjacent valleys ($\Delta h = 5$ μm).

4.6.3 Scratch methods

Scratch methods are based upon generating isolated scratches by straight surface grinding a single pass along a smooth flat workpiece. In order to obtain isolated scratches, instead of a ground surface consisting of overlapping scratches, the workpiece is moved relatively rapidly past a slowly turning wheel. Originally, the plane of the workpiece surface was inclined at a very small angle to the axis of the wheel to give a progressively increasing penetration depth across the width of the wheel [42], as in the taper print method, but a major difficulty appears to have been that of determining accurately the extremely small angles ($\approx 0.01°$) involved.

A similar scratch method was subsequently introduced, but without the need for tilting the workpiece [43]. With this latter method, the radial elevation of each cutting point is geometrically calculated from the length of the scratch it makes: cutting points protruding further from the wheel make longer scratches. Examples of the radial cutting-point distributions are shown in Figure 4-16 for conventional wheels of different grit sizes, and the same method has also been applied to diamond wheels [44]. This method gives a very high resolution (≈ 0.1 μm), but only measures the extreme outer portion of the wheel to a radial depth of 1-2 μm. The cumulative cutting-point densities in this region appear almost the same for the wheels of 30-, 46-, and

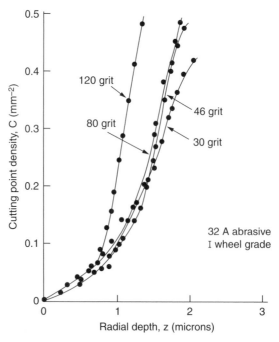

Figure 4-16 Cumulative cutting-point density versus radial depth obtained using scratch method for wheels of various grit sizes [43].

80-grit size, and slightly higher for the 120-grit size. These variations may seem too small, considering the relatively big differences in grain dimension. However, it should be borne in mind that it is the cumulative number of cutting points (scratches), not grains, which are given in Figure 4-16. There are fewer active grains per unit area with coarser-grained wheels, but this effect is offset by bigger grains each having more cutting points. This becomes evident by the higher incidence of multiple closely spaced side-by-side parallel scratches observed with coarser-grained wheels, which often can be seen to merge together when the scratches are deep enough.

Of the various measuring techniques, the scratch method provides the most detailed picture of the cross-sectional shape of a cutting point [43]. By measuring the width of an isolated scratch at various distances from one end, and in turn calculating the scratch depth from the distance, results are obtained for the width versus depth along the scratch. One example of the results obtained is shown in Figure 4-17. In this, as well as in numerous other cases over a range of grit sizes, the cross-sectional scratch shape was

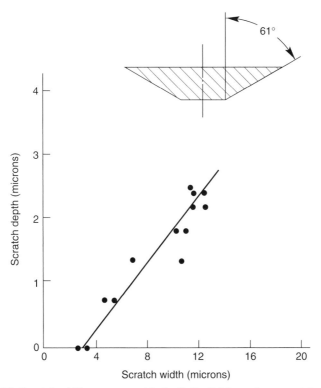

Figure 4-17 Scratch width versus scratch depth for 1.24 mm-long scratch obtained with 32A46I8VBE wheel. Also shown above is trapezoidal cross-sectional shape at maximum scratch depth [43].

found to be approximately trapezoidal with typically a base width of 1-3 μm and a semi-included angle of 55–70°.

4.6.4 Dynamometer and thermocouple methods

Dynamometer and thermocouple methods are based upon dynamically identifying the presence of cutting points during grinding from force pulses and thermal pulses, respectively. With the dynamometer method, forces are sensed while grinding an extremely small workpiece mounted in a high-frequency piezoelectric dynamometer. The grinding area, either a razor blade edge [45] or a tiny rectangular block [46, 47], is small enough to enable identification of individual force pulses corresponding to discrete interactions between abrasive grains and the workpiece. The thermocouple method works according to a similar principle, with cutting points identified from heat pulses generated as the abrasive grains interact with the small area of a thermocouple wire embedded in the workpiece [39, 48-50].

Using the dynamometer method with the razor blade, the force pulses are measured while grinding along the blade's length. Essentially, this corresponds to straight plunge grinding of a very narrow workpiece whose width equals the blade thickness (Figure 3-1(a)). With the tiny rectangular workpiece, the wheel rotating with a peripheral velocity v_s is fed down directly into the rectangular workpiece, which is held stationary under the wheel, at an infeed velocity v_r. The kinematic conditions provide a constant infeed angle ε along the grinding zone, as opposed to plunge grinding along the razor blade where the infeed angle ε varies along the grinding contact length (Chapter 3). Therefore, the force pulses can be measured as a function of the infeed angle ε (tan $\varepsilon = v_f/v_s$), which was shown in the previous chapter to be the kinematic parameter which uniquely controls the dynamic cutting-point density (C_{dyn}) for a given wheel topography.

Of the two dynamometer methods, measurements using the razor blade workpiece give a significantly higher cutting-point density than the rectangular workpiece. This may be attributed to differences in working conditions. In Figure 4-18, results for C_{dyn} obtained in the CIRP cooperative study using both methods are shown together, as a function of tan ε for the rectangular specimen, and tan $\bar{\varepsilon}$ corresponding to $(v_w/v_s)(a/d_e)^{1/2}$ (Eq. (3-35)) for plunge grinding of the razor blade edge [23]. These results suggest that the much lower values for C_{dyn} obtained with the rectangular specimen as compared with the razor blade can be mainly attributed to smaller infeed angles. With the rectangular specimen, test conditions have been generally limited to tan $\varepsilon < 20 \times 10^{-6}$, which is on the low end of typical values of $(v_w/v_s)(a/d_e)^{1/2}$ for most production grinding. Flattening of the curve at higher values of tanε at $C_{dyn} \approx 2.2$ mm^{-2} suggests that what is really being measured is not the number of cutting points per unit area, but rather the number of active grains per unit area G_{dyn}. It seems unlikely that multiple cutting

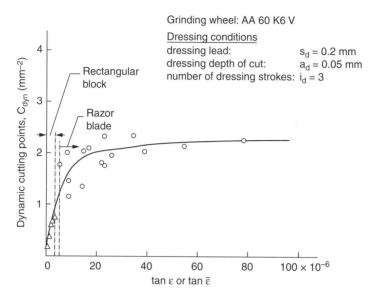

Figure 4-18 Dynamic cutting-point density obtained using dynamometer method [23]. Data for the rectangular block are shown versus tanε, and data for razor blade versus tan$\bar{\varepsilon}$

points, located more or less side by side on the leading edge of the same grain, would give force pulses which can be separately distinguished in these experiments. Furthermore, the value of $C_{dyn} = 2.2$ mm^{-2} would correspond to the grain density G_a in Eq. (4-8) and Figure 4-8 with 70% bond fracture during dressing (B = 0.7), which would be a reasonable value for this wheel.

4.6.5 Microscopic methods

Microscopic methods involve the use of an optical microscope or scanning electron microscope (SEM) to observe and measure topographic features of the wheel surface. The simplest method utilizes an optical metallurgical microscope with a built-in vertical illuminator [39, 51-55]. For convenience, the microscope may be mounted directly on the machine, thereby making it unnecessary to remove the wheel for observation. When viewing normal to the wheel surface (typical magnification 50-250X), flattened areas on the grain tips reflect light and appear shiny against a dark background. These flat areas are initially dressed onto the grains, but it is only after grinding that they become shiny and easily identifiable owing to adhering workpiece metal. Such areas are often referred to as 'wear flats', since after dressing they are also associated with attritious wear of the grain by rubbing (Chapter 5).

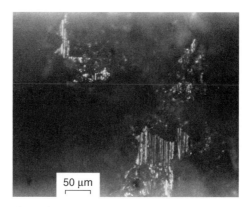

Figure 4-19 Optical micrograph showing shiny wear flats on grain tips.

An example of wear flats observed by optical microscopy on a used grinding wheel is shown in Figure 4-19. For the purpose of measurement and analysis, a number of sample viewing areas may be randomly selected. The number of distinct wear-flat areas counted per unit area of viewing on the wheel surface is sometimes interpreted as the cutting-point density [39, 40, 51, 53]. On the average, however, there is more than one wear-flat area per active grain. In order to count the number of active grains per unit area having wear flats, the focus on the wheel surface may be altered and side lighting added to observe whether adjacent wear flats belong to the same or to different grains [55]. The grain density G_a in Figure 4-8 was measured in this way after having taken one grinding pass.

The average number of cutting points per unit length can be estimated as the number of wear flats per unit length intersecting a straight line superposed in the viewing area. The inverse of this parameter is assumed to correspond to the average spacing between successive cutting points. The average grain spacing can also be determined in a similar manner.

The fraction of the total viewing area consisting of flattened areas is called the 'wear-flat area' or 'cutting-edge ratio'. This parameter may be determined with the aid of a fine grid superposed on the field of view [55]. It will be seen (Chapter 5) that the wear-flat area provides a useful quantitative measure of wheel dullness and is directly related to the grinding forces.

Since the microscopic measurements are made on wheel surfaces after grinding, the concentration of the cutting points and grain densities and their spacing represent dynamic values for the grinding conditions which were used. In distinguishing between static and dynamic conditions (Chapter 3), it was tacitly assumed in the analysis that successive prospective cutting points follow precisely the same track, one behind the other. This is not true, and some degree of overlapping must result depending upon the effective width of the cutting point or grain. Therefore, the difference between static and dynamic values might be much less than predicted from the analysis.

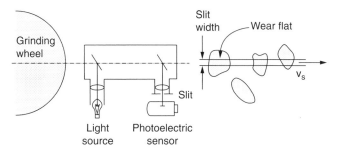

Figure 4-20 Illustration of an arrangement for counting wear flats and measuring the wear-flat area using a photoelectric sensor [57, 58].

Particularly in the case of active abrasive grains, which are relatively wide and widely spaced apart, the static and dynamic grain density may be very nearly the same. This could account for the flattering of the curve for C_{dyn} with the razor blade method (Figure 4-18) at a value which is reasonably consistent with that to be expected for G_a (Eq. (4-8)), as mentioned above. Most production grinding is carried out in this range.

These measurements using optical microscopy are extremely tedious, as they require counting the number of wear flats and measuring the combined areas. The measuring process can be automated by using an image processing system of the type normally used for automated quantitative microscopy [56]. A different approach allowing even for in-process measurement during grinding is based upon the use of a photoelectric sensor for detection of light pulses reflected from the wear-flat areas, as illustrated in Figure 4-20 [57, 58]. Owing to the rather slow response time of such a system, the technique usually provides more reliable results if it is applied post-process to a slowly rotating wheel rather than in-process to the rapidly spinning wheel.

The use of optical microscopy has also been applied in different ways. By viewing tangentially to the wheel surface at its periphery, it is possible to count abrasive grains and observe their profiles [53]. The radial distribution of abrasive grain tips can also be measured by viewing at a glancing angle to the wheel surface as illustrated in Figure 4-21 [43, 59]. While looking through the microscope, the wheel is very slowly rotated by hand until some point on a grain tip comes into sharp focus. Since the depth of focus is very shallow, all the grain tips can be considered to come into focus within the same vertical focal plane. The height h above a reference point where a grain tip comes into focus is measured, from which its relative radial distance z into the wheel is calculated. The resolution and maximum radial depth depend on the viewing angle α, which is analogous to the infeed angle ε (see Figure 3-3). With a smaller viewing angle α, the resolution is finer but the radial measuring depth is less because grains can be hidden from the field of view. With a viewing angle of $\alpha = 40°$, the radial distribution of abrasive grain tips can be measured to a depth of approximately 1 mm.

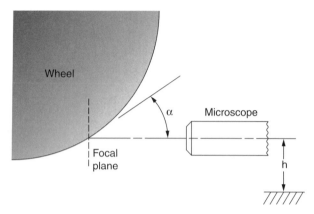

Figure 4-21 *Arrangement of grinding wheel and microscope for measuring radial distribution of abrasive grains. The radial depth is calculated from the height* **h** *at which the grain tip comes into sharp focus.*

More recently, optical microscopy has also been used to measure the packing density and grain tip height distribution on electroplated single-layer diamond and CBN wheels [60-62]. The effect of grain dimension on the areal packing density for CBN wheels is shown in Figure 4-22 for grain sizes ranging from 60 mesh (262 μm) to 270 mesh (52 μm). The number of grains per unit area decreases with grain dimension, but the dependence is less than

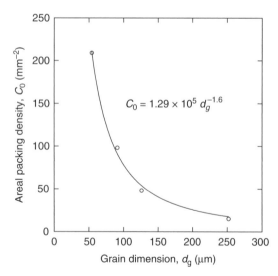

$$C_0 = 1.29 \times 10^5 \, d_g^{-1.6}$$

Figure 4-22 *Effect of grain dimension on areal packing density for electroplated CBN wheels [62].*

would be expected if this effect were due only to the grain size. The packing is relatively tighter with coarser grains. Assuming a simple square packing arrangement, the data in Figure 4-22 indicate a mean grain spacing of approximately 1.3 times the grain dimension for the finest 54 μm micron grain size and 1.1 times the grain dimension for the coarsest 262 μm grain size.

The height distributions of the grain tips on these same electroplated superabrasive wheels were measured using optical microscopy as illustrated in Figure 4-23. Viewing normal to the wheel surface, the heights where grain tips come into focus were measured relative to the nickel binder

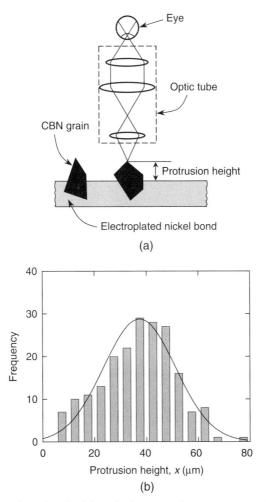

Figure 4-23 Illustration of method for wheel topography measurement in (a) and results for height distribution of new 180 grit electroplated CBN wheel in (b) [61, 62]

surface with the vertical micrometer in the microscope. Also included in this figure is an example showing measurements for a new 180 mesh electroplated CBN wheel. For this and other new CBN electroplated wheels with grain sizes ranging from 60 mesh (262 μm) to 270 mesh (50 μm), the height distributions were found to be normally distributed with a mean in each case of approximately half the grain dimension, a standard deviation one-sixth of the grain dimension, and maximum grain tip protrusion approximately equal to the grain dimension. Those grains protruding the most would be most weakly held in the nickel bond and be most likely to dislodge during subsequent grinding.

Scanning electron microscopy (SEM) is a powerful tool for observing the morphology of abrasive grains insofar as it provides for both high magnification and large depth of field. For this purpose, special wheels may be used with removable segments which are small enough to fit into the SEM chamber, or sections can be cut from a used wheel [53, 63-65]. Another possibility is to make a replica of the wheel surface [66], although this may be complicated by porosity at the wheel surface. The wheel topography can also be quantitatively described by topographical contour maps obtained from stereographic SEM photomicrographs of the wheel surface [67]. By comparing the contour maps with the corresponding SEM photographs, it appears that this method actually measures the radial distribution of grain tips. Much of the fine structural detail of the abrasive grain is lost in the mapping owing to the finite height differential (10 μm) between elevation contours.

REFERENCES

1. Verkerk, J. and Pekalharing, A. J., 'The Influence of the Dressing Operation on Productivity in Precision Grinding', *Annals of the CIRP*, 28/2, 1979, p. 385.
2. Anon., 'The Pre-grinding Preparation of Diamond Grinding Wheels', Diamond Information L24, De Beers Industrial Diamond Division, Ascot, England.
3. Saljé, E. and Mackensen, G. V., 'Dressing of Conventional and CBN Wheels with Diamond Form Dressers', *Annals of the CIRP*, 33/1, 1984, p. 205.
4. Pecherer, E. and Malkin, S., 'Grinding of Steels with Cubic Boron Nitride (CBN)', *Annals of the CIRP*, 33/1, 1984, p. 211.
5. Jacobuss, M., and Webster, J. A., 'Optimizing the Truing and Dressing of Vitrified-Bond CBN Grinding Wheels', *Abrasives Magazine*, Aug./Sept 1996, p. 23.
6. Ishikawa, T., and Kumar, K. V., 'Conditioning of Vitrified Bond Superabrasive Wheels', *Superabrasives '91*, SME, June 11–13, 1991, p. 7–91.
7. Pahlitzsch, G. and Appun, J., 'Effect of Truing Conditions on Circular Grinding', *Industrial Diamond Review*, 14, 1954, p. 185 and p. 212.
8. Vickerstaff, T. H., 'Diamond Dressing—Its Effect on Work Surface Roughness', *Industrial Diamond Review*, 30, 1970, p. 260.

9. Bhateja, C. P., Chisholm, A. W. J. and Pattinson, E. J., 'The Influence of Grinding Wheel Wear and Dressing on the Quality of Ground Surfaces', *Proceedings of the International Grinding Conference*, Pittsburgh, 1972, p. 685

10. Hahn, R. S. and Lindsay, R. P., 'The Influence of Process Variables on Material Removal, Surface Integrity, Surface Finish, and Vibration in Grinding', *Proceedings of the Tenth International Machine Tool Design and Research Conference*, 1969, p. 95.

11. Pattinson, E. J. and Lyon, J., 'The Collection of Data for the Assessment of Grinding Wheel Dressing Treatment', *Proceedings of the Fifteenth International Machine Tool Design and Research Conference*, 1974, p. 317.

12. Bhateja, C. P., 'The Intrinsic Characteristics of Ground Surfaces, *Proceedings of the Abrasive Engineering Society's International Technical Conference*, 1975, p. 139.

13. Verkerk, J., 'Kinematical Approach to the Effect of Wheel Dressing Conditions on the Grinding Process', *Annals of the CIRP*, 25/1 1976, p. 209.

14. Malkin, S. and Cook, N. H., 'The Wear of Grinding Wheels. Part II. Fracture Wear', *Trans. ASME, J. of Eng. for Ind.,* ,93, 1971, p. 1129.

15. Malkin, S. and Anderson, R. B., 'Active Grains and Dressing Particles in Grinding', *Proceedings of the International Grinding Conference*, Pittsburgh 1972, p. 161.

16. Vickerstaff, T. J., 'The Influence of Wheel Dressing on the Surface Generated in the Grinding Process', *Int. J. Mach. Tool Des. Res.*, 16, 1976, p. 145.

17. Pande, S. J. and Lal, G. K., 'Effect of Dressing on Grinding Wheel Performance', *Int. J. Mach. Tool Des., Res.*, 19, 1979, p. 171.

18. Bhateja, C. P., 'On the Mechanism of the Diamond Dressing of Grinding Wheels', *Proceedings of the International Conference on Production Engineering*, Tokyo, 1974, p. 733.

19. Sasaki, T. and Okamura, K., 'The Cutting Mechanism of Fine-grain Abrasive Stone', *Memoirs of Faculty, Kyoto University*, Oct. 1959, p. 367.

20. Malkin, S. and Murray, T., 'Comparison of Single Point and Rotary Dressing of Grinding Wheels', *Proceedings, Fifth North American Metalworking Research Conference*, 1977, p. 278.

21. Malkin, S. and Murray, T., 'Mechanics of Rotary Dressing of Grinding Wheels', *Trans. ASME, J. of Eng. for Ind.*, 100, 1978, p. 95.

22. Davis, C. E., 'The Dependence of Grinding Wheel Performance on Dressing Procedure', *Int. J. Mach. Tool Des. Res.*, 14, 1974, p. 33.

23. Verkerk, J., 'Final Report concerning CIRP Cooperative Work on the Characterization of Grinding Wheel Topography', *Annals of the CIRP*, 26/2, 1977, p.

24. Bruckner, K., 'Die Schneidflache der Schleifscheibe und ihr Einfluss auf die Schnittkrafte beim Aussenrundeinstechschleifen', *Industrie-Anzeiger*, 86 1964, No. 11, p. 173.

25. König, W. and Lortz, W., 'Properties of Cutting Edges Related to Chip Formation in Grinding', *Annals of the CIRP*, 24/1, 1975, p. 231.

26. McAdams, H. T., 'The Role of Topography in the Cutting Performance of Abrasive Tools', *Trans. ASME, J. of Eng. for Ind.*, 86, 1964, p. 75.

27. Baul, R. M. and Shilton, R., 'Mechanics of Metal Grinding with Particular Reference to Monte Carlo Simulation', *Proceedings of the Eighth International Machine Tool Design and Research Conference*, Part 2, 1967, p. 923

28. Hasegawa, M., 'Statistical Analysis for the Generating Mechanism of Ground Surface Roughness', *Wear*, 29, 1974, p. 31.

29. Story, R. W. and Keyes, E. J., 'Profile Measurement of Coated Abrasives', *Trans. ASME, J. of Eng. for Ind.*, 91, 1969, p. 597.

30. Kaliszer, H., Grieve, D. J. and Rowe, G. W., 'A Normal Wear Process Examined by Measurements of Surface Topography', *Annals of the CIRP*, 20, 1971, p. 76.

31. Bhattacharyya, S. K. and Hill, C. G., 'Characterization of Grinding Wheel Topography and Wear', *Proceedings of the Seventeenth International Machine Tool Design and Research Conference*, 1976, p. 171.

32. Hong, I. S., McDonald, W. J. and Duwell, E. J., 'A New Wheel Profilometer for the Analysis of Coated Abrasive Surfaces', *Proceedings of the International Grinding Conference, Pittsburgh*, 1972, p. 182.

33. Brough, D., Bell, W. F. and Rowe, W. B., 'A Re-examination of the Uncut Model of Grinding and its Practical Implications', *Proceedings of the Twenty Fourth international Machine Tool Design and Research Conference*, 1983, p. 261.

34. Pandit, S. M. and Sathyanarayan, G., 'A Model for Surface Grinding Based on Abrasive Geometry and Elasticity', *Trans. ASME, J. of Eng. for Ind.*, 104, 1982, p. 349.

35. Shaw, M. C. and Komanduri, R., 'The Role of Stylus Curvature in Grinding Wheel Surface Characterization', *Annals of the CIRP*, 25/1, 1977, p. 139.

36. Deutsch, S. 1., Wu, S. M. and Stralkowski, C. M., 'A New Irregular Surface Measuring System', *Int. J. Mach. Tool Des. Res.*, 13, 1973, p. 29.

37. Shah, G., Bell, A. C. and Malkin, S., 'Quantitative Characterization of Abrasive Surfaces using a New Profile Measuring System', *Wear*, 41, 1977, p. 315.

38. Backer, W. R., Marshall, E. R. and Shaw, M. C., 'The Size Effect in Metal Cutting', *Trans. ASME*, 74, 1952, p. 61.

39. Red'ko, S. G., 'The Active Grits on Grinding Wheels', *Machines and Tooling*, 31, No. 12, 1960, p. 11.

40. Peklenik, J., 'Untersuchen uber das Verschleisskriterium beim Schleifen', *Industrie-Anzeiger*, 80, 1958, p. 280.

41. Nakayama, K., 'Grinding Wheel Geometry', *Proceedings of the International Grinding Conference*, Pittsburgh, 1972, p. 197.

42. Nakayama, K. and Shaw, M. C., 'A Study of the Finish Produced in Surface Grinding, Part 2, Analytical', *Proc. Institution of Mechanical Engineers*, 182 pt. 3K, 1967–68, p. 179.

43. Thompson, D. L. and Malkin, S., 'Grinding Wheel Topography and Undeformed Chip Shape', *Proceedings of the International Conference on Production Engineering, Part 1*, Tokyo, 1974, p. 727.

44. Zelwer, O. and Malkin, S., 'Grinding of WC-Co Cemented Carbides', *Trans. ASME, J. of Eng. for Ind.*, 102, 1980, p. 173.

45. Brecker, J. N. and Shaw, M. C., 'Measurement of the Effective Number of Cutting Points in the Surface of Grinding Wheel', *Proceedings, International Conference on Production Engineering, Part I*, Tokyo, 1974, p 740.

46. Tigerström, L., A Model for Determination of Number of Active Grinding Edges in Various Grinding Processes', *Annals of the CIRP*, 24/1, 1975, p. 271.

47. Tigerström, L., 'Dynamic Measuring of Number of Grinding Edges and Determination of Chip Parameters, *Manufacturing Engineering Transactions*, SME, 1981, p. 1.

48. Peklenik, J. 'Ermittlung von geometrischen und physikalischen Kenngrossen fur die Grundlagenforschung des Schleifen', Dissertation, T. H. Aachen, 1957.

49. Peklenik, J., 'The Statistical Mechanism of Chip Formation in Grinding Process', *Proceedings, International Conference on Production Engineering, Part 2*, Tokyo, 1974, p. 51.

50. Verkerk, J., 'The Real Contact length in Cylindrical Plunge Grinding', *Annals of the CIRP*, 24/1, 1975, p. 259

51. Grisbrook, H., 'Cutting Points on the Surface of a Grinding Wheel and Chips Produced', *Proceedings of the Third International Machine Tool Design and Research Conference*, 1962, p. 155.

52. Yoshikawa, H., 'Criterion of Grinding Wheel Tool Life', *Bull. Japan Society of Grinding Engineers*, 3, 1963, p. 29.

53. Tsuwa, H., 'An Investigation of Grinding Wheel Cutting Edges', *Trans. ASME, J. of Eng. for Ind.*, 86, 1964, p. 371.

54. Hahn, R. S. and Lindsay, R. P., 'On the Effects of Real Area of Contact and Normal Stress in Grinding', *Annals of the CIRP*, 15, 1967, p. 197.

55. Malkin, S. and Cook, N. H., 'The Wear of Grinding Wheels. Part I, Attritious Wear', *Trans. ASME, J. of Eng. for Ind.*, 93, 1971, p. 1120.

56. Besuyen, A. M., 'The Measurement of the Grinding Wheel Wear with the Quantimet Image Analyzing Computer', *Annals of the CIRP*, 19, 1971, p. 619.

57. Sata, T., Suto, T., Waida, T. and Noguchi, H., 'In-process Measurement of the Grinding Process and its Application', *Proceedings of the International Grinding Conference*, Pittsburgh, 1972, p. 752.

58. Suto, T., Waida, T. and Sata, T., 'In-process Measurement of Wheel Surface in Grinding Operations', *Proceedings of the Tenth International Machine Tool Design and Research Conference*, 1969, p. 171.

59. Red'ko, S. G. and Korolev, A. V., 'Distribution of Abrasive Grains on Grinding Wheel Faces', *Machines and Tooling*, 41, No. 5, 1970, p. 64.

60. Hwang, T. W., Evans, C. J., Malkin, S., 'High Speed Grinding of Silicon Nitride with Electroplated Diamond Wheels, II. Wheel Topography and Grinding Mechanisms', *Trans. ASME, Journal of Manufacturing Science and Engineering*, 122, 2000, p. 42.

61. Shi, Z., and Malkin, S., 'Investigation of Grinding with Electroplated CBN Wheels', *Annals of the CIRP*, 53/1, 2003, p. 267.

62. Shi, Z., and Malkin, S., 'Wear of Electroplated CBN Grinding Wheels', *Trans. ASME, Journal of Manufacturing Science and Engineering*, 128, 2006, p. 110.

63. Besuyen, A. M. and Verkerk, J., 'Scanning Electron Microscope Observation of Grinding Grains', *Annals of the CIRP*, 21/1, 1971, p. 65.

64. Kirk, J. A., 'An Evaluation of Grinding Performance for Single and Polycrystal Grit Aluminum-oxide Grinding Wheels', *Trans. ASME, J. of Eng. for Ind.*, 98, 1976, p. 189.

65. Foerster, M. and Malkin, S., 'Wear Flats Generated during Grinding with Various Grinding Fluids', *Proceedings, Second North American Metalworking Research Conference*, 1974, p. 601.

66. Komanduri, R. and Shaw, M. C., 'A Technique for Investigating the Surface Features of a Grinding Wheel', *Annals of the CIRP*, 23/1, 1974, p. 95.

67. Matsuno, Y., Yamada, H., Harada, M. and Kobayashi, A., 'The Microtopography of the Grinding Wheel Surface with SEM', *Annals of the CIRP*, 24/1, 1975, p. 237.

BIBLIOGRAPHY

Carius, A. C., 'Preliminaries to Success-Preparation of Grinding Wheels Containing CBN', SME Paper No. MR84–547, 1984.

Farrago, F. T., *Abrasive Methods Engineering*, Volume II, Industrial Press, New York, 1980, Chapters 6-4 and 7-5.

Hitchiner, M. P., 'Dressing of Vitrified CBN Wheels for Production Grinding' *Ultrahard Materials Technical Conference*, Windsor, Ontario, Canada, May 28, 1998, p. 139.

Hughes, F. H., 'The Importance of Preparing a Diamond Wheel for Use', *Industrial Diamond Review*, 27, No. 319, June 1967, p. 244.

Marinescu, I. D., Hitchiner, M., Uhlmann, E., Rowe, W. B., and Inasaki, I., *Handbook of Machining with Grinding Wheels*, CRC Press, 2007, Chapter 7.

Metzger, J. L., *Superabrasive Grinding*, Butterworths, London, 1986, Chapter 12.

Tönshoff, H. K. and Grabner, T., 'Cylindrical and Profile Grinding with Boron Nitride Wheels', *Proceedings of the 5th International Conference on Production Engineering*, JSPE, Tokyo, 1984, p. 326.

Verkerk, J. and Pekelharing, A. J., 'The Influence of the Dressing Operation on Productivity in Precision Grinding', *Annals of the CIRP*, 28/2, 1979, p. 487.

Wiemann, H., 'Cutting Profiles in Ceramic Grinding Wheels: Report on a Significant Dressing Technique', *Progress in Industrial Diamond Technology*, Ed. J. Burls, Academic Press, London, 1966, p. 49.

Chapter **5**

Grinding Mechanisms

5.1 INTRODUCTION

Material removal during grinding occurs as abrasive grains interact with the workpiece. The penetration depths of the cutting points into the material being ground depend upon the topography of the wheel surface and the geometry and kinematic motions of the wheel and workpiece. These aspects of the grinding process were discussed in Chapters 3 and 4. The present chapter is concerned with what happens as abrasive grains interact with the workpiece.

A number of diverse methods can be used for gathering evidence to identify the mechanisms of abrasive-workpiece interactions. One possibility is to examine the grinding debris (swarf) produced by the process in order to theorize how it could have been produced. For this purpose, the scanning electron microscope (SEM) is an invaluable tool. Another approach is to measure the forces or power requirements of the process over a range of conditions. A fundamental parameter derived from these measurements is the specific grinding energy, which is defined as the energy expended per unit volume of material removal. The significance of this parameter lies in the fact that any proposed mechanism of abrasive-workpiece interaction must satisfy an energy balance which can account for the magnitude of the specific energy and its dependence on the processing conditions. A more direct observation of abrasive-workpiece interactions can be obtained from cutting experiments with single abrasive grains or with cutting tools shaped like grains. Of course, such results are valid only insofar as the cutting conditions accurately simulate those of abrasive grits in the grinding wheel.

5.2   GRINDING DEBRIS (SWARF)

For grinding of metals, it has been generally assumed that material removal occurs by a shearing process of chip formation, similar to that found with other machining methods such as turning or milling [1]. This idea was first suggested more than 90 years ago [2], and optical microscopy of grinding debris (swarf) more than 50 years ago revealed chip-like shapes [3-5].

With the advent of the electron microscope, the similarity between grinding chips and larger-scale metal-cutting chips became much more apparent [6-8]. A scanning electron micrograph of the swarf recovered after grinding a plain carbon steel is shown in Figure 5-1(a). Mostly curled chips are found, very much like those produced by turning or milling, although somewhat irregular in size and shape owing to the variability in cutting-point shape and penetration depth. These chips have a fine lamella structure, similar to what is generally found with other machining chips. Lamella formation during chip formation has been attributed to a thermal instability, whereby the shear resistance of the material decreases due to localized heating caused by intense plastic deformation [9]. At higher magnification in Figure 5-1(b), the lamella spacing can be estimated to be about 0.5 μm, which is somewhat finer than on chips produced by turning. The difference may be attributed to the typical rake angle being much more negative in grinding, as will be discussed later in this chapter. Another factor may be the much higher velocities and strain rates in grinding, which would result in the deformation during chip formation being more nearly adiabatic.

Two other types of particles found in the grinding swarf (Figure 5-1(a)) are short segmental blocky chips and spheres. Blocky-chip formation may occur by an extrusion-like bulging process with extreme negative rake angles [10]. This process is essentially one of plastic compression and bulging-up in front of the cutting edge in the cutting direction, together with shearing similar to that normally associated with chip formation.

An example of a spherical grinding particle is shown at higher magnification in Figure 5-1(c). It appears to be hollow and to have an extremely fine dendritic microstructure, which would indicate that it was once molten and solidified rapidly. Melting does not necessarily occur during grinding, but only afterwards by exothermic reaction of small hot chips with oxygen in the atmosphere. The round and hollow shape is a consequence of surface tension effects acting on the molten curled chip. Oxidation of grinding chips as they are emitted from the wheel is also responsible for the spark stream, which is actually glowing chips. Sparks are not observed when grinding in an oxygen-free environment. The color and intensity of the sparks depend on the particular workpiece material being ground.

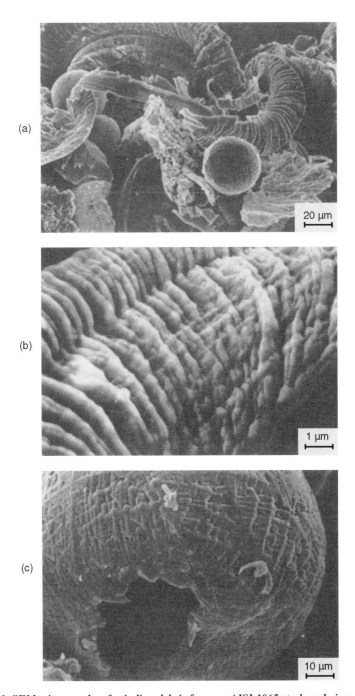

(a)

(b)

(c)

Figure 5-1 SEM micrographs of grinding debris from an AISI 1065 steel workpiece [1, 8].

5.3 GRINDING FORCES, POWER,
AND SPECIFIC ENERGY

Forces are developed between the wheel and the workpiece owing to the grinding action. For plunge grinding operations, as illustrated in Figure 5-2 for straight surface and external cylindrical grinding, the total force vector exerted by the workpiece against the wheel can be separated into a tangential component F_t and a normal component F_n. For shallow cut surface grinding with small depths of cut, the tangential force component is oriented very close to the horizontal direction and the normal force very close to the vertical direction, as shown in Figure 5-2 (a), because the line of action of the total force between the wheel and the workpiece are very close to the location on the wheel periphery directly below the wheel axis. However this 'approximation' may not apply with large depths of cut, such as typical of creep feed grinding, since the force line of action would be further moved along the contact length [11]. For grinding with non-symmetrical profiles or with traverse, there would also be an additional force component in a direction parallel to the wheel axis.

The grinding power P associated with the force components in Figure 5-2 can be written as

$$P = F_t(v_s \pm v_w) \qquad (5\text{-}1)$$

The plus sign is for up-grinding with the wheel and workpiece velocities v_s and v_w in opposite directions at the grinding zone as in Figure 5-2, and the minus sign is for down grinding with both velocities in the same direction. Since v_w is usually much smaller than v_s, the total power can be simplified to

$$P = F_t v_s \qquad (5\text{-}2)$$

This relationship is sufficient for most grinding situations. Additional power components associated with feed and traverse velocities are usually negligible.

A fundamental parameter derived from the power and machining conditions is the specific energy, which is defined as the energy per unit volume of material removal. (This parameter is identical to the specific power, which is the power per unit volumetric removal rate.) The specific grinding energy is obtained from the relationship

$$u = \frac{P}{Q_w} \qquad (5\text{-}3)$$

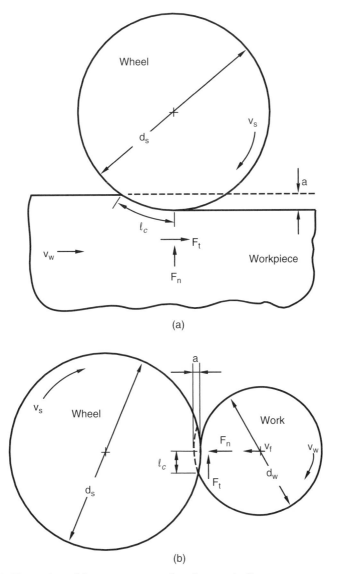

Figure 5-2 *Illustration of force components for plunge grinding.*

The numerator is the power (Eq. (5-2)), and the denominator Q_w is the volumetric removal rate given in terms of the grinding parameters in Figure 5-2 as:

$$Q_w = v_w ab = \pi d_w v_f b \qquad (5\text{-}4)$$

where b is the grinding width. From a fundamental perspective, the particular significance of the specific energy lies in the fact that any plausible mechanism of abrasive-metal interaction must be able to account for its magnitude and its dependence on the process parameters. The specific energy is also useful for estimating machine power requirements.

5.4 GRINDING MECHANISMS: CONVENTIONAL ABRASIVES

5.4.1 Size effect and energy considerations

Systematic measurements of grinding forces and specific energies in the early 1950s [12, 13] showed that the specific energies for grinding were much larger than for other metal-cutting operations. Furthermore, larger specific energies were found when the process parameters were adjusted so as to reduce the undeformed chip thickness, i.e., decrease v_w or a in Figure 5-2.

At the time of these studies, the classic model of chip formation formulated in 1945 [14] was beginning to be extensively applied to various metal-cutting processes. According to this model, chip formation occurs by an intense shearing process in an extremely thin zone followed by friction as the chip slides over the tool rake face. Typically, shearing accounts for about 75% of the total chip formation energy, and chip-tool friction the remaining 25%. Although many other secondary effects have been observed over the years, this same model is still considered to present a reasonably accurate description of chip formation.

As with other metal-cutting processes, an attempt was made to interpret the grinding forces in terms of this chip-formation model [13]. By invoking plausible assumptions for the typical cutting-point geometry, estimates of the shear stress for plastic deformation during chip formation were obtained. However, these calculated shear stresses greatly exceeded the known flow stress of the metal being ground. Moreover, larger shear stresses were generally obtained with finer grinding conditions, i.e., those giving smaller undeformed chip thicknesses corresponding to the higher specific energies found for this mode of grinding.

In order to account for these anomalous results, a 'size effect' theory was proposed, which attributes the apparent increase in flow stress with smaller undeformed chip thickness to a greater likelihood of shearing small volumes of metal free of strength-impairing dislocations. However, the application of dislocation theory to metal cutting predicts extremely high dislocation densities in the shear zone [15], and this has been experimentally verified by transmission electron microscopy of fine grinding chips [7]. This would seem to cast serious doubt on the 'size effect' theory.

Perhaps a more disturbing factor in applying the classic model of chip formation to grinding is found in the magnitude of the specific grinding energy. Virtually all the energy expended by grinding is converted to heat. Since the chip-formation process in grinding is extremely rapid, owing to the high cutting velocities and large strains, the process should be nearly adiabatic, which means that there is not sufficient time for any significant amount of the heat generated by plastic flow to be conducted away during deformation. Under adiabatic conditions, the plastic energy input per unit volume should be limited by the energy it takes to bring a unit volume of material from its ambient condition to its molten state. The melting energy per unit volume can be evaluated for metals from handbook data as the enthalpy difference between the liquid state at the melting point and ambient (room) temperature [16]. For iron, the melting energy per unit volume is 10.5 J/mm^3 and this value is generally representative of steels. Specific energies for production grinding of steels are bigger, typically ranging from 20 to 60 J/mm^3 and much larger values are not uncommon especially in fine grinding. It seems inconceivable that the plastic deformation energy associated with chip formation in grinding can be so much bigger than its melting energy.

5.4.2 Sliding forces and energy

Even though metal removal occurs mostly by chip formation, as seen from the swarf (Figure 5-1(a)), it would seem that much of the grinding energy must be expended by mechanisms other than chip formation. One such mechanism may involve the sliding of dulled flattened tips of the abrasive grains against the workpiece surface without removing any material [17]. Such 'wear flats' are generated by dressing prior to grinding as described in Chapter 4 (Figure 4-9). During grinding, the wear flats may become glazed and further enlarged through attritious wear and adhesion of metal particles from the workpiece (Figure 4-19). An isolated wear-flat area with adhering metal is seen in Figure 5-3, together with some grinding chips attached to its leading edge [18]. The growth of wear flats is offset, to a greater or lesser degree, by 'self-sharpening' due to bulk wear of the wheel, whereby some flats are wholly or partially removed by grit fracture or grit dislodgement from the binder (see Chapter 11).

The presence of the wear flats, with their characteristic striated markings in the grinding direction, indicates that part of the energy expended by grinding is due to their sliding against the workpiece. A direct relationship has been obtained between the grinding forces and the degree of wheel dullness as expressed in terms of the percentage of the wheel surface consisting of wear flats. (Wear-flat area measurement is described in Chapter 4.) With fixed machine settings, the normal and tangential forces F_n and F_t

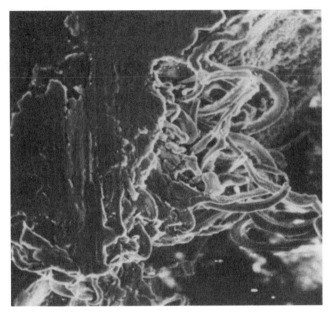

Figure 5-3 SEM micrograph of a wear flat with metal grinding chips adhering to its leading edge. AISI 52100 steel workpiece, 32A46K6VBE wheel.

increase with wear-flat area A, as seen in Figure 5-4. For a particular workpiece material, measured differences in forces and wear-flat area were obtained by changing the wheel grade, the dressing conditions, and the accumulated metal removed. With steel workpieces (Figure 5-4 (a)), the forces increase linearly with wear-flat area up to a critical point, beyond which the slopes become steeper and workpiece burn occurs (see Chapter 6). Linear relationships without discontinuities are obtained with many non-ferrous metals (Figure 5-4 (b)).

On the basis of the results in Figure 5-4 and other similar findings, it was proposed that the grinding forces, and hence the specific energy, can be considered to consist of cutting and sliding components [17]. The forces shown at the intercepts in Figure 5-4 ($A = 0$) are associated with cutting, and the additional forces in excess of the intercept values are associated with sliding, so that

$$F_t = F_{t,c} + F_{t,sl} \qquad (5\text{-}5)$$

and

$$F_n = F_{n,c} + F_{n,sl} \qquad (5\text{-}6)$$

where $F_{t,c}$ and $F_{n,c}$ are the tangential and normal forces for cutting, and $F_{t,sl}$ and $F_{n,sl}$ are those for sliding. The situation envisioned at a typical abrasive grain is illustrated in Figure 5-5.

The proportional relationship between the sliding forces and wear-flat area, only up to the discontinuities for steels (e.g. Figure 5-4 (a)), implies a constant average contact stress $\overline{p}$ and friction coefficient μ between the wear flats and the workpiece. Therefore, the forces in Eqs. (5-5) and (5-6) can be expressed as

$$F_t = F_{t,c} + \mu \overline{p} A_a \qquad (5\text{-}7)$$

and

$$F_n = F_{n,c} + \overline{p} A_a \qquad (5\text{-}8)$$

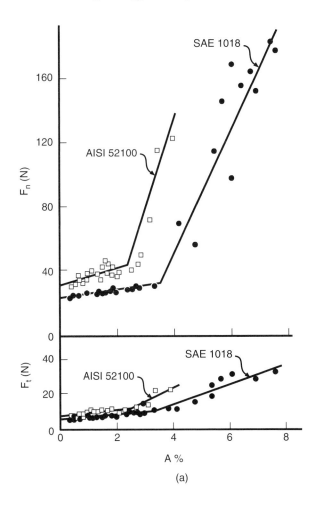

(a)

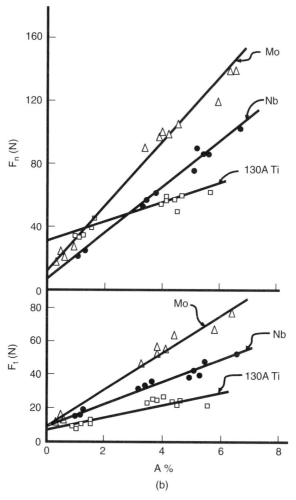

Figure 5-4 Tangential and normal force components, F_t and F_n, for straight surface plunge grinding of (a) two steels and (b) three non-ferrous metals: with vitrified aluminum oxide (32A46) wheels, Grinding conditions: v_s = 30 m/s, d_s = 200 mm, v_w = 4.6 m/min, a = 25 μm, b = 6.4 mm.

where A_a is the actual area of contact between the wear flats and the workpiece. The area A_a is obtained by multiplying the grinding zone area, which is the arc length of contact l_c multiplied by the grinding width b, by the fraction A of the wheel surface which consists of wear flats:

$$A_a = bl_c A \qquad (5\text{-}9)$$

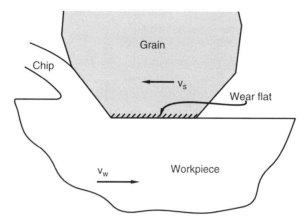

Figure 5-5 *Illustration of an abrasive grain cutting a chip with its dulled wear flat sliding against the newly generated surface.*

or substituting for l_c from Eq. (3.8):

$$A_a = b(d_e a)^{1/2} A \tag{5-10}$$

where d_e is the equivalent diameter (Eq. (3-9)) and a is the wheel depth of cut. Combining Eqs. (5-7), (5-8), and (5-10) leads to

$$F_t = F_{t,c} + \mu \bar{p} b (d_e a)^{1/2} A \tag{5-11}$$

and

$$F_n = F_{n,c} + \bar{p} A_a (d_e a)^{1/2} A \tag{5-12}$$

Within the framework of this model, the magnitudes of μ and $\bar{p}$, which characterize the contact process between the wear flats and the workpiece, can be evaluated. Combining Eqs. (5-7) and (5-8), we can show that

$$F_n = \frac{1}{\mu} F_t + \frac{\mu F_{n,c} - F_{t,c}}{\mu} \tag{5-13}$$

For given grinding conditions, $F_{t,c}$ and $F_{n,c}$ are constant, so a graph of F_n versus F_t should yield a straight line of slope μ^{-1}. An example of this behavior is shown in Figure 5-6 for grinding a plain carbon steel [19]. The big variation in forces shown here was obtained only by varying the dressing

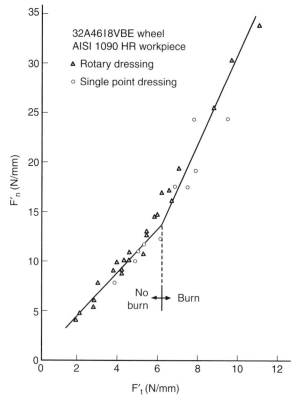

Figure 5-6 Normal versus tangential force per unit width for straight surface plunge grinding: $v_s = 30$ m/s, $d_s = 350$ mm, $v_w = 8.6$ m/min, $a = 25$ μm.

parameters. Finer dressing results in larger wear-flat areas, and so the forces are also larger. The initial slope up to the discontinuity implies a friction coefficient of $\mu \approx 0.4$. Beyond the discontinuity, workpiece burn occurs and the slope changes. Straight-line relationships such as these can also be drawn for the results in Figure 5-4.

The average contact pressure $\bar{p}$ between the wear flats and the workpiece can be obtained by differentiating Eq. (5-12) with respect to A, and solving for $\bar{p}$:

$$\bar{p} = \frac{dF_n/dA}{b(d_e a)^{1/2}} \tag{5-14}$$

From the experimental evidence available, it appears that the contact pressure $\bar{p}$ increases with the magnitude of the curvature difference Δ (see

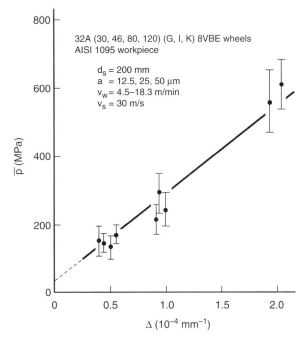

*Figure 5-7 Average contact pressure versus the magnitude of the curvature difference.
Adapted from Reference [20].*

Eq. (3-22)), as seen for example in Figure 5-7 [20]. With a bigger misfit or curvature difference between the wheel surface and the cutting path, the pressure needed to maintain contact between the wear flats and the workpiece is bigger, which would seem to make physical sense. It has also been found that materials having a higher hot hardness develop bigger contact pressures [17].

The sliding energy concept enables us to account quantitatively for the influence of wheel grade and dressing conditions on grinding forces. With harder-grade wheels and finer dressing, the wear-flat area is larger and so the sliding forces are proportionally larger. (The influence of wheel grade and dressing conditions on wheel dullness is discussed in Chapter 4.) Similarly, the effect of grinding fluids on the grinding forces can also be traced mainly to their influence on wear-flat area, as will be seen in Chapter 11.

5.4.3 Plowing and chip-formation energies

The specific energy for cutting, which is that portion of the specific grinding energy remaining after subtracting the contribution due to sliding, can now be calculated from the relationship

$$u_c = \frac{F_{t,c} v_s}{b v_w a} \qquad (5\text{-}15)$$

The numerator is the power associated with cutting, and the denominator is the volumetric removal rate (Eq. (5-4)). The tangential cutting force $F_{t,c}$ is equivalent to that of a perfectly sharp wheel ($A = 0$). In Figure 5-8, results are shown for the specific cutting energy versus the removal rate per unit width for grinding of a high-carbon steel [20-22]. All the results, even with different sizes of abrasive grains from 30 to 120 grit, fall on the same curve. At slow removal rates, the specific cutting energy is extremely big, but it decreases at faster removal rates tending towards a minimum value of approximately 13.8 J/mm^3 in the limit.

To obtain results such as those of Figure 5-8 entails a long and tedious experimental procedure. For each data point, it is necessary to develop a force versus wear-flat area relationship, as in Figure 5-4, in order to identify the tangential cutting-force component $F_{t,c}$. For this reason, values of u_c at different removal rates have been obtained for only one

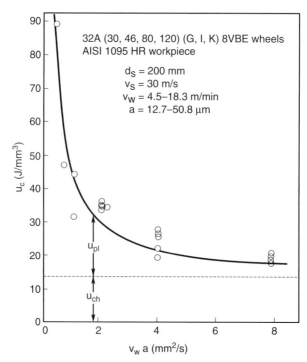

Figure 5-8 *Specific cutting energy versus volumetric removal rate per unit width ($v_w a$) in straight plunge grinding. Adapted from References [20-22].*

workpiece material. However, tests have been performed at one removal rate ($v_w a = 1$ mm^2/s) on diverse steels, ranging from a low-carbon steel to a hardened high-speed tool steel, and the specific cutting energies were all found to be nearly identical ($u_c \approx 40$ J/mm^3) [17]. This result is rather surprising, in view of the vast differences in workpiece hardnesses.

Even after subtracting the sliding energy, it is apparent from Figure 5-8 that there is still a 'size effect'. At slower removal rates, corresponding to finer undeformed chip thicknesses, the specific cutting energy becomes extremely large, and its magnitude cannot be reconciled with the classic chip-formation model. This would suggest that only part of the specific cutting energy is actually associated with chip formation, in which case there must be at least one other mechanism to account for the remaining energy.

Another mechanism associated with abrasive processes is plowing. Plowing energy is expended by deformation of workpiece material without removal. Plowing is usually associated with side flow of material from the cutting path into ridges, but it can also include plastic deformation of the material passing under the cutting edge [23].

Plowing deformation occurs as the abrasive initially cuts into the workpiece, as illustrated in Figure 5-9 [10, 20, 21, 24-27]. As the cutting point on the abrasive grain passes through the grinding zone, its depth of cut increases from zero to a maximum value h_m at the end of the cut. Initially the grit makes elastic contact, which is assumed to make a negligible contribution to the total energy and is not shown, followed by plastic deformation (plowing) of the workpiece. On the average, chip formation commences only after the cutting point has penetrated to some critical depth of cut h'. Factors which can affect the magnitude of h' include the sharpness of the cutting edge, its orientation, its rake angle, and the friction coefficient.

Even after chip formation begins, plowing may still persist, with some of the material from the cutting path being pushed aside into ridges rather than being removed as chips. This latter type of plowing has been extensively investigated, mostly by cutting experiments at fixed depths of cut with triangular-based or square-based pyramidal tools (Figure 5-10) to simulate abrasive cutting points [6, 28-33]. The tool is set orthogonally (Figure 5-10(a)) with one face perpendicular to the cutting direction, or obliquely (Figure 5-10(b)) with an edge between two adjacent faces at the front. The orientation of a tilted orthogonal tool is defined by an attack angle β (Figure 5-10 (a)), corresponding to a rake angle α:

$$\alpha = \beta - 90° \tag{5-16}$$

For $\beta < 90°$ as shown in Figure 5-10 (a), α is negative.

For orthogonal cutting tests with triangular-based pyramidal tools, a critical rake angle has been found, below which sideflow plowing occurs

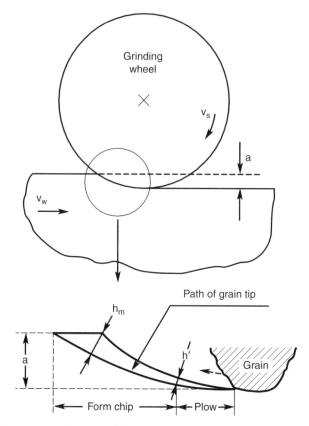

Figure 5-9 Illustration of plowing followed by transition to chip formation as a grain passes through the grinding zone [20].

and above which chip formation occurs [6, 28-30]. Subsequently, this transition was found to occur within a narrow range of rake angles, where both plowing and chip formation occur simultaneously [6, 30]. The material removed in this transition region resembles the blocky particles in Figure 5-1(a). Oblique tool orientations should increase the tendency for sideflow plowing.

Of the two types of plowing, hereafter referred to as critical-depth-controlled (Figure 5-9) and critical-rake-angle-controlled, only the former can account for the observed 'size effect'. One possible reason why the critical rake angle may not apply, at least with dressed conventional abrasive wheels, is because the cutting points are not shaped like pyramids. A typical active grit can have multiple cutting points, more or less side by side at its leading edge [34]. As seen in Figure 5-3, chip widths appear to

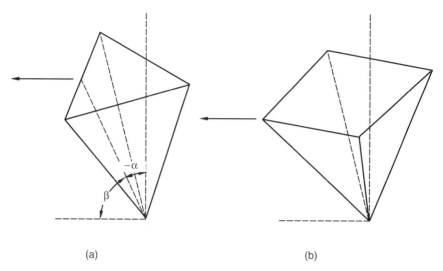

(a)   (b)

*Figure 5-10 Illustration of (a) triangular-based and (b) square-based pyramidal tools
and their orientation for cutting experiments to simulate abrasion.*

be considerably narrower than the plateau dimension across the width of a dressed and worn grit. In order for material to be plowed aside, it would have to flow around the whole extent of the grit plateau. This would be very unlikely except perhaps with cutting points towards the side of the grit. Furthermore, the cross-sections of the cutting points tend to be more nearly trapezoidal (Figure 4-17) rather than triangular, which would also decrease the tendency for sideflow.

   Critical-depth-controlled plowing can account for the observed 'size effect'. From Figure 5-9 and the undeformed chip geometry (Chapter 3), it can readily be shown that an increase in either v_w or a will enlarge the undeformed chip volume, so that the relative amount of material plowed aside before reaching h' decreases relative to that provided afterwards by chip formation [20]. Relatively less plowing at higher removal rates would decrease the specific plowing energy (Figure 5-8). In the limit, the specific plowing energy u_{pl} approaches zero, and the minimum specific cutting energy corresponds to the specific energy for chip formation u_{ch}, which is assumed to be constant. The results in Figure 5-8 imply a specific chip-formation energy of $u_{ch} = 13.8$ J/mm^3, and an inverse relationship between the specific plowing energy and removal rate with a constant tangential plowing force per unit width of $F'_{t,pl} = 1$ N/mm.

   Following this same line of reasoning, the use of down-grinding in place of up-grinding should reduce or even eliminate initial plowing, since each cutting point would now initially engage the workpiece at its maximum depth of cut. This can explain why the forces in down-grinding are usually slightly smaller than in up-grinding. However, the measured force differences appear to be somewhat less than the plowing force components, which would indicate that some plowing still persists.

In summary, the total specific grinding energy can be considered to consist of chip formation, plowing, and sliding components:

$$u = u_{ch} + u_{pl} + u_{sl} \tag{5-17}$$

Only the specific chip-formation energy is actually expended by material removal and, as such, is the minimum grinding energy required. For the results in Figure 5-8, the specific chip-formation energy $u_{ch} = 13.8$ J/mm^3 is still much bigger than the specific energy in larger-scale metal-cutting operations. Furthermore, unlike the energy or load requirements in other metal-working processes, the minimum specific grinding energy is insensitive to alloying and heat treatment. For example, a hot-worked steel requires nearly the same minimum grinding energy as a hardened alloy steel or even a hardened high-speed tool steel.

In order to explain the anomalous behavior in grinding, it is of interest to compare the magnitude of the minimum specific energy with the melting energy of the metal being ground. Since about 75% of the chip-formation energy is typically associated with shearing, and the remaining 25% with chip-tool friction, this would imply a specific energy for shearing of about 10.4 J/mm^3, which is virtually identical to the melting energy per unit volume for iron. This equality may be attributed to the large deformations and nearly adiabatic conditions which prevail during chip formation in grinding, as mentioned previously. Cutting points on abrasive grains have extremely large negative rake angles, which have been estimated to be about $-60°$ or even more negative [34, 35]. For such large negative rake angles, classic chip-formation theory predicts extremely large shear strains. This behavior has been experimentally verified for negative rake orthogonal cutting to simulate grinding [36]. Such large strains are obtained at extremely high cutting velocities so that the plastic deformation is virtually adiabatic. For adiabatic deformation to large strains, the plastic deformation energy should not exceed the melting energy of the workpiece material. This is schematically illustrated in Figure 5-11, where a hypothetical stress-strain curve is shown for adiabatic shear. The area under the curve is the plastic work per unit volume, which is virtually all converted to heat. Initial strain hardening is followed by strain softening at an increasing rate, with the plastic shear resistance dropping towards zero as the melting point is neared. Therefore, the total limiting area under the stress-strain curve is equal to the melting energy per unit volume. This does not imply that melting

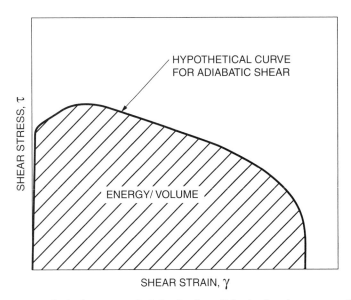

Figure 5-11 Hypothetical stress-strain behavior for adiabatic shearing up to melting [8, 21, 37].

occurs, but only that the shear energy during chip formation approaches the melting energy limit. Owing to the decreasing shear resistance of the material as the melting energy is approached, it seems unlikely that melting will actually occur.

This correlation between grinding and melting energies is not limited to steels, as can be seen in Figure 5-12 which presents experimental results for the minimum specific grinding energy versus the specific melting energy for a wide range of metals [8, 21, 37]. These experiments were conducted with sharp wheels under well lubricated conditions using a heavy-duty straight oil in order to minimize any friction contribution. The specific melting energy indicated for each metal is the enthalpy difference between the liquid state at its melting temperature and room temperature [16]. Considering the complexity of the process, it is remarkable to find a one-to-one relationship, with the minimum specific energy only slightly larger than the corresponding melting energy. According to the present theory, the minimum energy exceeds the melting energy because of chip-tool friction, which is expended in addition to shearing energy for chip formation, although it is likely that there is still a small plowing contribution. Heat treatment and alloying have little effect on the minimum specific energy because the melting energy is insensitive to these factors.

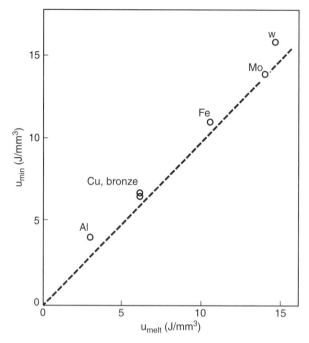

Figure 5-12 Minimum specific energy versus melting energy for various metallic materials [8, 21, 37].

5.5 GRINDING MECHANISMS: CBN WHEELS

The analysis of grinding mechanisms for metallic materials in the previous section was for conventional abrasive wheels. With wheels containing CBN abrasives, the grinding behavior is somewhat different. After wheel preparation, the initial grinding forces and energy may be extremely big, but with continued grinding they progressively decrease towards a steady-state value. Such transient grinding behavior has been observed with both resin- and vitreous-bonded CBN wheels [38-48]. An example is shown in Figure 5-13 for a resin-bond CBN wheel after brake-controlled truing and stick dressing, and similar results have also been found with multipoint diamond truing and stick dressing [38, 39]. Such high initial forces can cause thermal and mechanical damage to the workpiece, and can even break the wheel.

High initial grinding forces have been attributed to lack of protrusion of the abrasive grits and wheel clogging [42, 49-51]. However, SEM observations of the wheel surface suggest that the primary cause is dulling of the CBN grits during truing [39]. This phenomenon was discussed in Chapter 4,

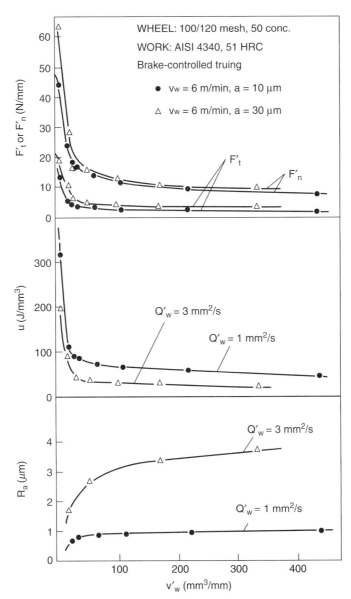

Figure 5-13 Force components, specific energy, and surface roughness versus volumetric removal per unit width for straight surface plunge grinding with a CBN wheel at two removal rates [38, 39].

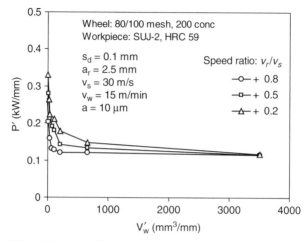

Figure 5-14 *Effect of diamond disk speed ratio (v_r/v_s) on initial power transient for grinding with a vitrified CBN wheel. Adapted from Reference [45].*

and an example was shown in Figure 4-10 for a resin-bonded CBN wheel. Stick dressing, after truing, tends to expose the grits and provides better protrusion, but the grits may still remain dull thereby causing high grinding forces. The steep decrease in the forces at the start of grinding (Figure 5-13) has been attributed mainly to grit dislodgement, whereby those grits flattened down the most by truing are preferentially removed. This is also accompanied by grit micro-fracture, which sharpens the remaining grits [38-40, 43]. Eventually, the wheel may become completely sharpened, with virtually no flattened areas left. In this condition, the grits appear very similar to those on a CBN wheel after a special 'sharpening' treatment (Figure 4-11).

With vitrified CBN wheels, more aggressive truing/dressing can be used to generate a sharper wheel, and thereby lessen or eliminate the need for a separate wheel sharpening treatment. This can be seen in Figure 5-14 which shows how the use of a bigger speed ratio with disk dressing reduces the magnitude and duration of the power transient at the start of grinding [45]. Increasing the dresser lead or depth also gives a sharper wheel with less of a transient.

The transient grinding behavior with both resin- and vitreous-bonded CBN wheels was further studied by incrementing the removal rate in steps and observing the time-dependent force behavior at each removal rate [38, 39, 48]. An example of what was obtained is shown in Figure 5-15 for straight surface plunge grinding using a resin-bonded CBN wheel with a fixed workpiece velocity. After dressing, grinding was performed with a depth of cut $a = 10$ μm until the approximate steady state was reached after

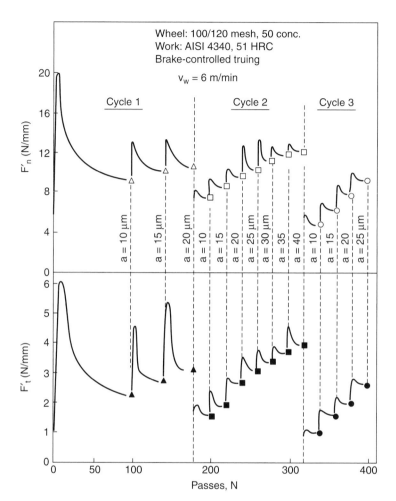

Figure 5-15 Transient CBN grinding behavior with incremental removal rates [38, 39].

100 grinding passes. Grinding was then continued with the force transients measured at larger depths of cut, $a = 15$ μm and $a = 20$ μm (Cycle 1). Thereafter, the depth of cut was reduced back to its initial value of $a = 10$ μm, and the procedure was repeated (Cycle 2), eventually reaching a larger depth of cut ($a = 40$ μm). This procedure was then repeated once more (Cycle 3).

With each subsequent cycle, the forces at the same removal rate became much smaller, and the force transient less pronounced. This is also seen in Figure 5-16 where the 'steady-state' forces for the three cycles are plotted against the corresponding removal rate per unit width ($Q'_w = v_w a$).

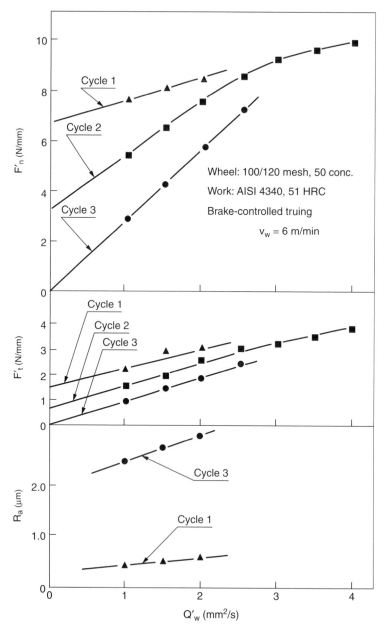

Figure 5-16 Steady-state force components from previous figure and surface roughness versus removal per unit width [38, 39].

Upon repeating the procedure, in a fourth cycle, the forces stayed the same as in the third cycle, which would indicate that the wheel had reached a fully stabilized condition. In this condition, both the normal and the tangential force components were virtually proportional to the removal rate (e.g., Cycle 3 in Figure 5-16). Essentially the same behavior was also obtained with a CBN wheel after a special sharpening treatment using a wire brush as described in Chapter 4 [39]. Since the forces very nearly intercept the origin at zero removal rate, it would appear that the stabilized wheel is almost perfectly sharp like the sharpened wheel in Figure 4-11, and this has been confirmed by SEM observations. It has been suggested that cyclically increasing the wheel depth of cut in steps as in these tests can be used as a sharpening treatment for vitrified CBN wheels [48].

The proportional relationship between the tangential force and the removal rate with a stabilized or sharpened CBN wheel means that the specific grinding energy is constant. The corresponding specific energies have been found to be approximately 25 to 30 J/mm^3 for grinding carbon and alloy steels, annealed or hardened, and about 60 J/mm^3 for hardened high speed steels [39]. The extent by which the specific energy exceeds these values can be taken as a relative indication of wheel dullness.

The observation that specific energy with a sharp CBN wheel does not depend on removal rate is contrary to what occurs with conventional abrasive wheels (e.g., Figure 5-8). The apparent absence of a 'size effect' with CBN might be due to the cutting edges being much sharper and more pointed. Sharper cutting edges favor an initial transition from plowing to cutting at a shallower initial depth, thereby reducing the initial plowing force which is responsible for the size effect with sharp conventional abrasives (Figure 5-8). However, the more-pointed cutting edges, which more closely resemble the pyramidal points used in simulated cutting tests (Figure 5-10) than the cutting points on conventional grits, would favor depth-independent sideflow plowing. Since the specific chip-formation energy component for grinding steels is limited to about 13.8 J/mm^3, as discussed in the previous section, this would imply that sideflow plowing contributes about 30 to 50% of the total specific energy for grinding carbon and alloy steels with a sharp CBN wheel and about 75% of the total for hardened tool steels.

While the grinding forces become progressively smaller during grinding as the wheel sharpens, the surface finish becomes progressively rougher, as seen in Figure 5-13 and Figure 5-16. The rougher finish with CBN has been is one of the main factors which has limited its more widespread use in place of conventional wheels. Surface finish is discussed in more detail in Chapter 10.

The transient behavior observed with resin and vitrified CBN wheels tends to be opposite to what is found with electroplated CBN wheels. Electroplated wheels are manufactured with only a single layer of abrasive and

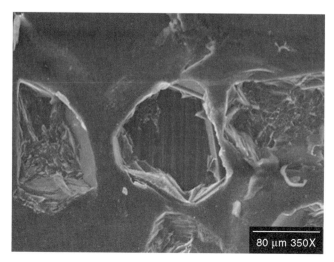

Figure 5-17 SEM of electroplated 270 mesh CBN wheel showing a grain with a smooth wear flat in the middle and grains with fragmented wear flats on either side [52].

are usually not dressed prior to use. Consequently they start out in an initially sharp state, and the forces and power tend to progressively increase with use [52-54]. Both smooth and fragmented wear flats develop and grow on the grain tips, as shown in Figure 5-17. This causes an increase in the forces and energy. This behavior is shown Figure 5-18 where the grinding power per unit

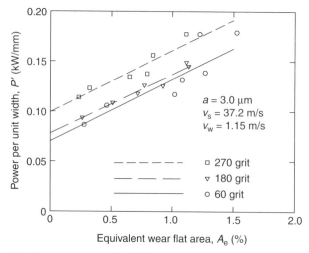

Figure 5-18 Grinding power per unit width versus equivalent wear flat area for grinding hardened AISI52100 steel with three electroplated CBN wheels.

width is plotted versus the "equivalent" wear flat area A_e for internal grinding of a hardened bearing steel with three electroplated CBN wheels with different grit sizes [52]. The same slopes for all three grit sizes were obtained when the "equivalent" wear flat area was taken as the sum of the smooth wear flat area and 70% of the fragmented wear flat area which makes only partial contact with the workpiece. The linear relationships in Figure 5-18 are analogous to what occurs with conventional abrasive wheels as described in the previous section. The intercept power values at zero wear flat area corresponding to "perfectly sharp wheels" indicate specific energies ranging from 20 J/mm^3 with the coarsest grit size up to 29 J/mm^3 with the finest.

5.6   CREEP-FEED GRINDING

Creep-feed grinding is characterized by the use of slow (creep) workpiece velocities and extremely large depths of cut which are hundreds or even thousands of times greater than those in regular grinding applications. With this process, it may be possible to grind complex profiles or deep slots in only a few or even a single pass. Applications of creep-feed grinding include the machining of drill flutes and the profiling of turbine blade roots for jet engines [55-57].

Because of the heavy wheel depths of cut in creep-feed grinding, which typically range from 1 to 10 mm, the wheel-work contact lengths and grinding zone areas are also very big. Therefore, we should expect much bigger specific energies than in regular grinding owing to much bigger sliding forces (Eq. (5-11)). An example of this behavior in Figure 5-19 illustrates the sensitivity of the specific grinding energy to the wear-flat area for straight creep-feed grinding of a nickel-base alloy with a high-porosity aluminum oxide wheel [58,59]. Whereas our grinding model predicts a linear relationship between specific energy and wear-flat area, these results suggests a discontinuous curve with two slopes. This discontinuity at a specific energy approaching 200 J/mm^3 may be associated with burn-out of the grinding fluid (see Chapter 7). The intercept at about 25 J/mm^3 in Figure 5-19 is comparable to that expected for the combined chip-formation and plowing energy components, and the very large specific grinding energies can be attributed to sliding of the wear flats against the workpiece.

Such large specific energies in creep-feed grinding necessitate special care to provide cooling so as to avoid thermal damage to the workpiece. (Heat transfer and cooling effects for creep-feed grinding are covered in Chapter 7.) Likewise, it is especially important to keep the wheel sharp in order to reduce the energy input. The tendency for wheel dulling in creep-feed grinding is promoted by the long sliding length per wheel revolution and less self-sharpening. Therefore, wheel dressing is of

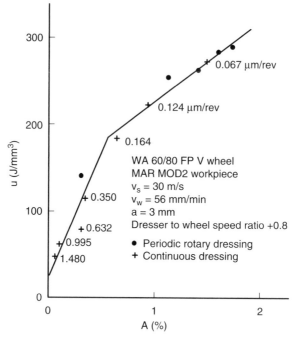

Figure 5-19 Specific energy versus wear-flat area for creep-feed grinding with periodic and continuous dressing. Numbers indicated are dresser infeed rates (μm/rev) for continuous dressing. Adapted from Reference [58].

critical importance (Chapter 4). The results in Figure 5-19 were obtained using two types of rotary diamond dressing: regular periodic dressing and continuous dressing during grinding. With regular dressing, the wheel is periodically resharpened. The different wear-flat areas indicated for regular rotary dressing were obtained after various amounts of grinding. Continuous dressing during grinding maintains a constant wear-flat area, with more severe dressing infeed rates giving smaller wear-flat areas as expected (see Chapter 4). Continuous Dress Creep Feed Grinding (CDCFG) is widely used for machining of turbine engine components.

5.7 CONTROLLED-FORCE GRINDING

Grinding processes considered up to this point are of the fixed-feed type, with the depth of cut or infeed rate being a controlled input parameter. Forces developed between the wheel and the workpiece cause elastic deflection of the grinding system, such that the true infeed is less than the input infeed. Changes in the controlled infeed rate during the grinding cycle, such as at the beginning and end of a grinding cycle (spark-out), are accompanied by grinding transients which depend mainly on the stiffness of the grinding system and the forces developed. Grinding transients of this type are discussed in Chapter 12.

While most grinding operations are of the fixed-feed type, some machines operate with controlled normal force instead. In this case, the normal force is input to the process, and the infeed or removal rate is an output or consequence of the process. Controlled-force grinding was originally introduced almost 50 years ago [60], and it has been extensively studied [60-67]. One particular advantage of controlled-force grinding is that it minimizes transient effects at the start of a grinding cycle, which is especially important for 'soft' grinding systems having low stiffnesses [62]. For this reason, controlled-force grinding was often selected for precision internal grinding, since the shaft supporting the grinding wheel in such machines limits the stiffness of the grinding system.

The effect of the normal force of the removal rate in controlled-force grinding with conventional abrasive wheels can be analyzed from the force relationships in section 5.4. The normal force per unit width for plunge grinding, F'_n, includes chip formation, plowing, and sliding components:

$$F'_n = F'_{n,ch} + F'_{n,pl} + F'_{n,sl} \tag{5-18}$$

Assuming that the normal chip-formation component is k, times the tangential component

$$F'_{n,ch} = k_1 F'_{t,ch} \tag{5-19}$$

it can be readily shown that

$$F'_{n,ch} = \frac{k_1 u_{ch} Q'_w}{v_s} \tag{5-20}$$

where Q'_w is the removal rate per unit width ($Q'_w = v_w a$). The normal plowing force component per unit width $F'_{n,pl}$ can be assumed to be constant, as was proposed for the tangential plowing component. From Eq. (5-12), the normal sliding force component per unit width can be written as

$$F'_{n,sl} = \overline{p}(d_e a)^{1/2} A \tag{5-21}$$

where $\overline{p}$ is the average contact pressure between the wear flats and workpiece, and A is the fractional wear-flat area. As seen in Figure 5-7, $\overline{p}$

depends on the curvature difference Δ, which in turn depends on the grinding parameters. Assuming that

$$\overline{p} = p_o \Delta \tag{5-22}$$

where p_o is a constant, and substituting for Δ from Eq. (3-22) (neglecting the sign)

$$F'_{n,sl} = \left(\frac{4p_o v_w^{1/2} A}{d_e^{1/2} v_s}\right) Q'_w{}^{1/2} \tag{5-23}$$

Combining Eqs. (5-18), (5-20) and (5-23), and dividing by $F'_{n,pl}$, we obtain the dimensionless normal force per unit width

$$\frac{F'_n}{F'_{n,pl}} = \frac{Q'_w}{Q'_{w,o}} + 1 + A_e\left(\frac{Q'_w}{Q'_{w,o}}\right)^{1/2} \tag{5-24}$$

where

$$Q'_{w,o} \equiv \frac{F_{n,pl} v_s}{k_1 u_{ch}} \tag{5-25}$$

and

$$A_e \equiv \left(\frac{16 p_o^2 v_w}{v_s d_e F'_{n,pl} k_1 u_{ch}}\right)^{1/2} A \tag{5-26}$$

The parameter $Q'_{w,o}$ can be interpreted as the removal rate per unit width obtained by chip formation for a force per unit width equal to the normal plowing component. The parameter A_e represents the effective wheel dullness, which is proportional to the actual wear-flat area A.

From Eq. (5-24), an expression for the removal rate per unit width as a function of the controlled normal force is obtained in dimensionless form as

$$\frac{Q'_w}{Q'_{w,o}} = \left\{\frac{-A_e + [A_e^2 + 4(F'_n/F'_{n,pl} - 1)]^{1/2}}{2}\right\}^2 \tag{5-27}$$

for $F'_n \geq F'_{n,pl}$ (If $F'_n < F'_{n,pl}$, the removal rate is zero.) This relationship is shown in Figure 5-20 as $Q'_w/Q'_{w,o}$ versus $F'_n/F'_{n,pl}$ for various values of effective dullness A_e. Each curve intercepts the abscissa at the normal

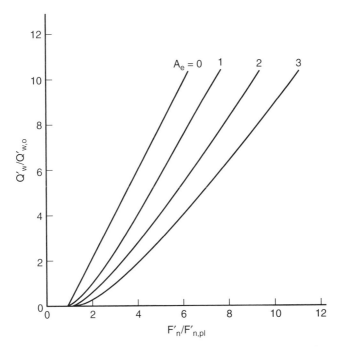

Figure 5-20 Dimensionless removal rate per unit width versus dimensionless normal force for various values of effective wheel dullness with controlled-force grinding.

plowing force component. For a perfectly sharp wheel ($A_e = 0$), the removal rate increases linearly with normal force. With dull wheels ($A_e > 0$) there is an initial non-linear region, beyond which the curves approximate straight lines.

Actual controlled-force grinding behavior shown in Figure 5-21 for four difficult-to-grind alloys [63] appears very similar to the curves in Figure 5-20. When obtaining results such as these, only the linear portion of the curve is often observed. The slope of Q'_w versus F'_n in the linear region is called the removal rate parameter Λ_w, and extrapolating the straight line to zero removal rate indicates an apparent threshold force, which would be equal to $F'_{n,pl}$ for $A_e = 0$ and greater than $F'_{n,pl}$ for $A_e > 0$. From Eq. (5-27) and Figure 5-20, it can be seen that an effectively sharper wheel (A_e smaller) should yield a bigger Λ_w. With softer-grade wheels or coarser dressing, the wear-flat area A is smaller, so Λ_w should be bigger. Also, a faster wheelspeed or higher ratio of wheelspeed to workspeed should reduce A_e, and thereby increase Λ_w. These effects are generally observed in controlled-force grinding tests [63-68].

A complication arises in analyzing controlled-force grinding from the dependence of wheel sharpness on removal rate. Each curve in

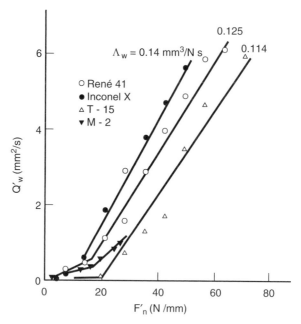

Figure 5-21 Controlled-force internal grinding behavior for four materials [63].

Figure 5-20 relates to a constant wheel dullness, but the actual degree of wheel dullness depends on the removal rate and grinding time [62, 63]. Higher removal rates tend to promote self-sharpening of the wheel. Some aspects of self-sharpening are discussed in the following section and in the discussion of wheel wear in Chapter 11.

Controlled-force grinding behavior with CBN wheels is similar to that observed with conventional abrasive wheels [68]. One particular advantage of controlled-force grinding with CBN is that the wheel can be 'broken-in' while avoiding high initial force peaks, as seen in Figure 5-19 [49, 69]. For conventional fixed-feed grinding at a constant infeed rate (Curve '1'), the normal force reaches a peak value followed by a progressive decrease towards a steady-state value as the wheel sharpens. If the wheel does not begin to sharpen readily or becomes clogged (Curve '2'), the normal force will continue to grow and possibly cause wheel breakage. The initial straight-line slope with conventional grinding (Curves '1' and '2') corresponds to the stiffness of the grinding system. With controlled-force grinding (Curve '3'), the normal force remains constant at its controlled value, while the removal rate progressively increases from zero towards a steady-state value as the wheel self-sharpens. Thus the danger of breaking the CBN wheel or damaging the workpiece is minimized.

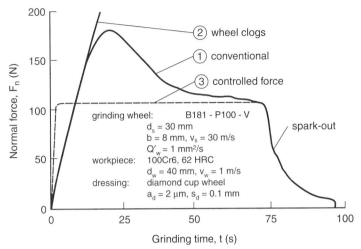

Figure 5-22 Normal grinding force versus time for conventional fixed-feed and controlled-force internal grinding with a CBN wheel [69].

One way to realize some of the advantages of controlled-force grinding with a fixed-feed machine is by controlled-power grinding. Modern grinding machines are often fitted with accurate power monitors, and in some cases the control systems allow for grinding at pre-set power levels. For a given wheel velocity, this is equivalent to controlled-force grinding, but with a given tangential force instead of normal force component. Some 'hybrid' machines can operate in either a controlled-force or a fixed-feed mode.

5.8   HEAVY-DUTY GRINDING

The objective in heavy-duty grinding is rapid material removal with only secondary concern for surface quality. Removal rates may be so big that the chips are readily visible to the naked eye. Unlike precision grinding, the wheel is usually not periodically dressed.

One important heavy-duty grinding operation is snagging, which is used to condition steel billets and slabs by removing surface defects and scale. Snagging is usually done with resinoid wheels under fixed normal force. The wheel may be either oriented with its plane normal to the workpiece surface or tilted as it is traversed across the billet or slab. Heavy-duty cut-off operations (Figure 3-14) may be either of the fixed-force or of the fixed-feed type. Other heavy-duty grinding processes fall into the fixed-feed category.

Heavy-duty grinding performance depends mainly on the removal rate and wheel-wear rate. A higher removal rate means faster production, and a higher wheel-wear rate indicates increased wheel costs. Wheel consumption with conventional abrasives is usually a minor cost factor in precision grinding, but it is very significant in heavy-duty grinding. There is generally a cost optimum removal rate which balances higher wheel costs at faster removal rates against lower stock removal costs.

The mechanics of heavy-duty fixed-force grinding have been described in somewhat different terms than those of controlled-force grinding (section 5.7), so it is being treated separately here. Material removal in heavy-duty grinding is often expressed in terms of weight instead of volume, but we will use volume for the purpose of comparison with other grinding results. A relationship for the volumetric removal rate in heavy-duty fixed-force grinding has been proposed as [70, 71]:

$$Q_w = \left(\frac{Q_s}{Q_s + q_s} \right) K F_n v_s \qquad (5\text{-}28)$$

where Q_s is the volumetric wheel-wear rate, K is a removal rate factor, and q_s is a parameter which characterizes the attritious wear susceptibility of the abrasive grain material. The quantity in parentheses can be regarded as a cutting efficiency, so that removal rate can also be written as a function of its maximum value:

$$Q_w = \left(\frac{Q_s}{Q_s + q_s} \right) Q_{w,m} \qquad (5\text{-}29)$$

where $Q_{w,m}$ is the maximum removal rate ($Q_{w,m} = K F_n v_s$). For an ideal abrasive with infinite resistance to dulling by attrition ($q_s = 0$), the efficiency is 100% and the removal rate reaches $Q_{w,m}$. For a finite value of q_s, a faster-wearing wheel (Q_s bigger) will self-sharpen to a greater degree, thereby also raising the efficiency. In terms of the previous force analysis (section 5.4.2), we can say that a higher efficiency indicates a sharper wheel with a smaller wear-flat area, so less of the fixed normal force is 'wasted' by sliding.

For the purpose of verifying the removal rate relationship and evaluating the parameters K and q_s, it is convenient to rewrite Eq. (5-28) as

$$\frac{F_n v_s}{Q_w} = \frac{q_s}{K Q_s} + \frac{1}{K} \qquad (5\text{-}30)$$

A plot of $F_n v_s / Q_w$ versus Q_s^{-1} should yield a straight line with intercept K^{-1} and slope $q_s K^{-1}$. Examples of such straight-line relationships are shown in Figure 5-23 for snagging of a plain carbon steel and a stainless steel on a relatively small laboratory grinder using resinoid wheels containing various alumina and alumina-zirconia abrasives [72], and similar results

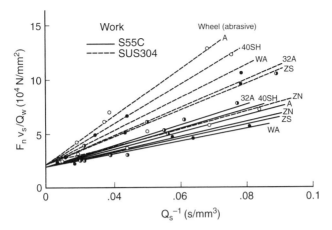

Figure 5-23 *Sagging results for a plain carbon steel (S55C) and a stainless steel (SUS 304) for resinoid wheels with various 20-grit abrasives: A (regular alumina), WA (white alumina), 32A (single-crystal alumina), 40SH (roasted regular alumina), ZS (alumina – 25% zirconia), ZN (alumina – 40% zirconia) [72].*

have been obtained on larger machines [71]. With narrower wheels or tilt, the results also fall on the same straight line [71].

An underlying feature of the present approach to the mechanics of heavy-duty grinding is the possibility, at least in theory, to differentiate between structural wheel characteristics and abrasive grain quality. The parameter K is related to geometrical and force factors, whereas the parameter q_s is regarded as an intrinsic property of the abrasive, which depends on its chemical composition but not on its size, shape, or microstructure. Alumina-zirconia abrasives with higher zirconia content should have a lower q_s value because zirconia has a higher melting point and is more chemically stable than alumina [70]. In Figure 5-23 it can be seen that the alumina-zirconia abrasive with the highest zirconia content indeed gave the lowest q_s value (smallest slope) for snagging of the stainless steel, and similar results were also found for snagging of an alloy steel [71]. However, a slightly lower q_s, value was obtained with the white alumina wheel when snagging the plain carbon steel.

A relationship for the power consumption in heavy-duty fixed-force grinding has been proposed as [70]

$$P = P_0 + mQ_w \qquad (5\text{-}31)$$

where m is a constant having the units of specific energy. Originally it was suggested that P_0 is proportional to the quantity $F_n v_s$, although this was

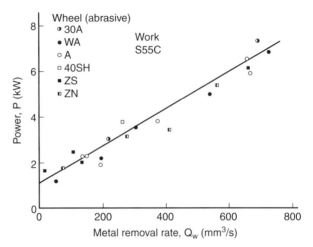

Figure 5-24 Grinding power versus removal rate corresponding to results in Figure 5-20 for a plain carbon steel. Adapted from Reference [72].

apparently never experimentally confirmed. The results in Figure 5-24 for snagging of a plain carbon steel with different abrasives obtained over a range of normal forces varying by about a factor of two all appear to fall on the same straight line, which would imply that P_0 is constant [72]. However, a proportional relationship between P_0 and F_n cannot be statistically excluded. The results for snagging of other steels all fell on the same straight line [72], which again shows the insensitivity of the grinding energy to the particular steel being ground.

In order to relate these results to the discussion in section 5.4, we can begin by repeating the specific energy equation:

$$u = u_{ch} + u_{pl} + u_{sl} \qquad (5\text{-}17)$$

Now because of the very high removal rates in heavy-duty grinding, the magnitude of the plowing component is negligible in comparison with the chip-formation component, so that

$$u \approx u_{ch} + u_{sl} \qquad (5\text{-}32)$$

In this case, the specific chip-formation energy is the specific grinding energy for a perfectly sharp wheel which removes material at the maximum removal rate $Q_{w,m}$. From Eq. (5-31), the power at the maximum removal rate is

$$P_m = P_0 + mQ_{w,m} \qquad (5\text{-}33)$$

and the corresponding specific energy u_{ch} is obtained by dividing Eq. (5-33) by $Q_{w,m}$

$$u_{ch} = \frac{P_m}{Q_{w,m}} = \frac{P_0}{Q_{w,m}} + m \qquad (5\text{-}34)$$

Also, by combining Eqs. (5-3), (5-29), (5-32) and (5-34), the sliding component is obtained as

$$u_{sl} = \frac{P_0}{Q_{w,m}} \left(\frac{q_s}{Q_s} \right) \qquad (5\text{-}35)$$

Since $Q_{w,m}$ is proportional to F_n, Eq. (5-34) implies that u_{ch} depends on F_n, but a proportional relationship between P_0 and F_n, as mentioned above, would result in u_{ch} being constant. For the results in Figure 5-23, it can be shown that the first term on the right-hand side of Eq. (5-23) amounts to only about 10 to 20% of m, in which case $u_{ch} \approx 1.1m - 1.2m$. The slope in Figure 5-24 is $m \approx 9.5$ J/mm^3, and other results for snagging of various steels yields slopes of 8 to 12 J/mm^3 [73]. These values are comparable to the specific melting energy for iron, and therefore imply specific chip-formation energies which are close to or slightly less than those for precision grinding of steels. Therefore, the grinding energy analysis in section 5.4 also provides a reasonable description of heavy-duty grinding.

For fixed-feed heavy-duty grinding, there is relatively little power and energy data available in the literature. Linear relationships between power and removal rate, like those in Figure 5-24, have been obtained for fixed-feed abrasive cut-off of steel bars [74] with slopes of 10 to 12 J/mm^3, which is again comparable to the melting energy of iron. However, with vertical-spindle rotary table surface grinding (Figure 1-1(c)) with a continuous downfeed, the specific energies were substantially bigger and found to increase with faster removal rates [75], which is contrary to what is normally observed. Such anomalous behavior may be due to excessive wheel wear and a high degree of continuous overlap between the wheel and the workpiece in this particular case, such that broken grain fragments from the wheel are trapped and roll around between the flat underside of the wheel and the workpiece surface. For vertical-spindle surface grinding with linear workpiece travel and discrete downfeed increments at the end of each stroke, most material removal occurs on the circular rim area of the wheel, rather than on its flat underside. In this case, the specific energy progressively decreases with faster removal rates [76].

Linear relationships between power and removal rate are not necessarily limited to heavy-duty grinding, but are also found with some precision grinding operations [74]. For controlled-force internal precision grinding of a bearing steel with finely dressed wheels [66, 67], the slopes obtained were about three times as big as those for heavy-duty grinding. Such large values reflect the high values of u_{sl} associated with dull wheels and long contact lengths (Eq. (5-11)). With harshly dressed, sharp wheels, the slopes were close to the specific melting energy for iron.

5.9 EMPIRICAL RELATIONSHIPS

Our objective in this chapter has been to present a physical description of the grinding process, which can account for the magnitude of the forces and energy. Other more empirical approaches have attempted to correlate grinding behavior with 'basic' process parameters. With cylindrical plunge grinding, for example, it has been shown that the tangential and normal force components per unit width may be fairly well approximated by power function relationships [77]:

$$F'_t = F_1 \left(\frac{v_w a}{v_s} \right)^f = F_1 \left(\frac{Q'_w}{v_s} \right)^f \tag{5-36}$$

and

$$F'_n = F_2 \left(\frac{v_w a}{v_s} \right)^f = F_2 \left(\frac{Q'_w}{v_s} \right)^f \tag{5-37}$$

where F_1, F_2 and f are constants. The exponents are typically in the range 0.4–0.9. The corresponding specific energy from Eq. (5-3) is

$$u = F_1 \left(\frac{v_w a}{v_s} \right)^{f-1} = F_1 \left(\frac{Q'_w}{v_s} \right)^{f-1} \tag{5-38}$$

The quantity within the parentheses in Eqs. (5-36)–(5-38) is called the equivalent chip thickness

$$h_{eq} \equiv \frac{v_w a}{v_s} = \frac{Q'_w}{v_s} \tag{5-39}$$

the name being given because it would correspond to the thickness of a continuous layer of material (chip) being removed at a volumetric rate per unit width Q'_w and cutting velocity v_s. This parameter is also equal to the volumetric removal rate per unit area of wheel surface passing through the grinding zone. As a relative measure of grinding severity, the equivalent chip thickness correlates fairly well not only with the grinding forces and energy, but also with other performance characteristics including surface roughness and wheel wear, as will be seen in Chapters 10 and 11. However, these and other empirical relationships tend to be of limited practical use for predicting grinding performance because the constants F_1, F_2 and f depend on the particular wheel, workpiece, grinding fluid, and dressing conditions, as well as on the accumulated stock removal.

REFERENCES

1. Malkin, S., 'Grinding of Metals: Theory and Application', *J. Applied Metalworking*, 3, 1984, p. 95.
2. Alden, G. I., 'Operation of Grinding Wheels in Machine Grinding', *Trans. ASME*, Paper No. 1446, 1914, p. 451.
3. Dall, A. H., 'A Review of the Grinding Theories', *Modern Machine Shop*, November 1939.
4. Ernst, H., *Metal Cutting: Art to Science', Machining—Theory and Practice*, American Society for Metals, Metals Park, Ohio, 1950, p. 1.
5. Tarasov, L. P., 'Some Metallurgical Aspects of Grinding', *Machining - Theory and Practice,* American Society for Metals, Metals Park, Ohio, 1950, p. 409.
6. Doyle, E. D. and Dean, S. K., 'An Insight into Grinding Viewpoint', *Annals of the CIRP*, 29/2, 1980, p. 571.
7. Doyle. E. D and Aghan, R. L., "Mechanism of Metal Removal in the Polishing and Grinding of Hard Materials," *Metall. Trans. B.*, 6B, 1975, p. 143.
8. Malkin, S., 'Grinding Mechanisms for Metallic and Non-metallic Materials, *Proceedings, Ninth North American Manufacturing Research Conference*, SME, 1981, p. 235.
9. Black, J. T., 'Flow Stress Model in Metal Cutting', *Trans. ASME, J. of Eng. for Ind.*, 101, 1979, p. 403.
10. Rowe, G. W. and Wetton, A. G., 'Theoretical Considerations in the Grinding of Metals', *J. Inst. Met.*, 97, 1969, p. 193.
11. Guo, C., Krishnan, N., and Malkin, S., 'Matching Forces and Power Measurements in Creep Feed Grinding', *Fifth International Grinding Conference*; SME, Cincinnati, Ohio, October 26 - 28, 1993.
12. Marshall, E. R. and Shaw, M. C., 'Forces in Dry Surface Grinding', *Trans. ASME*, 74, 1952, p. 51.
13. Backer, W. R., Marshall, E. R. and Shaw, M. C., 'The Size Effect in Metal Cutting', *Trans. ASME*, 74, 1952, p. 61.
14. Merchant, E., 'Mechanics of the Metal Cutting Process', *J. Appl. Phys.*, 16, 1945, p. 207.
15. Von Turkovich, B. F., 'Shear Stress in Metal Cutting', *Trans. ASME, J. of Eng. for Ind.*, 94, 1970, p. 151.
16. Hultgren, R., Desai, P. D., Hawkins, D. T., Gleiser, M., Kelly, K. K. and Wagman, D. D., *Selected Values of the Thermodynamic Properties of the Elements*, American Society for Metals, Metals Park, Ohio, 1973.
17. Malkin, S. and Cook, N. H., 'The Wear of Grinding Wheels, Part 1, Attritious Wear', *Trans. ASME, J. of Eng. for Ind.*, 93, 1971, p. 1129.

18. Foerster, M. and Malkin, S., 'Wear Flats Generated during Grinding with Various Grinding Fluids', *Proceedings, Second North American Metalworking Research Conference*, SME, 1974, p. 601.
19. Malkin, S. and Murray, T., 'Comparison of Single Point and Rotary Dressing of Grinding Wheels', *Proceedings, Fifth North American Metalworking Research Conference*, SME, 1977, p. 278.
20. Kannapan, S. and Malkin, S., 'Effects of Grain Size and Operating Parameters on the Mechanics of Grinding', *Trans. ASME, J. of Eng. for Ind.*, 94, 1972, p. 833.
21. Malkin, S., 'Specific Energy and Mechanisms in Abrasive Processes', Proceedings, *Third North American Metalworking Research Conference*, Carnegie Press, 1975, p. 453.
22. Malkin, S., 'Selection of Operating Parameters in Surface Grinding of Steels', *Trans. ASME., J. of Eng. for Ind.*, 98, 1976, p. 56.
23. Abebe, M. and Appl, F. C., 'A Slip Line Field for Negative Rake Angle Cutting', *Proceedings, Ninth North American Manufacturing Research Conference*, SME, 1981, p. 341.
24. Shonozaki, T. and Shigematu, H., 'Mechanism of Rubbing and Biting of Cutting Edge on Work Surface in Grinding Process', *Bull. Japan Soc. of Prec. Eng.*, 2, 1966, p. 8.
25. Takenaka, N., 'A Study on the Grinding Resistance Force on Single Grits', S., *Proceedings of the International Conference on Manufacturing Technology*, Ann Arbor, 1967, p. 617.
26. Okamura, K. and Nakajima, K., 'The Surface Generation Mechanics in the Transitional Cutting Process', *Proceedings of the International Grinding Conference*, Pittsburgh, 1972, p. 305.
27. Hahn, R. S., 'On the Nature of the Grinding Process', *Proceedings of the Third International Machine Tool Design and Research Conference*, 1962, p. 129.
28. Sedricks, A. J. and Mulhearn, T. O, 'The Mechanics of Cutting and Rubbing in Simulated Abrasives Processes,' *Wear*, 6, 1963, p. 457.
29. Mulhern, T. O. and Samuels, T. O, 'The Abrasion of Metals: a Model of the Process', *Wear*, 5, 1962, p. 478.
30. Samuels, L. E., *Metallographic Polishing by Mechanical Methods*, 3rd edn., Chapter 3, ASM, 1982.
31. Maan, N. and Broese Van Groenou, A., 'Low Speed Scratch Experiments on Steels', *Wear*, 42, 1977, p. 365.
32. Gilormini, P. and Felder, E., 'Theoretical and Experimental Study of the Ploughing of a Rigid-plastic Semi-infinite Body by a Rigid Pyramidal Indentor', *Wear*, 88, 1983, p. 195.
33. Torrance, A. A., 'A New Approach to the Mechanics of Abrasion', *Wear*, 67, 1981, p. 233.
34. Thompson, D. L. and Malkin, S., 'Grinding Wheel Topography and Undeformed Chip Shape', *Proceedings of the International Conference on Production Engineering*, Tokyo, 1974, p. 727.
35. Werner, G., 'Kinematik und Mechanik des Schleifprozesses', Doctoral Dissertation, TH Aachen, 1971.
36. Malkin, S., 'Negative Rake Cutting to Simulate Chip Formation in Grinding', *Annals of the CIRP*, 28/1, 1979, p. 209.
37. Malkin, S. and Joseph, N., 'Minimum Energy in Abrasive Processes', *Wear*, 32, 1975, p. 15.
38. Malkin, S., 'Current Trends in CBN Grinding Technology, *Annals of the CIRP*, 34/2, 1985, p. 557.

39. Percherer, E. and Malkin, S., 'Grinding of Steels with Cubic Boron Nitride (CBN), *Annals of the CIRP*, 33/1, 1984, p. 211.

40. Bhattacharyya, S. K. and Hon, K. K., 'Grindability Study of CBN Wheels', *Proceedings of the Nineteenth international Machine Tool Design and Research Conference*, 1978, p. 645.

41. Hon, K. K., 'Characteristics for Force Patterns in CBN Grinding', *Proceedings of the Twenty Fourth International Machine Tool Design and Research Conference*, 1973, p. 671.

42. Smith, L. I. and Tsujigo, Y., 'An Analysis of CBN Grinding', *Cutting Tool Engineering*, May/June 1977. p. 49.

43. Kishi, K. and Ichida, Y., 'Grindability of High Carbon—High Vanadium Steel with CBN Wheel', *Proceedings of the 5th International Conference on Production Engineering*, Tokyo, 1984, p. 679.

44. Saljé, E. and Heidenfelder, H., 'Comparison of CBN and Conventional Grinding Processes', *Proceedings, Thirteenth North American Manufacturing Research Conference*, SME, 1985.

45. Ishikawa, T., and Kumar, K. V., 'Conditioning of Vitrified Bond Superabrasive Wheels', *Superabrasives, 91*, SME, June 11–13, 1991, p. 7.

46. Hitchiner, M. P., 'Dressing of Vitrified CBN Wheels for Production Grinding', *Ultrahard Materials Technical Conference*, Windsor, Ontario, Canada, May 28, 1998, p. 139.

47. Jacobuss, M. and Webster, J. A., 'Optimizing the Truing and Dressing of Vitrified-Bond CBN Grinding Wheels', *Abrasives Magazine*, Aug./Sept 1996, p. 23.

48. Klocke, F. and König, W., 'Appropriate Conditioning Strategies Increase the Performance of Vitrified-Bond CBN Grinding Wheels', *Annals of the CIRP*, 44/1, 1995, p. 305.

49. Tönshoff, H.K. and Grabner, T., "Cylindrical and Profile Grinding with Boron Nitride Wheels', *Proceedings of the 5th International Conference on Production Engineering*, JSPE, Tokyo, 1984, p. 326.

50. Carius, A. C., 'Preliminaries to Success—Preparation of Grinding Wheels Containing CBN', SME Paper No. MR84-547, International Grinding Conference, Fontana, Wisconsin 1984.

51. Saljé, E. and von Mackensen, H. G., 'Dressing of Conventional and CBN Grinding Wheels with Diamond Form Rollers', *Annals of the CIRP*, 33/1, 1984, p. 205.

52. Shi, Z. and Malkin, S., 'Wear of Electroplated CBN Grinding Wheels', *Trans. ASME Journal of Manufacturing Science and Engineering,* 128, 2006, p. 110.

53. Upadhyaya, R. P. and Malkin, S, 'Thermal Aspects of Grinding with Electroplated CBN Wheels', *Trans. ASME, Journal of Manufacturing Science and Engineering*, 126, 2004, p. 107.

54. Shi, Z. and Malkin, S., 'Investigation of Grinding with Electroplated CBN Wheels', *Annals of the CIRP*, Vol. 53/1, 2003, p. 267.

55. Andrew, C., Howes, T. D. and Pearce, T. R. A., *Creep-feed Grinding*, Holt, Rinehart, and Winston, London, 1985.

56. Werner, G. and Schlingensiepen, R., 'Creep feed—an Effective Method to Reduce Work Surface Temperature in High Efficiency Grinding, *Proceedings, Eighth North American Manufacturing Research Conference*, SME, 1980, p. 312.

57. Saljé, E., 'Creep Feed Grinding', *Proceedings of the 5th International Conference on Production Engineering*, JSPE Tokyo, 1984, p. 37.

58. Howes, T. D., 'The Technique of Dressing during Grinding', *Proceeding of the International Conference on Creep-feed Grinding,* Bristol, 1979, p. 184.
59. Pearce, T. R. A. and Howes, T. D., 'Applications of Continuous Dressing in Grinding Operations', *Proceedings of the Twenty Third International Machine Tool Design and Research Conference*, 1982, p. 203.
60. Hahn, R. S., 'Controlled Force Grinding—a New Technique for Precision Internal Grinding', *Trans. ASME, J. of Eng. for Ind.*, 86, 1964, p. 287.
61. Hahn, R. S. and Lindsay, R. P., 'Principles of Grinding', 5-part series, *Machinery*, New York, 1971.
62. Hahn, R. S., 'Some Characteristics of Controlled Force Grinding', *Proceedings of the Sixth International Machine Tool Design Research Conference*, 1965, p. 597.
63. Lindsay, R. P. and Hahn, R. S., 'On the Basic Relationships between Grinding Parameters', *Annals of the CIRP*, 19, 1971, p. 657.
64. Hahn, R. S. and Lindsay, R. P., 'On the Effects of Real Area of Contact and Normal Stress in Grinding, *Annals of the CIRP*, 15, 1967, p. 197.
65. Lindsay, R. P., 'The Effect of Parameter Variations in Precision Grinding', *Trans. ASME, J. of Eng. for Ind.*, 92, 1970, p. 683.
66. Lindsay, R. P., 'Factors Affecting Forces and Power during High-speed Precision Grinding', High Speed Machinery Symposium, Winter Annual Meeting, ASME, 1984.
67. Lindsay, R. P. 'The Effect of Contact Time on Forces, Wheelwear Rate and G-ratio during Internal and External Grinding', *Annals of the CIRP*, 33/1, 1984, p. 193.
68. Lindsay, R. P., 'Characteristics of Precision Production Grinding with Borazon (CBN) Wheels', *Proceedings, Second North American Metalworking Research Conference*, 1974, p. 610.
69. Tönshoff, H. K., Grabner, T. and Zinngrabe, M., 'A Breakthrough in Production Grinding with BN Wheels, *Proceedings of the International Conference on Manufacturing Science and Technology of the Future*, MIT, Cambridge, 1984.
70. Coes, L., Jr., *Abrasives*, Springer-Verlag, New York, 1971, Chapter 12.
71. Reichenbach, G. S. and Coes, L., Jr., 'The Mechanics and Economics of Steel Conditioning', *Proceedings of the International Grinding Conference*, Pittsburgh, 1972, p. 538.
72. Matsuo, T. and Sonoda, S., 'The Rating of Wheels in Laboratory Snag Grinding', *Annals of the CIRP*, 29/1, 1986, p. 221.
73. Coes, L., Jr., private communication.
74. Lindsay, R. P., 'Similarities among Grinding Operations', SME Paper No. MR76-257, 1976.
75. Pollock, C., 'Significant Factors in Vertical Spindle Abrasive Machining Tests', *Trans. ASME, J. of Eng. for Ind.*, 87, 1965, p. 365.
76. Lal, G. K., 'Forces in Vertical Surface Grinding', *Int. J. Mach. Tool Des. Res.*, 8, 1968, p. 33.
77. Snoeys, R., Peters, I. and Decneut, A., 'The Significance of Chip Thickness in Grinding', *Annals of the CIRP*, 23/2, 1974, p. 227.

Chapter 6

Thermal Aspects: Conventional Grinding[1]

6.1 INTRODUCTION

The grinding process requires an extremely high energy expenditure per unit volume of material removed. Virtually all of this energy is converted to heat which is concentrated within the grinding zone. The high temperatures produced can cause various types of thermal damage to the workpiece, such as burning, phase transformations, softening (tempering) of the surface layer with possible rehardening, unfavorable residual tensile stresses, cracks, and reduced fatigue strength [2-3]. Furthermore, thermal expansion of the workpiece during grinding contributes to inaccuracies and distortions in the final product. The production rates which can be achieved by grinding are often limited by grinding temperatures and their deleterious influence on workpiece quality.

From metallurgical examinations of ground hardened steel surfaces reported in 1950 [4], it was conclusively shown that most grinding damage is thermal in origin. In the first attempt to correlate actual grinding temperatures with structural metallurgical changes in the workpiece five years later [5], the temperature distribution in the subsurface was measured during grinding of a hardened bearing steel by means of a thermocouple embedded in the workpiece. Numerous other methods have also been developed to measure grinding temperatures using either thermocouples and radiation sensors [6, 7]. While considerable difficulties may arise in interpreting such measurements due to the extreme temperature gradients in time and space near to the surface, embedded thermocouples and infrared radiation sensors utilizing fiber optics have been shown to provide a reasonably good indication of the workpiece temperature near the ground

---

[1] *Parts of this chapter were adapted from Reference [1].*

surface [2, 7-13]. Both of these temperature-measuring techniques have been found to give results which are consistent with each other, and also with measurements of the surface temperature using a thin foil thermocouple [13].

Temperatures generated during grinding are a direct consequence of the energy input to the process. In general, the energy or power consumption is an uncontrolled output of the grinding process, which may vary considerably and is sensitive to the wheel condition. Consequently, the temperature generated is also uncontrolled and varying. Temperature-measuring methods do not provide a practical means to identify and control grinding temperatures, as they are generally restricted to the laboratory and cannot be applied in a production environment. In-process monitoring of the grinding power, when coupled with a thermal analysis of the grinding process, offers a better approach for estimating grinding temperatures and controlling thermal damage.

Thermal analyses of grinding processes are usually based upon the application of moving heat source theory to the workpiece being ground. For this purpose, the grinding zone is usually modeled as a band source of heat which moves along the surface of the workpiece. All the grinding energy expended is considered to be converted to heat at the grinding zone where the wheel interacts with the workpiece. A critical parameter needed for calculating the temperature response is the energy partition to the workpiece, which is the fraction of the total grinding energy transported to the workpiece as heat at the grinding zone. The energy partition depends on the type of grinding, the wheel and workpiece materials, and the operating conditions. For conventional shallow cut grinding with conventional aluminum oxide wheels, the energy partition is usually generally bigger than for creep-feed grinding or for grinding with CBN wheels.

The present chapter is concerned with the thermal aspects of conventional shallow cut grinding process, which is mainly directed towards calculating temperatures and controlling thermal grinding damage. Thermal aspects of creep-feed grinding will be addressed in Chapter 7, and thermal aspects of grinding with CBN abrasive wheels in Chapter 8. The present chapter begins with a relatively simple heat transfer analysis of the grinding process to establish the grinding zone temperature model for cylindrical and straight surface plunge grinding processes in terms of the power consumption, energy partition, and the other grinding parameters. By inverting the heat transfer solution, the allowable power corresponding to a critical surface temperature can then be specified in terms of the grinding parameters. It is demonstrated how this result can be applied to predicting and controlling the onset of thermal damage for plunge grinding of steels. Thermal analyses are also presented for face grinding and abrasive cut-off processes.

6.2 HEAT TRANSFER ANALYSIS: PLUNGE GRINDING

Grinding occurs by the interaction of discrete abrasive grains on the wheel surface with the workpiece. According to the analysis of grinding mechanisms in Chapter 5, the total grinding energy input includes chip formation, plowing, and sliding energy components. Peak 'flash' temperatures are generated which approach the melting point of the material being ground [14-16]. However, these extreme temperatures are of extremely short duration and highly localized on the shear planes of microscopic grinding chips. Just beneath the surface, the workpiece 'feels' nearly continuous heating owing to the multiplicity of interactions with the abrasive grits passing quickly through the grinding zone. Therefore, the temperature associated with 'continuous' heating over the grinding zone, rather than the peak 'flash' temperature, is found to be responsible for most thermal damage. Also of interest is the temperature in the bulk of the workpiece, which causes thermal expansion leading to distortions and dimensional inaccuracies [17-20]. The average bulk temperature rise is generally much smaller than the grinding zone temperature.

In order to calculate the grinding zone temperature rise, consider the cylindrical plunge-grinding situation illustrated in Figure 6-1(a). The grinding energy is dissipated over the rectangular grinding zone of length l_c along the arc of contact and width b normal to the plane of the figure. For simplicity, the grinding heat flux q_w entering the workpiece is assumed for now to be uniformly distributed over the grinding zone [14, 15, 21, 22], although a different distribution (e.g. triangular) can also be used [2]. Since the cylindrical workpiece is generally much bigger than the dimensions of the grinding zone, the heated area can be likened to a plane band source of heat which moves along the surface of a semi-infinite solid (the workpiece) at the workpiece velocity. For this two-dimensional heat transfer model illustrated in Figure 6-1(b), the temperature rise can be written as [23]:

$$\theta = \frac{1}{\pi k}\int_{-l}^{l} q_w(u)e^{-\frac{v(x-u)}{2\alpha}}K_0\left[\frac{V}{2\alpha}\{(x - u)^2 + z^2\}^{\frac{1}{2}}\right]du \qquad (6\text{-}1)$$

where $K_0(x)$ is the modified Bessel function of the second kind of order zero, $q_w(u)$ is the heat flux distribution at the surface of the semi-infinite body, k is the thermal conductivity of the workpiece, α is the thermal diffusivity of the workpiece, l is the half of the heat source length, and V is the heat source velocity.

In many practical cases, the maximum temperature rise θ_m, which occurs on the workpiece surface towards the trailing edge of the heat source, is of particular interest. If the heat source is moving sufficiently fast so that the heat condition in the direction of motion is much slower than

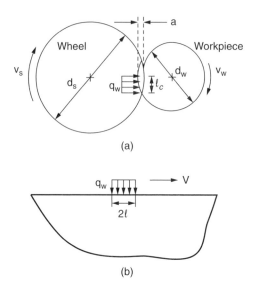

(a)

(b)

Figure 6-1 Illustration of (a) external cylindrical plunge grinding and (b) thermal model for grinding zone temperature.

heat source velocity, the maximum temperature rise for a uniform heat source ($q_w(u) = q_w$) can be approximated as [21, 23]

$$\theta_m = \frac{1.595 q_w \alpha^{1/2} l^{1/2}}{k V^{1/2}} \qquad (6\text{-}2)$$

This 'high speed' condition is satisfied provided that the dimensionless thermal number (Peclet number) $L \equiv Vl/2\alpha$ is bigger than 5, although this equation can also be practically applied to estimating the maximum surface temperature in grinding down to $L = 1$, which includes most actual grinding situations [21]. The average surface temperature θ_a over the source is 2/3 θ_m for $L > 5$.

   An expression for θ_m can now be obtained by making the appropriate substitutions into Eq. (6-1). Referring in Figure 6-1, the velocity V in the heat transfer model corresponds to the workpiece velocity v_w:

$$V = v_w \qquad (6\text{-}3)$$

and the half-length l of the heat source is half the arc length of contact:

$$l = l_c/2 \qquad (6\text{-}4)$$

From geometrical considerations (Chapter 3), the arc length of contact is given by

$$l_c = (ad_e)^{1/2} \qquad (6\text{-}5)$$

The parameter d_e is the equivalent diameter given by Eq. (3-9) for either straight, external cylindrical, or internal cylindrical grinding, so this temperature solution is not limited to external cylindrical plunge grinding (Figure 6-1) but is also applicable to straight and internal grinding. This contact-length relationship neglects elastic deformation, which would tend to make l_c bigger, in which case the calculated temperature would be less. On the other hand, a more realistic energy distribution, rather than the uniform one which has been assumed, might lead to a higher temperature. It will be seen that the use of a uniform heat distribution and neglecting elastic deflection is adequate for predicting some types of thermal damage.

Combining Eqs. (6-2)-(6-5), the maximum grinding zone temperature rise becomes

$$\theta_m = \frac{1.13 q_w \alpha^{1/2} a^{1/4} d_e^{1/4}}{k v_w^{1/2}} \tag{6-6}$$

With a triangular heat source, instead of the uniform rectangular one considered here, the factor 1.13 in Eq. (6-6) reduces slightly to 1.06 [2, 24]. For calculating θ_m, the parameter which remains to be determined is the heat flux q_w, which is the energy input rate to the workpiece per unit area over the grinding zone. Of the total grinding energy generated, only the fraction ε is conducted as heat to the workpiece at the grinding zone. For a specific grinding energy u or power P (see Chapter 5), the heat flux to the workpiece can be written as

$$q_w = \frac{\varepsilon u v_w ab}{l_c b} = \frac{\varepsilon P}{l_c b} = \varepsilon q \tag{6-7}$$

where the numerator is that portion of the grinding power entering the workpiece and the denominator is the area of the grinding zone.

In order to proceed with the thermal analysis, it is necessary to specify the energy partition to the workpiece ε [9-13, 24-31]. The first direct measurement of the energy partition, obtained by calorimetric methods, was reported to be 84% [30], although subsequent calorimetric measurements for regular shallow cut grinding of steels indicated values ranging from about 60 to 90% [31]. More recent studies have shown that the energy partition can vary widely, depending on the type of grinding, fluid application conditions, and wheel composition. For example, it will be seen in the following chapters that creep feed grinding with porous aluminum oxide wheels can give extremely low energy partitions of only 3 to 7%, and that comparably low values may be obtained for grinding with vitrified and electroplated CBN wheels.

For regular shallow cut grinding of steels with aluminum oxide wheels, the energy partition typically varies from 60 to 90% [31]. These results have been rationalized in terms of the grinding energy model

(Chapter 5), according to which the total specific energy includes contributions from chip formation, plowing, and sliding:

$$u = u_{ch} + u_{pl} + u_{sl} \tag{6-8}$$

where u_{ch}, u_{pl}, and u_{sl} are the chip-formation, plowing, and sliding components, respectively. From heat transfer considerations, it can be shown that almost all the sliding energy generated at the interface between the wear flats and the workpiece is conducted as heat to the workpiece. Likewise, virtually all the plowing energy is in the workpiece, as plowing involves workpiece deformation without material removal. Using calorimetric methods, it has been found that approximately 55% of the chip-formation energy is conducted to the workpiece, which is reasonably consistent with expectations from a heat transfer analysis of the chip-formation process [31]. Therefore, all the grinding energy except for about 45% of the chip-formation energy is conducted as heat to the workpiece, so the overall fraction of the grinding energy entering the workpiece is:

$$\varepsilon = \frac{u_{pl} + u_{sl} + 0.55u_{ch}}{u} = \frac{u - 0.45u_{ch}}{u} \tag{6-9}$$

Substituting this result into Eq. (6-7) and combining with Eq. (6-6) leads to the maximum temperature rise:

$$\theta_m = \frac{1.13\alpha^{1/2}a^{3/4}v_w^{1/2}(u - 0.45u_{ch})}{kd_e^{1/4}} \tag{6-10}$$

Various techniques have been used to measure the energy partition in grinding. As stated above, early investigations were based upon calorimetric methods whereby the heat content in the workpiece is obtained by placing it in a bath or measuring its average temperature rise immediately after grinding [30, 31]. The energy partition is then obtained as the ratio of heat content in the workpiece to the total grinding energy obtained from measurements of forces or power. Calorimetric methods have been applied only to dry grinding, since bulk cooling of the workpiece by the applied fluid would make such experiments much more difficult.

The energy partition can also be obtained using temperature matching and inverse heat transfer methods [12, 28, 29]. Both of these methods utilize measurements of the temperature response in the workpiece subsurface, rather than its average temperature rise. The temperature response in the workpiece subsurface can be measured during straight surface plunge using either an embedded thermocouple or infrared detector with an optical fiber. The thermocouple or optical fiber is fed into a blind hole from the underside of the workpiece toward the surface being ground. The temperature response is measured at the bottom of the blind hole as the grinding wheel passes over the workpiece. With each successive grinding pass, the temperature response is measured closer to the surface being ground.

With the temperature matching method [12], it is necessary to find the energy partition ε for which the measured temperature response most closely matches the temperature computed according to the thermal model (Eq. (6-1)). For example, it is seen in Figure 6-2 that the individual

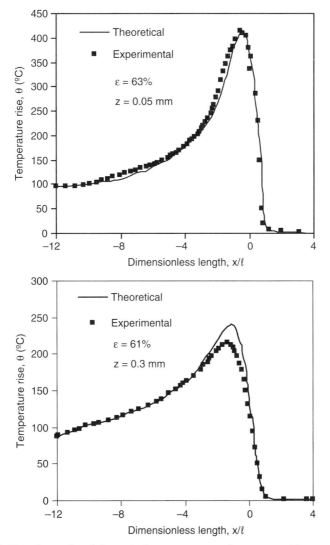

*Figure 6-2 **Experimental and theoretical temperatures and energy partition at two depths: AISI 1020 steel workpiece, aluminum oxide wheel, triangular heat source.***

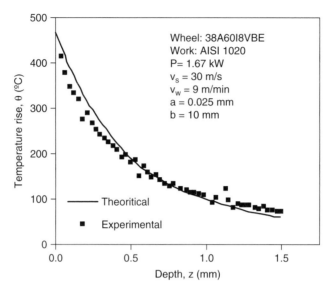

Figure 6-3 Experimental and theoretical maximum temperature rise versus depth for energy partition ε = 65%: triangular heat source.

temperature responses at various depths for grinding of AISI 1020 steel with an aluminum oxide wheel matches the analytically computed temperature response quite well for $\varepsilon = 65\%$. Virtually the same energy partition value can be obtained by matching the maximum temperature obtained at various depths to the analytically computed temperature as seen in Figure 6-3.

With the inverse heat transfer method [28, 29], the heat flux to the workpiece surface is calculated from the measured temperature distribution within the workpiece. An example of a heat flux distribution obtained in this way is shown in Figure 6-4 for grinding of a hardened bearing steel with an aluminum oxide wheel. It can be seen that the heat flux distribution on the workpiece surface consists of both positive and negative values. Positive heat flux implies heat flow into the workpiece, and negative heat flux indicates localized cooling by the applied grinding fluid. The positive heat flux region would correspond to the grinding zone, and the area under the heat flux curve at this location is the total energy to the workpiece per unit width of grinding. Integrating along the heat source gives the total heat flux per unit width to the workpiece, which is about 129 W/mm for this example. This corresponds to an energy partition of about 75%.

Energy partition values obtained from temperature measurements tend be comparable to but smaller than predicted from Eq. (6-9). The specific energy for the example in Figure 6-2 and Figure 6-3 was 45 J/mm^3, which

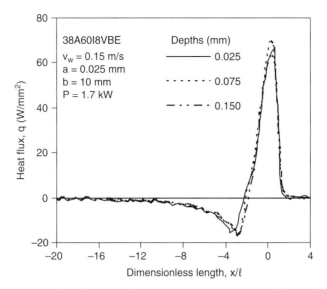

Figure 6-4 Estimated heat flux distribution using inverse heat transfer method for grinding AISI 52100 steel with an aluminum oxide wheel

would result in a predicted energy partition of 86%. Likewise for the example in Figure 6-4, the specific energy is about 40 J/mm^3, which would lead to an energy partition of about 85%. These discrepancies between predicted and measured values may be attributed to a number of factors. First of all, the thermal model assumes that all the energy expended by grinding is converted to heat. However it has been shown that up to 10% of the energy expended by plastic deformation may not be converted to heat, although for the large strains encountered in machining this may be only 1–3% [32]. The model also assumes that all the sliding energy due to rubbing between the wear flats and the workpiece is conducted as heat into the workpiece. A closer examination of the sliding energy partition using a friction-slider thermal model for aluminum oxide on steel indicates that 90–93% of the sliding energy, rather than all of it, should be conducted to the workpiece. Likewise, all the plowing energy associated with deformation of the workpiece material was also assumed to be retained as heat in the workpiece, but the abrasive grains in contact with the workpiece may also conduct away some of this heat. Furthermore, cooling by the fluid would also tend to remove some heat from the workpiece. Cooling by the fluid at the grinding zone will be considered in the following chapters.

As with most thermal analyses, a linear heat transfer model was assumed with the assumption of thermal properties independent of temperature. However, the calculated grinding zone temperatures are often so large that the assumption of constant thermal properties may no longer be justified. One approach to account for the influence of temperature on thermal properties is to use an iterative procedure whereby the constant-property solution is used but with the thermal properties evaluated at the average surface temperature over the heat source [33]. However, the validity of this method is questionable, since the temperature gradients beneath the heat source are very steep and most material through which heat conduction occurs is much cooler than the average surface temperature. A numerical analysis of this moving-heat-source problem with temperature-dependent thermal properties for a plain carbon steel, which would be comparable to those of other plain carbon and low-alloy steels, has shown that the increase in specific heat and decrease in thermal conductivity with temperature partially offset each other, so that the linear constant-property solution (Eq. (6-10)) only slightly underestimates the actual temperature up to a maximum temperature of about 1000°C [34]. With nickel-base alloys, however, it has been shown using a finite element analysis that the constant-property thermal model may significantly overestimate the grinding temperature [35].

    Thermal damage control may require that the temperature rise be kept below a critical value θ^*. For conventional grinding with aluminum oxide wheels, the allowable specific grinding energy corresponding to a maximum temperature rise θ_m can be obtained by rearranging Eq. (6-10) [21]:

$$u = u_0 + Bd_e^{1/4}a^{-3/4}v_w^{-1/2} \tag{6-11}$$

where

$$u_0 \equiv 0.45u_{ch} \tag{6-12}$$

and

$$B \equiv \frac{k\theta_m}{\beta\alpha^{1/2}} \tag{6-13}$$

Therefore, straight-line plots of specific energy u versus the quantity $d_e^{1/4}a^{-3/4}v_w^{-1/2}$ with intercept u_0, as seen in Figure 6-5, should each correspond to a constant maximum grinding zone temperature [36]. The slope B is proportional to the maximum temperature. Since $u_{ch} \approx 13.8$ J/mm³ for steels, u_0 would be approximately 6.2 J/mm³. The allowable grinding power, P, for a maximum grinding zone temperature θ_m is obtained by multiplying (6-11) by the volumetric removal rate ($Q_w = bv_wa$):

$$P = u_0bv_wa + Bd_e^{1/4}a^{1/4}v_w^{1/2}b \tag{6-14}$$

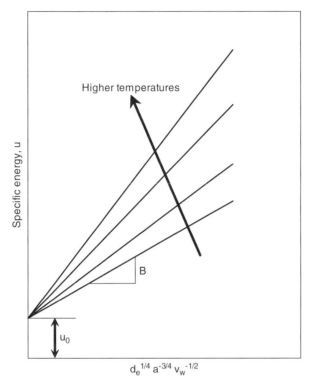

Higher temperatures

Specific energy, u

B

u_0

$$d_e^{1/4}\, a^{-3/4}\, v_w^{-1/2}$$

Figure 6-5 Lines of constant maximum grinding zone temperatures.

where b is the grinding width and u_0 and B are as defined in Eqs. (6-12) and (6-13).

Eq. (6-14) provides a practical basis for in-process identification of thermal damage in excess of a critical maximum grinding zone temperature. For this purpose, an acceptable limiting value of the parameter B must be specified, which can be obtained from Eq. (6-11) by in-process monitoring of the grinding power and post-grinding inspection of workpiece quality. The actual critical temperature θ^* need not be specified. After calibrating the thermal damage limit in this way, the measured grinding power can be compared with the allowable threshold power (Eq. (6-14) with $\theta_m = \theta^*$) to identify whether thermal damage is occurring. This is illustrated in the following section for thermal damage of steels by workpiece burn.

The foregoing analysis considers the grinding zone temperature at the workpiece surface. Thermal damage also occurs in the subsurface, in which case it is of interest to consider the temperatures reached beneath the surface. For the moving-band heat source (Figure 6-1(b)), it has been shown that the maximum dimensionless temperatures reached at depth z beneath the surface can be approximated by [37]

$$\bar{\theta}_m \equiv \left(\frac{\pi k V}{2\alpha q_w}\right)\theta_m = 3.1 L^{0.53}\exp[-0.69 L^{-0.37}Z] \qquad (6\text{-}15)$$

where L is the thermal number and Z is the dimensionless depth defined from the actual depth z as

$$Z \equiv \frac{V z}{2\alpha} \qquad (6\text{-}16)$$

At the surface ($Z = 0$), the exponential term in Eq. (6-15) would become unity and the dimensionless temperature should be identical to Eq. (6-2). There is, however, a slight discrepancy because the constants in Eq. (6-15) were obtained by fitting the results over the range of values $0.5 < L < 10$ and $0 < Z < 4$. More accurate approximations covering a wider range of L and Z have also been obtained [2]. From Eq. (6-15) it can be readily shown, for a given maximum surface temperature, that a steeper gradient of maximum temperature, and hence shallower heat penetration, is obtained with a faster velocity V and, to a lesser extent, with a shorter heat source half-length l.

Up to this point, the heat transfer analysis neglects any influence of cooling. The workpiece has been modeled as a semi-infinite body (Figure 6-1(b)) with its surface thermally insulated except at the moving heat source. Most grinding operations are performed using a grinding fluid which cools the workpiece, and an analysis of the moving-heat-source problem has been presented which also takes into account the effect of cooling [22]. In order for cooling to lower the grinding zone temperature to any significant degree, it is necessary that heat be removed from within the grinding zone area. In most practical cases, cooling by grinding fluids at the grinding zone is ineffective, a notable exception being found in creep-feed grinding (see Chapter 7). However, grinding fluids do provide bulk cooling of the workpiece, thereby helping to control dimensional inaccuracies due to thermal deformations [38]. The use of grinding fluids lowers the grinding zone temperatures mainly by providing lubrication, which reduces wheel dulling and lowers the energy input (Chapter 11). Straight oils are generally better lubricants and more effective in reducing grinding energies than water-based soluble oils and emulsions. However, water-based fluids are much better coolants than oils, since their specific heats are typically two to three times and their thermal conductivities about four times those of oils.

Another factor which has been neglected up to this point is repetition of grinding over the same nominal area. With cylindrical plunge grinding, for example, the heat source passes over the same location on the work-piece periphery once per workpiece revolution. The temperature rise in Eq. (6-10) is for one pass over the workpiece, which should be superposed on the residual temperature from previous grinding passes. In most practical cases the residual temperature is a negligible portion of the maximum temperature rise for a single pass, and it is neglected.

Although the thermal analysis has been developed for the case of plunge grinding of straight cross-sections without profiles, it can also be applied to grinding of gentle profiles having moderate variations in equivalent diameter across the width and shallow profile angles (see Chapter 3). For example, the present heat transfer analysis appears to be sufficient for grinding of typical profiles on ball-bearing races. With more severe profiles, there should be significant temperature variations across the grinding width. From heat transfer considerations, the highest temperatures should be developed near the crests of convex protrusions ('peaks') in the profile [2] where there is less material to conduct away the heat, which explains why these areas are usually more prone to thermal damage and why the lowest temperatures tend to occur at concave locations ('valleys').

The present two-dimensional heat transfer analysis can also be applied to traverse grinding with cross-feed provided that the radial wheel wear towards the leading edge is much less than the depth of cut (see Figure 3-7). In this case, the grinding action and, thus, the heat source can be considered to be concentrated over the leading edge of the wheel of width s_t corresponding to the cross-feed per workpiece revolution. By analogy with plunge grinding, the cross-feed s_t would correspond to the grinding width b. A correction factor in the thermal analysis may also be necessary if the cross feed (source width) is comparable in size or smaller than the arc length of contact (source length) [23]. If the low wheel-wear condition is not satisfied, then the energy is distributed in steps across the grinding width (Figure 3-7), and it is necessary to know the radial wheel wear on each step and to estimate the energy distribution among the steps in order to calculate the grinding zone temperature.

6.3   THERMAL DAMAGE

Excessive grinding temperatures cause thermal damage to the work-piece. In this section, a few common types of thermal damage will be considered. By establishing a direct relationship between the heat transfer analysis of the previous section and some types of thermal damage, it becomes practically feasible to predict and control thermal damage by in-process monitoring of the grinding power.

6.3.1 Workpiece burn

One of the most common types of thermal damage is workpiece burn. This phenomenon has been investigated mainly for grinding of plain carbon and alloy steels, although it is also a problem with some other metallic materials [3]. Visible workpiece burn with steels is characterized by bluish temper colors on the workpiece, which are a consequence of oxide-layer formation [4]. The temper colors are usually removed by spark-out at the end of the grinding cycle (Chapter 12), especially with cylindrical grinding, but this effect is cosmetic and the absence of temper colors on the ground surface does not necessarily mean that workpiece burn did not occur.

At the onset of burning, there is a tendency for increased adhesion of metal workpiece particles to the abrasive grains, thereby causing the forces to grow, the workpiece surface to deteriorate, and the rate of wheel wear to increase [39]. A discontinuity in the force versus wear-flat-area relationship also occurs (Figure 5-4a), which indicates an abrupt change in the grinding mechanisms possibly related to a metallurgical transformation. From microhardness distributions in the subsurface of hardened steels, visible burn has been found to be accompanied by reaustenitization of the workpiece [4, 14]. For a hardened steel ground without any burning, there is generally some softening due to tempering close to the surface [2-4, 14, 37]. Some examples of such tempering behavior are shown for grinding of a hardened tool steel at four depths of cut in Figure 6-6 [37, 40]. Starting with

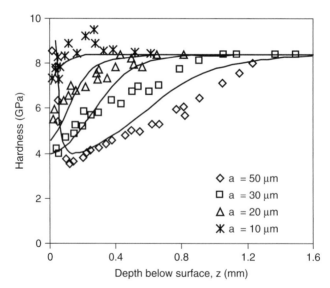

Figure 6-6 Microhardness versus depth beneath surface after grinding SK7 tool steel at four depths of cut.

an initial hardness of about 8 GPa, a greater degree of tempering can be seen with increasing wheel depth of cut due to higher temperatures to a greater depth below the surface. With the onset of burning, rehardening of the steel workpiece also occurs towards the surface as shown by the hardness curve at the biggest depth of cut in Figure 6-6. Rehardening is a consequence of reaustenitization followed by the formation of untempered martensite, which can be identified after etching as a white phase in the surface layer or in patches. Workpiece burn and austenitization by grinding heat of soft steels, even hardable types, is not necessarily accompanied by surface hardening. The metallurgical evidence and microhardness measurements suggest that the visible burn threshold is virtually co-incident with the onset of that for austenitization [14].

Burning might be expected to occur when a critical grinding zone temperature is exceeded. The critical specific energy at the burning threshold, u^*, should behave according to Eq. (6-11), so that a graph of u^* plotted against the quantity $d_e^{1/4}a^{-3/4}v_w^{-1/2}$ should yield a straight line. This is indeed what has been found, as seen in Figure 6-7 for straight and external cylindrical grinding of a wide variety of steels having similar thermal

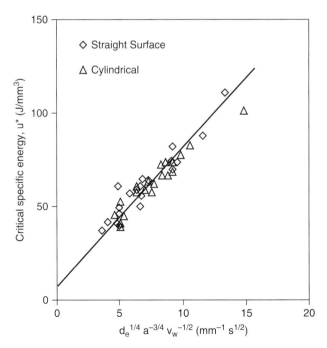

Figure 6-7 Specific energy at workpiece burn threshold for straight surface and external cylindrical grinding of carbon and low-to-medium-alloy steels. The results fall on a constant temperature line as in Figure 6-5.

properties. The particular straight line at the workpiece burn threshold in Figure 6-7 corresponds to Eq. (6-11) with a critical slope $B^* = 7.2$ J/mm$^2 \cdot$ s$^{1/2}$. The intercept value is $u_0 = 6.2$ J/mm^3, which corresponds to $0.45u_{ch}$ and, according to the thermal model, is that portion of the specific energy not entering the workpiece. This same approach has been applied to controlling thermal damage for grinding of hardened alloy steels for helicopter gears [41].

The maximum grinding zone temperature θ_m at the burning threshold in Figure 6-7 can be calculated from Eq. (6-13). Using room temperature values for the thermal properties k and α, the magnitude of the slope corresponds to a maximum temperature rise $\theta_m = 650°$C. Adding to this a typical initial temperature of 40 to 70°C, and taking into account the inaccuracies in the thermal model associated with the assumption of constant thermal properties as discussed in the previous section, the maximum temperature becomes comparable to the eutectoid temperature of 723°C for plain carbon steels. This would be the minimum temperature required for austenitization, although a slightly higher temperature of about 800°C might be necessary in order to form untempered martensite by rehardening at the ground surface [2].

These findings offer a practical means for in-process identification and control of thermal damage. During grinding, the measured specific energy can be compared with the critical specific energy u^*, in order to predict whether workpiece burn is occurring. Likewise the allowable specific energy can be set higher or lower than u^*, depending on the practical situation. For grinding of high strength critical components, it may be important to completely avoid any thermal damage, whereas non-critical components may be more efficiently ground if some workpiece burn is allowed at least during initial rough grinding. Laboratory monitoring of specific energy is usually based upon force measurements, but this is generally not feasible in a production environment. Machine power can be accurately measured in production using inexpensive solid-state power transducers. In order to determine the net grinding power, it is usually sufficient to subtract the measured idling power from the total power. The net grinding power can then be compared with the threshold burning power as given by Eq. (6-14) using the values of u_0 and B^* as above. The use of in-process power monitoring in this way also facilitates adaptive optimal control of the grinding process while satisfying surface quality requirements (Chapter 13).

Workpiece burn of bearing steels has been found to have an adverse effect on fatigue life, which might be attributed mainly to the formation of untempered martensite. This has been dramatically demonstrated in an investigation of the rolling contact fatigue lives of hardened bearing rings finished under various grinding conditions [42]. In each case, the specific energy during grinding was obtained from measurements of the grinding

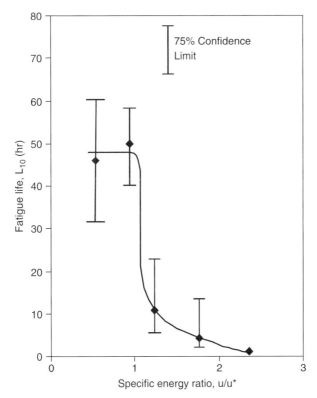

Figure 6-8 L_{10} fatigue life versus specific energy ratio u/u* for bearing rings.

force components. The results summarized in Figure 6-8 show a direct relationship between the L_{10} fatigue life and the ratio of the specific energy to the critical specific energy at the burning threshold, u/u^*. (L_{10} is the fatigue life exceeded by 90% of the specimens.) Below the predicted burning limit ($u/u^* < 1$) the L_{10} fatigue life is almost 50 hours, whereas just above the burning limit it drops to only about 10 hours. Therefore, avoiding workpiece burn when grinding bearing steels can prevent a catastrophic reduction in the L_{10} fatigue life.

6.3.2 Tempering and Rehardening

Steels are often ground in the hardened state. Transformations which may occur due to excessive grinding temperatures include tempering (softening) of the hard martensite phase, and also the formation of brittle untempered marensite (rehardening) if the temperature is high enough and persists long enough for reaustenitizeation to occur. The formation of untempered martensite is the consequence of rapid cooling of the reaustenitized material mainly by heat conduction to the workpiece bulk after the grinding zone (heat source) passes.

Tempering is a complex phenomenon which is mainly due to carbon diffusion and is dependent upon both temperature and time. In general, the hardness H obtained after tempering at temperature θ for time t can be expressed a single-valued function of the time-modified temperature parameter, $\theta(C + log\ t)$, or [43]

$$H = H[\theta(C + \log t)] \tag{6-17}$$

where C is an experimentally determined constant for each steel.

As a practical matter, it would be useful to be able to predict the reduction in hardness due to grinding heat by combining the tempering behavior (Eq. (6-17)) with the thermal analysis. The application of this approach to grinding necessitates determining an effective temperature and corresponding effective time at that temperature at varying depths below the surface. While the temperature history at any point in the subsurface is very complex, it might be expected that the maximum temperature could be taken as the effective temperature, particularly in view of the very strong influence of temperature on tempering. The maximum temperature rise at a depth z below the surface can be obtained from Eq. (6-15). It has been proposed that the effective time might be conveniently expressed by the parameter (l_c/v_w) which is how long it takes for the grinding zone (heat source) to pass any given point on the workpiece surface. Experimental results for the microhardness distribution in the subsurface of a hardened steel due to tempering by grinding heat show the expected trends, whereby a higher temperature and/or longer time parameter (l_c/v_w) causes a greater degree of softening [37]. However, these and other results cannot be fitted to a single master tempering curve (Eq. (6-17)).

A more fundamental 'differential' analysis of the tempering process has been developed which couples the reaction kinetics with the thermal analysis [40]. Based upon an analysis of existing tempering data [43], a third order reaction rate equation was found to describe the tempering phenomena of the form:

$$\frac{d\psi}{dt} = h_1(1 - \psi)^3 \tag{6-18}$$

where ψ is the probability (fraction) of transformation and h_1 is a thermally activated parameter:

$$h_1 \equiv A_1 \exp\left\{-\frac{U_1}{R\theta_a}\right\} \tag{6-19}$$

Here U_1 is the activation energy for diffusion of carbon in alpha iron ($U_1 = 80$ kJ/mol), R is the universal gas constant, θ_a is the absolute temperature, and A_1 is a constant for a particular steel. Furthermore the hardness H was assumed to have a linear dependence on ψ:

$$H = H_3 - (H_3 - H_1)\psi \tag{6-20}$$

where H_1 and H_3 are the minimum and maximum hardnesses of the fully tempered and fully quenched materials, respectively.

In addition to tempering, a similar 'differential' analysis was also developed for rehardening. Rehardening causes an increase in hardness or, in other words, a decrease the probability ψ of tempering. The rate controlling step in rehardening of steels is reaustenitization, which can be described by the first order rate equation [44, 45]

$$\frac{d\psi}{dt} = -h_2(1 - \psi) \tag{6-21}$$

where analogous to tempering

$$h_2 \equiv A_2 \exp\left\{-\frac{U_2}{R\theta_a}\right\} \tag{6-22}$$

Here the activation energy U_2 is for diffusion of carbon in gamma iron ($U_2 = 135$ kJ/mol).

In order to predict the effect of the grinding temperatures on the metallurgical state and hardness distribution in the workpiece subsurface after grinding, the reaction rate kinematics (Eqs. (6-18)-(6-22)) were coupled with a thermal analysis, similar to the one presented above, which also took into account the effect of temperature on the thermal properties and of multiple grinding passes over the same area [40]. Some examples of the results obtained are shown by the solid lines for the subsurface hardness distributions in Figure 6-6. Considering the complexity of the tempering and rehardening phenomena, the results obtained from the coupled kinetic and thermal analysis are in very good agreement with the experimental measurements.

Tempering commonly occurs near the workpiece surface during grinding of hardened steels, and it may be accompanied in severe cases by rehardening. The depth of the thermally affected layer may be reduced by the use of faster workpiece velocities which results in shallower heat penetration and shorter heating times. Some or even all of a shallow thermally affected layer produced during aggressive rough grinding at high removal rates may be removed by gentler finish grinding and spark-out at the end of the grinding cycle.

6.3.3 Residual stresses

The grinding process invariably leads to residual stresses in the vicinity of the finished surface, which can significantly affect the mechanical behavior of the material. Residual stresses are induced by non-uniform plastic deformation near the workpiece surface [2, 3, 46-65]. Mechanical interactions of abrasive grains with the workpiece result in predominantly residual compressive stresses by localized plastic flow. The effect may be likened to that of shot peening. Residual tensile stresses are caused mainly by thermally induced stresses and deformation associated with the grinding temperature and its gradient from the surface into the workpiece. At the grinding zone, the thermal expansion of hotter material closer to the surface is partially constrained by cooler subsurface material. This generates compressive thermal stresses near the surface which, if sufficiently big, cause plastic flow in compression. During subsequent cooling, after the grinding heat passes, the plastically deformed material wants to be shorter than the subsurface material, so the requirement for material continuity causes tensile stresses to develop near the surface. In order to ensure mechanical equilibrium, residual compressive stresses must also arise deeper in the material, but these are much smaller in magnitude than the residual tensile stresses. The formation of thermally induced residual stresses is further complicated by any solid phase transformations which may occur during the heating and cooling cycle, since these generally involve volume changes.

Some examples of the distribution of the residual stress component along the grinding direction are shown in Figure 6-9 for an alloy steel [2]. Residual stress measurements, which are usually based upon X-ray methods, typically reveal a biaxial stress state in the surface layer, with the stress along the grinding direction approximately equal to the stress across the grinding direction [46, 52]. In much production grinding, the residual stresses are predominantly tensile, which would indicate that they are mainly thermal in origin. Residual compressive stresses are considered to have a beneficial effect on mechanical strength properties, whereas residual tensile stresses have an adverse effect.

The influence of residual stresses is relatively more pronounced with higher strength brittle materials for which strength considerations are often critical. More severe grinding conditions on high-strength steels and aircraft alloys generally cause larger residual tensile stresses, thereby leading to reduced fatigue strength and cracking. The situation may be further aggravated in steels by hydrogen embrittlement, owing to significantly higher levels of hydrogen being released as a result of grinding fluid breakdown [60]. Abusively ground hardened steel components exposed to hot acid develop surface cracks, which can be attributed to the presence of residual tensile stresses acting on brittle untempered martensite formed by workpiece burn. Cracks induced by acid etching and abusive grinding are usually oriented normal to the grinding direction [55, 56], which would suggest that the residual

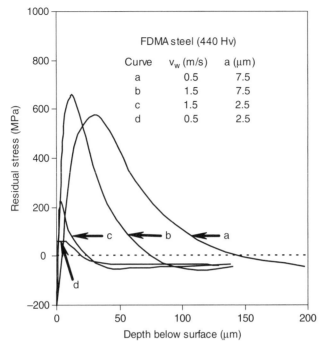

Figure 6-9 Residual stress distributions induced by grinding of an alloy steel. Residual tensile stresses observed are considered to be thermally induced.

tensile stress component along the grinding direction is the predominant one. The tendency for microcracking in high strength nickel base alloys due to residual tensile stresses may be further promoted by the onset of non-equilibrium constitutional melting at elevated grinding temperatures [35].

It is generally desirable to control the grinding conditions so as to induce residual compressive stresses or, at least, to limit the magnitude of the peak residual tensile stress. As a practical matter, demands for more efficient production and faster removal rates result in higher residual tensile stresses, such as are seen in Figure 6-9. In order to obtain residual compressive stresses, it is usually necessary to maintain extremely low removal rates. However, the introduction of CBN abrasive wheels in place of aluminum oxide has been shown to induce compressive instead of tensile residual tensile stresses when grinding hardened bearing races [61, 62]. This would suggest reduced temperatures for grinding with CBN, due to lower specific energies. Another factor is the very high thermal conductivity of CBN, which promotes conduction to the wheel, thereby lowering the energy partition to the workpiece (Chapter 8).

In principle, it should be possible to analytically predict the thermally induced residual stress distribution from the grinding temperature solution coupled with thermal stress and strain calculations. Such analyses have been performed using the finite-element method [35, 63, 64], taking into account the initial elastic-plastic stresses and deformation during thermal loading followed by elastic unloading during cooling. An extensive amount of computational time is usually required, especially when also taking into account the influence of temperature on the mechanical and thermal properties of the particular workpiece material, and the non-linear nature of the phenomenon does not allow for generalization of the results. Significant simplification in the calculation of the residual stresses may be achieved in some cases by the use of an approximate one-dimensional analysis [65]. Perhaps of more practical interest is the experimental observation of a direct relationship between the peak residual tensile stress and the maximum grinding zone temperature, as seen in Figure 6-10 for three different steels [2]. The discontinuity near temperature A_{c1} ($\theta_m = 721°C$) is due to austenite formation, and this corresponds to the visible burning threshold (section 6.3.1). These results suggest that the magnitude of the peak residual tensile stress can be controlled by keeping the grinding temperature θ_m below a certain value. Therefore, it should be possible to specify an allowable specific energy or power (Eqs. (6-11) or (6-14)) with the temperature θ_m in the parameter B corresponding to a limiting residual stress condition.

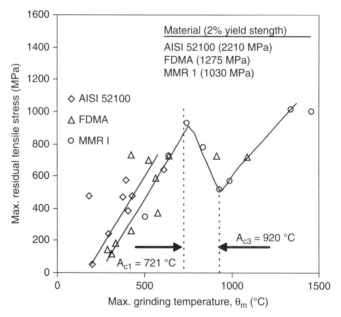

Figure 6-10 Peak residual tensile stress versus maximum grinding zone temperature for three steels.

6.4   FACE GRINDING

Consider the face-grinding situation illustrated in Figure 3-12. In this particular case, the wheel surface is wider than the workpiece and the grinding zone extends over the entire cross-section of the workpiece. In other cases, especially with cup wheels, the wheel (rim) width is narrower than the workpiece, and the workpiece cross-section is ground by cross-feeding the wheel back and forth with intermittent infeed.

The heat transfer situation for continuous face grinding over the entire workpiece cross-section can be simply approximated by one-dimensional heat conduction into the workpiece (Figure 6-11). For a workpiece dimension T and sufficiently big and slow infeed velocity v_f, the workpiece can be approximated as extending to infinity in the z-direction from the surface being ground. Neglecting any cooling effects, the maximum temperature rise reached uniformly over the grinding area for a workpiece of arbitrary cross-section is given by [66]

$$\theta_m = \frac{2q_w t^{1/2}}{\pi^{1/2} \alpha^{1/2} \rho c} \qquad (6\text{-}23)$$

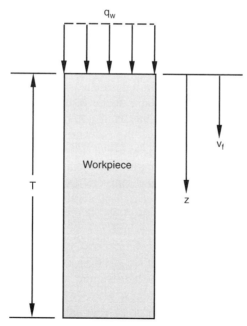

Figure 6-11 Thermal model for face grinding with continuous contact over the workpiece cross-section.

where q_w is the heat flux to the workpiece, α is the thermal diffusivity, ρc is the volumetric specific heat, and t is the time from the start of grinding. The requirement of an 'infinite' workpiece in the z-direction is that $T > (\alpha t)^{1/2}$. Analogous to the heat transfer analysis in section 6.2, the heat flux q_w over the cross-sectional area A_c of the workpiece can be expressed in terms of the grinding power P as

$$q_w = \frac{\varepsilon P}{A_c} \tag{6-24}$$

where ε is the energy fraction conducted into the workpiece, as given by Eq. (6-9). For this face-grinding situation of stationary continuous heating with a small infeed velocity, such that the infeed motion of the ground surface into the workpiece can be neglected, the temperature does not reach a steady state but continually rises. (The situation in which the infeed velocity has also been taken into account is considered in the following section.) One way to prevent this undesirable situation is to apply grinding fluid to cool around the workpiece periphery. However, in practice, many face-grinding operations, especially for tool and cutter grinding applications, are carried out dry, in which case the grinding operation should be periodically interrupted in order to allow for some cooling. This is particularly important when using resin-bonded superabrasive-wheels, which are sensitive to temperature.

    With a narrow-rimmed-cup wheel, the heat transfer situation is more difficult to analyze. In this case, the grinding zone moves back and forth across the workpiece surface, similar to the situation in straight surface grinding. However, the actual shape of the grinding zone is not generally known and would depend upon the wear pattern of the wheel across the rim. If the traverse distance across the workpiece is small, as is often the case, grinding is repeated over the same area after a very short time duration. In this case the workpiece would 'feel' nearly continuous heating from the surface being ground, and the temperature would progressively rise with time. Therefore, special precautions should be taken to provide for periodic cooling, as in the previous case of continuous grinding over the entire workpiece cross-section.

    More complex face-grinding heat transfer analyses are required when the workpiece is not of uniform cross-section, is not of 'infinite' length, or is cooled by a grinding fluid. Each particular situation would require the development of a particular thermal analysis which takes into account the heated grinding areas and their motions, the particular part geometry, and possible grinding fluid interactions with the workpiece.

6.5   ABRASIVE CUT-OFF

In abrasive cut-off operations, as illustrated in Figure 6-12, a thin rotating wheel of diameter d_s and thickness b_s is fed down into the workpiece at a velocity v_f. One approach to modeling the thermal situation for this operation has been to assume that all the grinding energy is generated at the grinding zone between the wheel periphery and the workpiece, part being conducted as heat down into the workpiece and the balance conducted up into the wheel [67, 68]. There is no energy generated or heat transfer at the sidewalls between the wheel with the workpiece. Since the wheel diameter is usually substantially bigger than the workpiece width W, the grinding zone arc of contact can be approximated as a chord and the heat source as planar. The workpiece thickness T to be cut through is assumed to be infinite, which is justified for practical infeed velocities except near the end of the cut. Heat conduction within the wheel is considered to occur only radially with uniform cooling around the wheel periphery. Heat entering the workpiece is assumed to be conducted only straight down in the z-direction within the material to be removed, with no side flow to the remainder of the

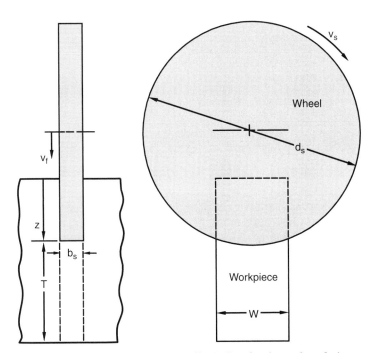

Figure 6-12 Illustration of cut-off grinding for thermal analysis.

workpiece. The numerical solution to this problem indicates a transient behavior, with the grinding zone temperature rising from zero at the initial contact between the wheel and the workpiece towards a steady-state condition. The transient is primarily controlled by conduction in the wheel, which is much slower than the transient in the workpiece.

For this heat transfer model, the partition of the grinding energy between the fractions conducted to the wheel and workpiece is imposed by a condition that the temperature rise at the wheel periphery is equal to that of the workpiece over the entire grinding zone. This situation would imply that the wheel and workpiece are in complete mutual sliding contact, which is certainly not the case. Heat transfer to the wheel is mainly by its contact with the grinding chips at discrete abrasive cutting points.

More realistically, the fraction of the grinding heat entering the workpiece is controlled by the grinding mechanisms, as discussed in the foregoing thermal analysis (section 6.2). Therefore, the thermal analysis can be performed, as before, by considering heating of the workpiece. For this purpose, the grinding zone can be approximated as a uniform planar heat source moving directly down into the workpiece at the infeed velocity v_f with the material crossing the grinding zone being removed, as in the inclined source creep-feed analysis. The solution to this heat transfer problem can be derived from the aforementioned cut-off analysis [67,68] for the limiting case where the conductivity of the wheel and its cooling are both set equal to zero. For a heat flux q_w over the grinding zone area, the temperature rise is

$$\theta_m = \frac{q_w}{\rho c v_f}\left[\left(1 + \frac{\tau}{2}\right)erf\left(\frac{\sqrt{\tau}}{\pi}\right) - \frac{\tau}{2} + \sqrt{\frac{\tau}{\pi}}e^{-\tau/4}\right] \qquad (6\text{-}25)$$

where 'erf' refers to the error function and τ is a dimensionless time parameter:

$$\tau \equiv \frac{v_f^2 t}{\alpha} \qquad (6\text{-}26)$$

where the time t is measured from initial wheel engagement when $z = 0$. For a fraction ε of the total grinding energy entering the workpiece, the heat flux q_w can be given either in terms of the grinding power P as

$$q_w = \frac{\varepsilon P}{W b_s} \qquad (6\text{-}27)$$

where the denominator is the grinding zone area, or in terms of the specific energy u as

$$q_w = \varepsilon u v_f \qquad (6\text{-}28)$$

Introducing the parameter

$$\theta_{ad} = \frac{\varepsilon u}{\rho c} \tag{6-29}$$

and combining with Eq. (6-28), the temperature rise of Eq. (6-25) can also be written as

$$\theta_m = \theta_{ad}\left[\left(1 + \frac{\tau}{2}\right)erf\left(\frac{\sqrt{\tau}}{\pi}\right) - \frac{\tau}{2} + \sqrt{\frac{\tau}{\pi}}e^{-\tau/4}\right] \tag{6-30}$$

where θ_{ad} is the adiabatic temperature rise due to uniform heating by the grinding energy entering the workpiece. The complicated function in the brackets of Eqs. (6-25) and (6-30) progressively increases with τ at a decreasing rate from zero to unity, so θ_{ad} is the steady-state temperature rise. It can be shown from these equations that the transient temperature reaches 90% of the steady-state value when $\tau \approx 1.5$ or

$$\frac{v_f^2 t}{\alpha} = 1.5 \tag{6-31}$$

Introducing the dimensionless parameter

$$L_1 \equiv \frac{v_f b_s}{\alpha} \tag{6-32}$$

analogous to the thermal number, the significance of which will be considered in the continuing discussion, and noting that the depth $z = v_f t$, 90% of the steady-state temperature rise is reached when

$$z = 1.5\frac{b_s}{L_1} \tag{6-33}$$

For $L_1 = 1$, which is about the lower limit for heavy-duty cut-off, the near-steady state condition is reached after the wheel has progressed into the workpiece only to about 1.5 times its thickness b_s, so the transient is very quick. For an extremely slow infeed velocity, the transient becomes much slower, and in the limit as $v \to 0$ the temperature in Eq. (6-25) is identical to the one-dimensional face-grinding situation of Eq. (6-23) and continually rises.

In formulating this heat transfer analysis, it has been assumed that heat flow is one-dimensional down into the material to be removed. Actually, there is always some side flow such that part of the heat is conducted out of the path of the advancing wheel and is not removed with the chips. The degree of side flow heat transfer is less with larger values of the parameter L_1. On the basis of a finite element heat transfer analysis, which takes into account both downwards and sideways conductions, it appears that almost all the grinding heat is removed with the chips provided that $L_1 > 1$, which would include most heavy-duty abrasive cut-off operations [2]. Under such conditions, the temperature would quickly rise towards its steady-state value θ_{ad}.

The particular heat transfer situation with cut-off grinding can account for the relatively small specific energies which are obtained, as mentioned in the previous chapter. With heavy-duty cut-off at fast removal rates, the plowing and sliding energy components are negligible, in which case $\varepsilon \approx 0.5$ (Eq. (6-9)). In most grinding operations, the main workpiece motion is tangential to the wheel surface (e.g. Figure 6-1), so that the heat source (grinding zone) is continually moving into cool material and the chip-formation energy u_{ch} can be related to the melting energy per unit volume starting from near room temperature (Chapter 5). For cut-off operations, however, the grinding zone moves directly down into hot material. After the initial transient, the material being ground away can be considered to have been 'preheated' to θ_{ad}. Therefore, the melting energy limitation should be based upon heating from θ_{ad}, and not from room temperature. For $\varepsilon \approx 0.5$, this would then imply a specific grinding energy for heavy-duty cut-off which is only about half of the chip-formation energy for other grinding processes. For abrasive cut-off of steels, we should expect specific energies down to about $u \approx 6.9$ J/mm^3, which is very close to what is actually obtained. Using a room temperature value of ρc in Eq. (6-29), this would imply an adiabatic temperature rise of $\theta_{ad} \approx 1850°C$, which exceeds the melting point. A more realistic temperature rise of $\theta_{ad} \approx 1250°C$ is obtained when taking into account the influence of temperature on the volumetric specific heat ρc, and this is comparable to grinding zone temperatures measured with thermocouples [69].

With no sideways heat conduction, the temperature rise across the grinding zone for cut-off of steel would be about 1250°C, or even higher. Such high temperatures should cause severe burn on the sidewalls of the workpiece being cut. However, it has been shown even with very large values of L_1, that sideways conduction at the extreme edge of the grinding zone can reduce the local steady-state temperature to only 35–45% of θ_{ad} [2]. This local edge temperature corresponds to the maximum sidewall temperature on the cut surface of the workpiece. With such low temperatures, thermal damage should not be a problem. Thermal damage is more likely to occur at the end of the cut, as the wheel is about to break through, since there is virtually no workpiece material remaining to conduct heat downward.

REFERENCES

1. Malkin, S. and Guo, C., 'Thermal Analysis of Grinding', *Annals of the CIRP*, 57/2, 2007, p. 760.
2. Snoeys, R., Maris, M. and Peters, J., 'Thermally Induced Damage in Grinding', *Annals of the CIRP*, 27/2, 1978, p. 571.
3. Torrance, A. A., 'Metallurgical Effects Associated with Grinding', *Proceedings of the Twelfth International Machine Tool Design and Research Conference*, 1978, p. 637.
4. Tarasov, L. P., 'Some Metallurgical Aspects of Grinding', *Machining Theory and Practice,* ASM, 1950, p. 409.
5. Littman, W. E. and Wulff, J., 'The Influence of the Grinding Process on the Structure of Hardened Steels', *Trans. ASM*, 47, 1955, p. 692.
6. Wetton, A. G., 'A Review of Theories of Metal Removal in Grinding', *J. Mech. Eng. Sci.*, 11, 1969, p. 412.
7. Ueda, T., Hosokawa, A. and Yamamoto, A., 'Studies of Temperature of Abrasive Grains in Grinding - Application of Infrared Radiation Pyrometer', *Trans. ASME, J. of Eng. for Ind.*, 107, 1984, p. 127.
8. Maris, M. and Snoeys, R., 'Heat Affected Zone in Grinding', *Proceedings of the Fourteenth International Machine Tool Design and Research Conference*, 1973, p. 659.
9. Guo, C. Wu, Y, Varghese, V., and Malkin, S., 'Temperature and Energy Partition for Grinding with Vitrified CBN Wheels', *Annals of the CIRP*, 48/1, 1999, p. 247.
10. Kim, N., Guo, C., and Malkin, S., 'Heat Flux and Energy Partition in Creep-Feed Grinding,' *Annals of the CIRP*, 46/1, 1997, p. 227.
11. Zhu, B., Guo, C., Sunderland, J. E., and Malkin, S., 'Energy Partition to the Workpiece for Grinding of Ceramics,' *Annals of the CIRP*, 44/1, 1995, p. 267.
12. Kohli, S., Guo, C. and Malkin, S., 'Energy Partition to the Workpiece for Grinding with Aluminum Oxide and CBN Abrasive Wheels,' *Trans. ASME, J. of Eng. for Ind.*, 117, 1993, p. 160.
13. Xu, X. and Malkin, S., 'Comparison of Methods to Measure Grinding Temperature', *Trans. ASME, J. of Manufact. Sci. and Eng.,* 123, 2001, p. 191.
14. Malkin, S., 'Thermal Aspects of Grinding, Part 2 - Surface Temperatures and Workpiece Burn', *Trans. ASME, J. of Eng. for Ind.*, 96, 1974, p. 484.
15. Des Ruisseaux, N. R. and Zerkle, R. D., 'Thermal Analysis of the Grinding Process', *Trans. ASME, J. of Eng. for Ind.*, 92, 1970, p. 428.
16. Loladze, T. N., Bokuchava, G. V. and Siradze, A. M., 'The Problems of Grinding Process Tribology and Abrasive Tool Improvement', *Proceedings of International Conference on Production Engineering, Part 1,* Tokyo, 1974, p. 713.
17. Masuda, M. and Shiozaki, S., 'Out-of-straightness of Workpiece, Considering the Effect of Thermal Work Expansion on Depth of Cut in Grinding', *Annals of the CIRP*, 23/1, 1974, p. 91.
18. Yokoyama, K. and Ichimiya, R., 'Thermal Deformation of Workpiece in Surface Grinding', *Bull. Japan Soc. Prec. Engg.*, 11, 1977, p. 195.
19. Thé, J. H. L. and Scrutton, R. F., 'Thermal Expansions and Grinding Forces Accompanying Plunge-cut Surface Grinding', *Int. J. Mach. Tool Des. Res., 13*, 1973, p. 287.

20. Kops, L. and Hucke, M., 'Simulated Thermal Deformation in Surface Grinding', *Proceedings, International Conference on Production Engineering, Pt. 1,* Tokyo, 1974, p. 683.

21. Malkin, S., 'Burning Limits for Surface and Cylindrical Grinding of Steels', *Annals of the CIRP,* 27/1, 1978, p. 233.

22. Des Ruisseaux, N. R. and Zerkle, R. D., 'Temperatures in Semi-infinite and Cylindrical Bodies Subjected to Moving Heat Sources and Surface Cooling', *Trans. ASME, J. of Eng. for Ind.,* 70, 1970, p. 456.

23. Jaeger, J. C., 'Moving Sources of Heat and the Temperature at Sliding Contacts', *Proceedings of the Royal Society of New South Wales,* 76, 1942, p. 203.

24. Guo, C. and Malkin, S., 'Analytical and Experimental Investigation of Burnout in Creep-Feed Grinding,' *Annals of the CIRP,* 43/1, 1994, p. 283

25. Guo, C., and Malkin, S., 'Energy Partition and Cooling During Grinding,' *Journal of Manufacturing Processes,* 2, 2000, p. 151.

26. Guo, C. and Malkin, S., 'Cooling Effectiveness in Grinding,' *Trans. NAMRI/SME,* 23, 1996, p. 111.

27. Guo, C. and Malkin, S., 'Analysis of Energy Partition in Grinding,' *Trans. ASME, J. of Eng. for Ind.,* 117, 1995, p. 55.

28. Guo, C. and Malkin, S., 'Inverse Heat transfer Analysis of Heat Flux to the Workpiece: I. Model,' *Trans. ASME, J. of Eng. for Ind.,* 118, 1996, p. 137.

29. Guo, C. and Malkin, S., 'Inverse Heat Transfer Analysis of Heat Flux to the Workpiece: II. Application,' *Trans. ASME, J. of Eng. for Ind.,* 118, 1996, p. 143.

30. Sato, K., 'Grinding Temperature', *Bull. Japan Soc. Grind. Engns.,* 1, 1961, p. 31.

31. Malkin, S. and Anderson, R. B., 'Thermal Aspects of Grinding, Part 1 - Energy Partition', *Trans. ASME, J. of Eng. for Ind.,* 96, 1974, p. 1177.

32. Trigger, K. J. and Chao, B. T., 'An Analytical Evaluation of Metal Cutting Temperatures', *Trans. ASME,* 73, 1951, p. 57.

33. Outwater, J. O. and Shaw, M. C., 'Surface Temperatures in Grinding', *Trans. ASME,* 74, 1952, p. 73.

34. Isenberg, J. and Malkin, S., 'Effects of Variable Thermal Properties on Moving Band Source Temperatures', *Trans. ASME, J. of Eng. for Ind.,* 96, 1974, p. 1184.

35. Kovach, J. A. and Malkin, S., 'Thermally Induced Grinding Damage in Superalloy Materials', *Annals of the CIRP,* 37/1, 1988, p. 309.

36. Malkin, S., 'In-process Control of Thermal Damage during Grinding', SME Paper No. MR2-532, SME, 1984.

37. Takazawa, K., 'Effects of Grinding Variables on Surface Structure of Hardened Steels', *Bull. Japan Soc. Prec. Engg.,* 2, 1966, p. 14.

38. Yamamoto, Y., Horike, M. and Hoshina, N., 'A Study of the Temperature Variation of Workpieces during Cylindrical Plunge Grinding', *Annals of the CIRP,* 25/1, 1977, p. 151.

39. Malkin, S. and Cook, N. H., 'The Wear of Grinding Wheels, Part I, Attritious Wear', *Trans. ASME, J. of Eng. for Ind.,* 93, 1971, p. 1120.

40. Fedoseev, O, and Malkin, S., 'Analysis of Tempering and Rehardening for Grinding of Hardened Steels'. *Trans. ASME, J. of Eng. for Ind.,* 113, 1991, p. 388.

41. Mayer, J. E., Purushothaman, G., and Gopalakrishnan, S., 'Model of Grinding Thermal Damage for Precision Gear Materials', *Annals of the CIRP,* 48/1, 1999, p. 251.

42. Torrance, A. A., Stokes, R. J. and Howes, T. D., 'The Effect of Grinding Conditions on the Rolling Contact Fatigue Life of Bearing Steel', *Mechanical Engineering,* October 1983, p. 63.

43. Hollomon, J. H. and Jaffe, L. D., 'Time-temperature Relation in Tempering Steel', AIME Technical Publication 1831, February 1945.

44. Ashby, M. F., and Eaterling, K. E., 'The Transformation Hardening of Steel Surfaces by Laser Beams-I. Hypoeutectoid Steels,' *Acta Metal.*, 32, 1984, p. 1935.

45. Li, W. D., Eaterling, K. E., and Ashby, M. F., 'The Transformation Hardening of Steel Surfaces by Laser beams-II. Hypereutectoid Steels,' *Acta Metal.*, 34, 1986, p. 1533.

46. *Grinding Stresses - Cause, Effect, and Control, Collected Papers*, Grinding Wheel Institute, Cleveland, Ohio, 1960.

47. Littman, W. E., 'Control of Residual Stresses in Metal Surfaces', *Proceedings of the International Conference on Manufacturing Technology*, ASTME, 1967, p. 1303.

48. Littman, W. E., 'The Influence of Grinding on Workpiece Quality', ASTME Paper MR67-593, 1967.

49. Wakabayashi, M. and Nakayama, M., 'Experimental Research on Elements Composing Residual Stresses in Surface Grinding', *Bull. Japan Soc. Prec. Engg.*, 13, 1979, p. 75.

50. Hahn, R. S., 'On the Loss of Surface Integrity and Surface Form due to Thermoplastic Stress in Grinding Operations', *Annals of the CIRP*, 25/1, 1976, p. 203.

51. Lenning, R. L., 'Controlling Residual Stresses in Cylindrical Grinding', *Abrasive Engineering*, December 1968, p. 24.

52. Winter, P. M. and McDonald, W. J., 'Biaxial Residual Surface Stresses from Grinding and Finish Machining 304 Stainless Steel Determined by a New Dissection Technique', *Trans. ASME, J. of Basic Eng*, 91, 1969, p. 15.

53. El-Helieby, S. O. A. and Rowe, G. W., 'A Quantitative Comparison between Residual Stresses and Fatigue Properties of Surface-Ground Bearing Steel (En 31)', *Wear*, 58, 1980, p. 155.

54. Field, M., Koster, W. P. and Kahles, J. F., 'Effect of Machining Practice on the Surface Integrity of Modern Alloys', Proceedings, *International Conference on Manufacturing Technology*, ASTME, 1967, p. 1319.

55. Hahn, R. S., 'Survey of Technical Factors in Grinding High Strength Heat Resistant Alloys', *Annals of the CIRP*, 17, 1969, p. 107.

56. Hahn, R. S. and Lindsay, R. P., 'The Production of Fine Surfaces while Maintaining Good Surface Integrity at High Production Rates in Grinding'. *Proceedings, International Conference on Surface Technology*, SME, 1973, p. 402.

57. Field, M. and Koster, W. P., 'Surface Integrity in Grinding', *Proceedings, International Grinding Conference*, Pittsburgh, 1972, p. 666.

58. 'Surface Integrity', *Machinability Data Handbook*, Vol. 2, Section 18.3, 3rd edition, Machinability Data Center, Cincinnati, 1980, p. 18.

59. Tarasov, L. P., Hyler, W. S. and Letner, U. R., 'Effects of Grinding Conditions and Resultant Residual Stresses on the Fatigue Strength of Hardened Steel', *Proc. ASTM*, 57, 1957, p. 601.

60. Das, K. B. and Parrish, P. A., Influence of Grinding Fluids on the Introduction of Hydrogen in High Strength Steel', *Corrosion, NACE*, 31, 1975, p. 72.

61. Tönshoff, H. K. and Grabner, T., 'Cylindrical and Profile Grinding with Boron Nitride Wheels', *Proceedings of the 5th International Conference on Production Engineering,* Tokyo, 1984, p. 326.

62. Tönshoff, H. K. and Hetz, F., 'Influence of Abrasive on Fatigue in Precision Grinding', *Milton C. Shaw Grinding Symposium*, PED-16, ASME, 1985, p. 191.

63. Mishra, A., Rao, U. R. K. and Natarajan, R., 'An Analytical Approach to the Determination of Residual Stresses in Surface Grinding', *Proceedings, International Conference on Production Engineering*, Inst. Mech. Engrs. (India) and CIRP, New Delhi, 1977, p. VI-40.

64. Skalli, N., Turbatt, A. and Flavenot, J. F., 'Prevision of Thermal Stresses in Plunge Grinding of Steels', *Annals of the CIRP*, 31/1, 1982, p. 451.

65. Xiao, G., Stevenson, R., Hanna, I. M., and Hucker, S. A., 'Modeling of Residual Stress in Grinding of Nodular Cast Iron, *Trans. ASME, J. of Manufact. Sci. and Eng.*, 124, 2002, p. 833.

66. Jakob, M., *Heat Transfer*, Volume 1, Wiley, New York, 1949, p. 258.

67. Eshghy, S., 'Thermal Aspects of the Abrasive Cut-off Operation. Part 1 - Theoretical Analysis', *Trans. ASME, J. of Eng. for Ind.*, 89, 1967, p. 356.

68. Eshghy, S., 'Thermal Aspects of the Abrasive Cutoff Operation. Part 2 - Partition Functions and Optimal Cutoff', *Trans. ASME, J. of Eng. for Ind.*, 90, 1968, p. 360.

69. Korolev, V. V., 'Temperature Distribution Calculations during Abrasive Machining', *Machines and Tooling*, 42, 1971, p. 47.

Chapter 7

Thermal Aspects: Creep-Feed Grinding[1]

7.1 INTRODUCTION

Creep-feed grinding utilizes very slow (creep) workpiece velocities and extremely large depths of cut. Straight surface grinding under creep feed conditions allows for much faster removal rates than can be reached with regular shallow cut grinding without causing thermal damage to the workpiece, even though creep-feed grinding generally requires much bigger specific energies. For creep-feed grinding of steels, the specific energy may substantially exceed the threshold value for workpiece burn as obtained for regular grinding in Chapter 6, but with no evidence of any thermal damage.

It has been suggested that the improved thermal situation in creep-feed grinding can be attributed to the extremely large depths of cut, such that much of the heat input to the workpiece is removed together with the grinding chips before it can be conducted out of the path of the advancing wheel [2]. This is somewhat analogous to the thermal situation with abrasive cut-off as described in the previous chapter. In order to evaluate this effect for creep feed grinding, the heat transfer analysis (section 6.2) was modified to take into account the large depth of cut by using an inclined heat source as illustrated in Figure 7-1 [3]. For this thermal analysis, the circular arc of the grinding zone was approximated by a chord AB oriented at an angle ϕ to the direction of workpiece motion. The heat flux was considered to be uniformly distributed over the source length AB, with the material from the workpiece crossing AB being removed during grinding. Owing to the inclination ϕ and workpiece motion, part of the heat entering the workpiece at the grinding zone is not conducted down into the workpiece

---

[1] *Parts of this chapter were adapted from Reference [1].*

189

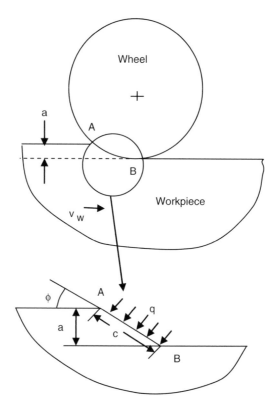

Figure 7-1 Inclined heat-source model for creep-feed grinding.

below B, but is convected out across the advancing boundary AB together with the material (chips) being removed, thereby reducing the maximum temperature where the newly ground surface is generated at B. However, for inclination angles typical of creep-feed grinding, which range from 5 to 10 degrees, the calculations using a finite element method showed only a very moderate reduction in the maximum temperature at the trailing edge B, as compared with zero inclination, and cannot account for the ability to creep-feed grind steels without workpiece burn.

The main factor which enhances the thermal situation in creep-feed grinding is cooling by the fluid at the grinding zone. Creep-feed grinding requires a copious flow of fluid delivered at high pressure to the grinding zone in order to remove heat by forced convection. Cooling by the grinding fluid is effective only up to a critical burnout temperature associated with film boiling. At this point, considerable vapor is generated, thereby making it difficult for the fluid to wet and cool the heated surface at the grinding zone [4-13]. The burnout transition at a critical threshold temperature

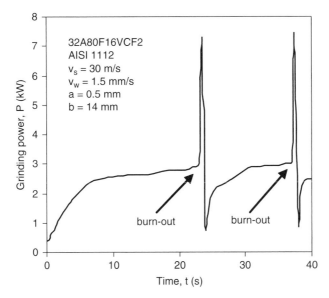

Figure 7-2 Power surges due to burnout.

corresponds to a critical burnout heat flux at the grinding zone. The critical temperature at the burnout threshold is about 130 °C with water-based soluble oils and 300 °C with straight oils. However, the critical burnout heat flux with water-based fluids is much higher than with straight oils [9], owing to the higher thermal conductivity of water-based fluids. It should be noted that these critical temperatures are usually exceeded for regular shallow cut grinding, as considered in the previous chapter, so cooling by the fluid at the grinding zone was neglected.

When burnout occurs, the grinding zone temperature may jump to about 1000°C or more. This thermal instability may be accompanied by a surge phenomenon whereby the grinding power periodically builds up and drops off, owing to cyclical metal build-up on the wheel followed by self-sharpening [5, 6]. An example of this cyclical burnout behavior is shown in Figure 7-2 [12]. In one investigation the workpiece burn threshold for creep-feed grinding of a bearing steel with a water based fluid over a wide range of conditions was found to occur at nearly the same heat flux (power per unit area of grinding zone) of $q^* = 7$–8 W/mm^2 [13]. Much bigger burnout heat fluxes up to 50 W/mm^2 have also been reported [6-11]. These values are typical of those reported for burnout of water in forced convection [4]. More extensive investigation indicates that the burnout heat flux depends on the grinding conditions and the location along the grinding path as shown in Figure 7-3 [12, 13]. For creep-feed grinding of workpieces

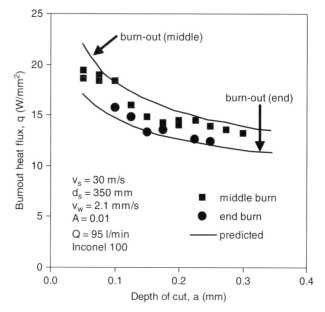

Figure 7-3 Influence of depth of cut on burnout heat flux for a nickel base alloy.

which are long enough for the quasi-steady state temperature to be reached, burnout is usually observed to occur either in the middle portion of the pass where the quasi-steady state prevails, or at the end during cut out as the wheel disengages from the workpiece [14].

Fluid burnout in creep-feed grinding leads to a significant grinding zone temperature rise and the likelihood of catastrophic thermal damage, so it should be avoided. Therefore, a main focus in this chapter is on the prediction of fluid burnout for control of thermal damage in creep-feed grinding.

7.2 WORKPIECE TEMPERATURE AND BURNOUT

The thermal analysis for calculating the temperatures and predicting the burnout limit for creep-feed grinding is similar to that for regular shallow-cut grinding, but with some notable differences. For creep-feed grinding, cooling by the fluid at the grinding zone needs to be taken into account, which can be done through its influence on the energy partition. Furthermore the slow workpiece velocities may lead to situations whereby heat conduction along the grinding direction becomes significant $(L \equiv v_w l_c / 4\alpha < 5)$. In this case, Eq. (6-6) would tend to over predict the maximum temperature. On the other hand, the workpiece may experience an additional temperature rise during the thermal transient at the end of a grinding pass as the wheel disengages from the workpiece (cut out) due to the sudden absence of material to conduct away the grinding heat. For this reason, localized thermal damage often occurs during cut out [14].

By analogy with Eq. (6-6), the maximum temperature rise for a triangular heat source can be written as [12, 14]:

$$\theta_m = \frac{\beta \varepsilon q}{(k\rho c)^{1/2}} \left(\frac{l_c}{v_w}\right)^{1/2} = \frac{\beta \varepsilon q}{(k\rho c)^{1/2}} \frac{(ad_s)^{1/4}}{v_w^{1/2}} \tag{7-1}$$

where the parameter β considers the effect of the thermal number L and the transient behavior during cut out as the wheel disengages. For the quasi-steady state [12, 14]:

$$\beta = \begin{cases} 1.02L^{0.03} & 0.1 \leq L \leq 5 \\ 1.06 & L > 5 \end{cases} \tag{7-2}$$

and for the transient during cut out at the end of the grinding pass:

$$\beta = \begin{cases} 1.02L^{0.03} + 0.23L^{-0.37} & 0.1 \leq L \leq 5 \\ 1.06 + 0.23L^{-0.37} & L > 5 \end{cases} \tag{7-3}$$

The second term on the right side of Eq. (7-3) in each case is the additional maximum temperature rise above the quasi-steady state due to cut out [14]. Therefore burnout during cut out should occur at a lower heat flux because of higher β values than during the quasi-steady state.

Because the fluid burnout is a critical temperature phenomenon, the heat flux at burnout can be readily obtained from Eq. (7-1) as

$$q^* = \frac{(\theta^* - \theta_0)(k\rho c)_w^{1/2}}{\varepsilon \beta} \left(\frac{v_w}{l_c}\right)^{1/2} \tag{7-4}$$

where θ^* is the critical burnout temperature (about 130 °C for water-based fluid and 300 °C for oil) and θ_0 is the initial temperature of the fluid. The corresponding critical grinding power is:

$$P_c^* = q^*[l_c b] = q^*[(ad_s)^{1/2}b] \tag{7-5}$$

In order to use Eq. (7-4) to predict the critical burnout heat flux q^* and Eq. (7-5) to predict the corresponding burnout power P^*, it is necessary to specify the energy partition to the workpiece ε. The energy partition model developed in Chapter 6 (Eq. 6-9) for regular shallow-cut grinding is based on the grinding energy model. The fluid influence was not considered in the model because the grinding zone temperature is generally high enough to cause fluid burnout under regular grinding conditions. However, cooling by the fluid at the grinding zone is usually essential under creep-feed conditions and greatly reduces the energy partition.

7.3 ENERGY PARTITION: SIMPLE MODEL FOR CREEP-FEED GRINDING

Grinding energy is dissipated as heat at the grinding zone where the wheel interacts with the workpiece. For creep-feed grinding, it is convenient to envision the grinding zone as the interface between the workpiece and the wheel. The wheel surface is impregnated with fluid in its pores which can cool the workpiece if the temperature is below the burnout limit. A fraction of the grinding energy equal to the energy partition is transported to the work-piece, which moves relative to the grinding zone at the workpiece velocity. Most of the remaining grinding energy is carried away by the wheel/fluid composite, which moves relative to the grinding zone at the wheel velocity. (It will be seen in Chapter 9 that the applied fluid quickly accelerates up to the peripheral wheel velocity before it enters the grinding zone). Some energy would also be carried away with the grinding chips, although this is usually a negligible fraction of the total heat generated in creep-feed grinding [15].

One relatively simple way to estimate the energy partition is to equate the maximum temperature rise on the workpiece at the grinding zone to the maximum temperature rise on the wheel/fluid composite surface [16]. The maximum workpiece temperature rise is given by Eq. (7-1). The maximum temperature rise at the wheel-fluid composite surface can be written in an analogous manner by likening the grinding heat input at the wheel-fluid com-posite to a heat source which moves along its surface at the wheel velocity v_s. In this case, the thermal number L_c for the wheel/fluid composite becomes

$$L_c = \frac{v_s l_c}{2\alpha_c} \tag{7-7}$$

where the workpiece velocity is now replaced by the wheel velocity and α_c is the thermal diffusivity of the wheel/fluid composite. Since $v_s \gg v_w$ it is generally found that $L_c > 5$, so the maximum temperature rise becomes:

[AQ3]

$$(\theta_m)_s = \frac{1.06(1 - \varepsilon)q\alpha_c^{1/2}a^{1/4}d_e^{1/4}}{k_c v_s^{1/2}} \tag{7-8}$$

where the subscript c refers to the composite and $(1 - \varepsilon)q$ is the heat flux to the composite. By equating the composite surface temperature (Eq. (7-8)) to the workpiece surface temperature (Eq. (7-1), the energy partition for $L > 5$ and $\beta = 1.06$ becomes:

$$\varepsilon = \frac{1}{1 + \left(\dfrac{v_s}{v_w}\right)^{1/2}\left[\dfrac{(k\rho c)_c}{(k\rho c)_w}\right]^{1/2}} \tag{7-9}$$

In order to calculate the energy partition from Eq. (7-9), it is necessary to estimate the composite thermal properties. Assuming that the surface porosity is completely filled with grinding fluid and that the thermal properties of the composite can be approximated by a weighted volumetric average of the thermal properties of the abrasive grain material and grinding fluid, the composite thermal conductivity and volumetric specific heat is written as:

$$k_c = \phi_a k_f + (1 - \phi_a)k_g$$
$$(\rho c)_c = \phi_a (\rho c)_f + (1 - \phi_a)(\rho c)_g \tag{7-10}$$

where k is the thermal conductivity, ρc the volumetric specific heat, and ϕ_a the near surface wheel porosity. The subscripts g and f refer to the grain and fluid, respectively.

The bulk porosity of a vitrified aluminum oxide wheel for creep-feed grinding is typically about 50%, but the near surface porosity should be much bigger. In order to take this effect into account, the grinding wheel has been modeled as equally spaced spheres and the average near surface porosity calculated to a radial depth equal to the estimated thermal boundary layer thickness [16]. For creep-feed grinding with water-based fluids, the average surface porosity ϕ_a was estimated to be about 87%. Similar values were obtained for creep-feed vitrified wheels when considering their ability to pump fluid through the grinding zone, as will be seen in Chapter 9. The estimated thermal properties for an aluminum oxide wheel for water-based fluids with $\phi_a = 87\%$ would be $k_c = 9.7$ W/mK and $(\rho c)_c = 4.3 \times 10^6$ J/m³k. For grinding a plain carbon steels under typical creep-feed grinding conditions ($v_s = 30$ m/s, $v_w = 1$–20 mm/s), the energy partition predicted from Eq. (7-9) would be only 1.3%–5.4%, and even lower values would be obtained for grinding of nickel base alloys. These predictions agree with measurements reported for down creep-feed grinding of a plain carbon steel [17].

7.4 ENERGY PARTITION: VARIATION ALONG THE GRINDING ZONE

In the previous section, the energy partition was obtained by equating the maximum interface temperatures for the workpiece and the wheel/fluid composite. However, it should be noted that the temperature distribution along the grinding zone interface on the workpiece side may differ significantly from that on the wheel side if a constant energy partition to the workpiece is assumed. The difference is much greater for up grinding than for down grinding. The temperatures should match everywhere along the grinding zone at the interface between the workpiece and the composite, which requires an energy partition which varies along the grinding zone.

In order to analyze how the energy partition varies along the grinding zone, consider the down grinding situation as shown in Figure 7-4. The

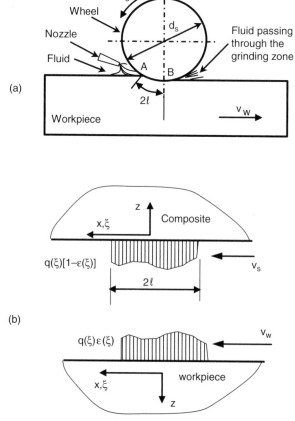

Figure 7-4 Moving heat sources on the workpiece surface and wheel/fluid composite surface for down grinding

wheel/fluid composite moves relative to the grinding zone at the wheel velocity, and the workpiece moves relative to the grinding zone at the workpiece velocity. The grinding energy expended is transported away by the workpiece passing beneath the grinding zone and by the wheel/fluid composite passing above the grinding zone. As the wheel/fluid composite and the workpiece pass through the grinding zone, they should each be heated up to the same temperature at each point along their common interface.

For down grinding where the workpiece moves in the same direction as the wheel/fluid composite at the grinding zone, the leading edge of the heat source on the workpiece coincides with the leading edge of the heat source on the wheel/fluid composite as shown in Figure 7-4(b). In this case, the energy partition to the workpiece should be almost constant along the grinding zone if the heat source moves sufficiently fast relative to both the workpiece and the composite so that conduction along the direction of motion can be neglected ($L > 5$ and $L_c > 5$). But for up grinding where the grinding wheel moves opposite to the workpiece at the grinding zone, the composite surface of lower temperature comes into contact with the workpiece surface of higher temperature at point B. This could even cause local heat flow from the workpiece to the composite in the region near point B (negative energy partition). At the other end of the grinding zone near point A, the workpiece of lower temperature comes into contact with the composite of higher temperature, which could cause heat flow from the composite to the workpiece. Therefore, the energy partition to the workpiece should deviate more significantly from a constant value for up grinding than for down grinding.

Now let us develop a generic model to predict the variation of the energy partition to the workpiece along the grinding zone both up and down grinding. For a total heat flux distribution $q(\xi)$ along the grinding zone as in Figure 7-4, the heat flux distribution to the workpiece is $q_w(\xi)$ and the remaining heat flux to the composite is $q(\xi) - q_w(\xi)$. The fraction of the total grinding energy to the workpiece at the grinding zone can then be denoted by $\varepsilon(\xi)$, which is also the ratio of the heat flux to the workpiece $q_w(\xi)$ to the total heat flux $q(\xi)$:

$$\varepsilon(\xi) = \frac{q_w(\xi)}{q(\xi)} \tag{7-11}$$

The workpiece moves relative to the grinding zone (heat source) at the workspeed v_w and the composite moves at the wheel speed v_s. The quasi-steady state temperature distribution on the workpiece surface $\theta_w(x)$ and that on the composite surface $\theta_c(x)$ can both be calculated analogous to

Eq. (6-1) as:

$$\theta_w(x) = \frac{1}{\pi k_w} \int_{-l}^{l} q(\xi)\varepsilon(\xi)e^{-\frac{v_w(x-\xi)}{2\alpha_w}} K_0\left[\frac{v_w}{2\alpha_w}\{(x-\xi)^2 + z^2\}^{\frac{1}{2}}\right]d\xi \quad (7\text{-}12)$$

$$\theta_c(x) = \frac{1}{\pi k_c} \int_{-l}^{l} q(\xi)[1-\varepsilon(\xi)]e^{\mp\frac{v_s(x-\xi)}{2\alpha_c}} K_0\left[\frac{v_s}{2\alpha_c}\{(x-\xi)^2 + z^2\}^{\frac{1}{2}}\right]d\xi \quad (7\text{-}13)$$

The energy partition to the workpiece $\varepsilon(\xi)$ must satisfy the compatibility requirement that the temperature on the workpiece surface $\theta_w(x)$ equals the temperature on the composite surface $\theta_c(x)$ everywhere along the grinding zone:

$$\theta_w(x) = \theta_c(x) \quad (7\text{-}14)$$

The numerical solution of this problem gives us the energy partition distribution along the grinding zone [14].

Two examples will be presented to illustrate how the energy partition varies along the grinding zone, one for down grinding and one for up grinding. For down grinding, the calculated energy partitions are shown in Figure 7-5(a) for a triangular heat flux distribution under creep-feed grinding conditions. Also included in this figure is the constant energy partition value obtained using Eq. (7-9). It can be seen that the energy partition distribution deviates significantly from the constant energy partition value near the trailing and the leading edges of the grinding zone. The energy partition becomes negative very close to the leading edge of the grinding zone ($x/l = 1$). However, the temperature distribution is nearly the same in both cases as seen in Figure 7-5(b).

The situation for up grinding is different from down grinding as stated above. For up grinding, the wheel/fluid composite moves opposite to the workpiece at the grinding zone, so the leading edge on the workpiece corresponds to the trailing edge on the composite and the trailing edge on the workpiece corresponds to the leading edge on the composite. In this case, the energy partition to the workpiece deviates much more drastically from a constant value than for down grinding as can be seen in Figure 7-6(a). The energy partition exceeds unity at the heat source leading edge ($x/l = 1$) indicating net heat flow from the composite to the workpiece, and decreases along the grinding zone becoming negative near the trailing edge ($x/l = -1$) indicating net heat flow from the workpiece to the composite. From the corresponding temperature distributions in Figure 7-6(b), it can be seen that the constant partition model predicts a lower temperature near the heat source leading edge, a higher temperature near the trailing edge, and a lower maximum temperature.

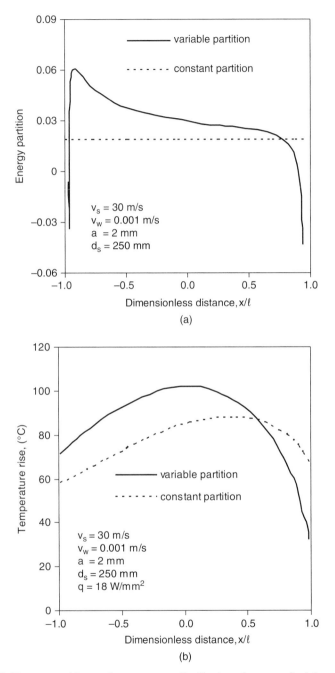

Figure 7-5 Energy partition and temperature distributions for creep-feed down grinding with triangular total heat flux distribution

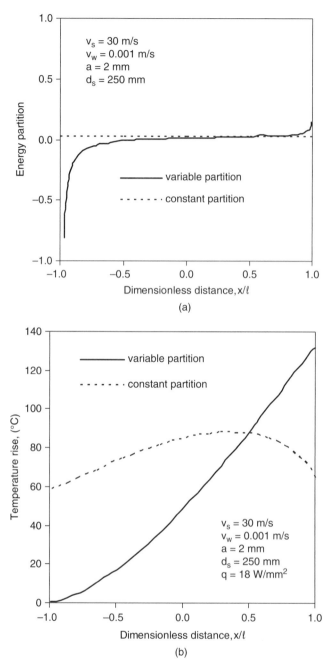

***Figure 7-6 Energy partition and temperature distributions for creep-feed up grinding
with triangular total heat flux distribution***

7.5 ENERGY PARTITION: SINGLE GRAIN MODEL

Another approach to estimating the energy partition analyzes the heat transfer to the abrasive grains (wheel), fluid, and workpiece by considering a single grain surrounded by fluid interacting with the workpiece [12, 15] as shown in Figure 7-7. Each single active grain is modeled as a truncated cone of radius r_o at its tip moving along the workpiece surface at the wheel velocity v_s. All of the grinding energy is considered to be uniformly dissipated as heat at the grain-workpiece interface of area $A_0 = \pi r_0^2$. The maximum temperature rise at the grain-workpiece interface is

$$\theta_{maxg} = 1.13 \frac{(1 - \varepsilon_{dry})q}{(k\rho c)_g^{1/2}} \left(\frac{l_c}{v_s}\right)^{1/2} \frac{1}{f(\zeta)A} \tag{7-15}$$

where A is the fraction of wheel surface consisting of truncated grain tips (wear flats), ε_{dry} is the 'initial' energy partition to the workpiece, k the thermal conductivity, ρc the volumetric specific heat, and the subscript 'g' refers to the grain material. The function $f(\zeta)$ is [15]:

$$f(\zeta) = \frac{2}{\pi^{1/2}} \frac{\zeta}{1 - \exp(\zeta^2)\mathrm{erfc}(\zeta)} \tag{7-16}$$

where

$$\zeta \equiv \left(\frac{l_c\gamma 2\pi\alpha_g}{A_0 v_s}\right)^{1/2} \tag{7-17}$$

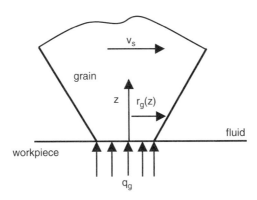

Figure 7-7 Single grain model for estimating the energy partition

γ is a geometric grain shape factor defined as

$$\gamma \equiv \frac{dr_g}{dz} \tag{7-18}$$

and r_g is the grain radius. Part of the initial energy partition ε_{dry} is then immediately transferred from the workpiece to the fluid, thereby leaving the energy partition ε in the workpiece.

Cooling by the fluid in this way is taken into account by considering the temperature at the fluid-workpiece interface within the grinding zone. Grinding fluid entering the wheel pores is quickly accelerated and can be considered to be moving at the wheel velocity within the grinding zone [4]. For a thermal number $L \equiv v_w l / 2\alpha > 5$, the maximum temperature rise of the fluid at its interface with the workpiece can be written as [18]:

$$\theta_{maxf} = 1.06 \frac{q(\varepsilon_{dry} - \varepsilon)}{(k\rho c)_f^{1/2}} \left(\frac{l_c}{v_s}\right)^{1/2} \frac{1}{(1 - A)} \tag{7-19}$$

where ε is the fraction of the total energy not removed by the fluid (energy partition) and subscript f refers to the fluid. The maximum workpiece temperature rise can be expressed as

$$\theta_{max} = \beta \frac{\varepsilon q}{(k\rho c)_w^{1/2}} \left(\frac{l_c}{v_w}\right)^{1/2} \tag{7-20}$$

If the maximum temperature at the workpiece-fluid interface is the same as at the workpiece-grain interface ($\theta_{max} = \theta_{maxf} = \theta_{maxg}$), the overall energy partition to the workpiece is finally obtained by combining Eqs. (7-15), (7-19) and (7-20):

$$\varepsilon = \frac{1}{1 + \Omega \left(\dfrac{v_s}{v_w}\right)^{1/2} \beta} \tag{7-21}$$

where β is defined in Eqs. (7-2) and (7-3) and

$$\Omega \equiv 0.94 \frac{(k\rho c)_g^{1/2}}{(k\rho c)_w^{1/2}} A f(\zeta) + \frac{(k\rho c)_f^{1/2}}{(k\rho c)_w^{1/2}} (1 - A) \tag{7-22}$$

The burnout heat flux predicted by substituting Eqs. (7-21) and (7-22) into Eqs. (7-4) or (7-5) agrees quite well with the measured burnout heat flux under various conditions as shown in Figure 7-3 [12, 13]. In this figure, the upper line is the model prediction for burnout at the middle of the grinding (quasi-steady state) and the lower line is the result for burnout during cut out.

This single grain energy partition model can be used to account for the differences in energy partition to the workpiece under various grinding and fluid application conditions. For regular grinding with conventional aluminum oxide wheels and water-based fluids, the grinding zone temperatures are often much higher than the burnout limit of 130 °C, so cooling by the fluid is not effective at the grinding zone. In this case, the term $(k\rho c)_f$ in Eq. (7-21) is essentially zero, which leads to an energy partition of about 60–70%. For creep-feed grinding, the temperature at the grinding zone is below 130 °C. The term $(k\rho c)_f$ is approximately 2.72×10^6 J^2/m^4K^2s which leads to an energy partition less than 5%, comparable to actual measurements [17]. In Chapter 8, this single grain model will also be used to account for the energy prediction for grinding with CBN abrasive wheels.

7.6   TRANSIENT TEMPERATURE

In the previous sections, we have considered the situation where the workpiece is long enough for the temperature within the grinding zone to reach a quasi-steady state. Temperatures generated in the workpiece during straight surface plunge grinding follow a transient behavior as the grinding wheel engages with and disengages from the workpiece [19]. Transient conditions also prevail throughout the entire grinding pass for workpieces which are shorter than needed to reach a quasi-steady state condition [14], which is often the case for creep-feed grinding. The temperature rises rapidly during initial wheel-workpiece engagement (cut in), subsequently reaches a quasi-steady state value if the workpiece is sufficiently long, and increases still further during final wheel-workpiece disengagement (cut out) as workpiece material is suddenly unavailable to dissipate heat. In this section, models are developed for the transient temperature distribution under creep-feed conditions. Thermal damage to the workpiece often occurs near the end of a grinding pass as the wheel disengages from the workpiece (cut out) [12], which also suggests that the highest temperature might be reached at this location.

For analyzing transient grinding temperatures, the workpiece can be categorized as either long or short in terms of both the geometric conditions and the thermal conditions. A workpiece is geometrically long if its length

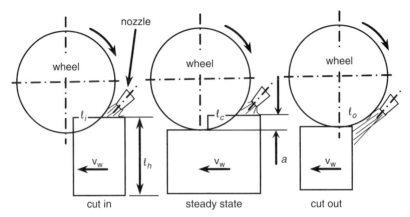

***Figure 7-8** Illustration of cut in, steady state, and cut out for grinding a geometrically long workpiece ($l_w > l_c$).*

exceeds the wheel-workpiece contact length at the full depth of cut ($l_w > l_c$) as shown in Figure 7-8, and it is geometrically short if its length is less ($l_w < l_c$) as shown in Figure 7-9. A geometrically long workpiece may be further classified as thermally long or short according to whether it is

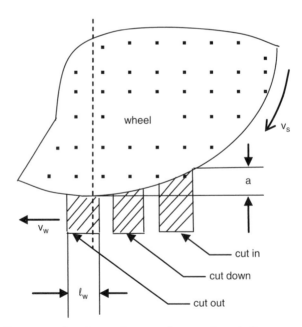

***Figure 7-9** Illustration of cut in, cut down, and cut out for grinding a geometrically short workpiece ($l_w < l_c$). For illustration purposes, the workpiece is shown at different locations relative to the wheel.*

sufficiently long for the quasi-steady state surface temperature to be reached. Geometrically short workpieces are also thermally short.

For analyzing the temperatures generated during grinding of a geometrically long workpiece, it is useful to consider three sequential regimes as illustrated in Figure 7-8: cut in, steady state, and cut out. Heat input to the workpiece in each regime occurs over the wheel-workpiece arc length of contact which can be approximated by the chord length. During cut in, the contact length l_i increases approximately linearly from zero to its geometrical steady state value l_c as the depth of wheel engagement with the workpiece in the downfeed direction increases from zero to the specified wheel depth of cut a. The corresponding workpiece temperature at the grinding zone should rapidly rise during cut in as the metal removal rate increases from zero to its steady state value. Whether the workpiece is thermally long enough to reach the quasi-steady state temperature depends on its thermal properties and the grinding conditions. Cut out occurs during disengagement at the end of the grinding pass as the wheel-workpiece contact length l_o decreases from its steady state value l_c back to zero. During cut out the temperature may exceed the quasi-steady value insofar as workpiece material is suddenly unavailable to conduct heat away.

For grinding a geometrically short workpiece as shown in Figure 7-9, the wheel depth of engagement in the downfeed direction is smaller than the specified wheel depth of cut a throughout the grinding pass [20]. The grinding pass for the short workpiece can also be divided into three regimes: cut in, cut down, and cut out. The wheel-workpiece arc length of contact, again approximated here by the chord length as for the geometrically long workpiece, increases nearly linearly from zero to its maximum value during cut in. Cut down begins after the workpiece has traveled a distance from initial engagement equal to the workpiece length l_w at which point the workpiece top surface becomes completely covered by the grinding zone. Since l_w is generally much longer than the maximum depth of engagement, the contact length in this regime can be approximated as being equal to l_w. During cut out which begins when the workpiece has traveled a distance l_c, the contact length decreases from its maximum value (workpiece length) back to zero.

The thermal model for analyzing the transient temperature distribution during straight surface grinding of a rectangular block (workpiece) of height l_h and length l_w is illustrated in Figure 7-10. The heat flux to the workpiece at the grinding zone is modeled as a continuously distributed planar band source of intensity $\varepsilon q(x)$ on the top surface of the workpiece which moves at the workpiece velocity v_w. The length of the heat source corresponds to the wheel-workpiece contact length which varies during cut in and cut out as described above. Wheel depths of cut, even for creep-feed grinding, are usually sufficiently small such that the inclination and curvature of the heat source can be neglected [3]. Crosswise heat transfer is

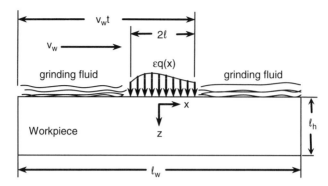

Figure 7-10 Thermal model for straight surface grinding.

neglected, so the problem is two-dimensional. The dimensionless form of the governing heat transfer equation can be written as

$$\frac{\partial \bar{\theta}}{\partial \tau} = \frac{1}{4L}\left(\frac{\partial^2 \bar{\theta}}{\partial X^2} + \frac{\partial^2 \bar{\theta}}{\partial Z^2}\right) \qquad (7\text{-}23)$$

where

$$\tau \equiv \frac{v_w t}{l_c}, \quad X \equiv \frac{x}{l_c}, \quad Z \equiv \frac{z}{l_c}$$

$$L_w \equiv \frac{l_w}{l_c}, \quad \bar{\theta} \equiv \frac{\theta k}{q_w l_c}, \quad L \equiv \frac{v_w l_c}{4\alpha} \qquad (7\text{-}24)$$

The temperature distribution $\theta(x, z, t)$ within the workpiece can be obtained by solving Eq. (7-23) with appropriate initial and boundary conditions. Since the temperature rise is our main concern, the initial temperature can be taken as zero everywhere within the workpiece. The bottom of the workpiece is assumed to be sufficiently remote from the top surface so as to remain at its initial temperature throughout the grinding pass. At any location x on the workpiece top surface as in Figure 7-10, the intensity of the heat input to the workpiece is time dependent due to the motion of the workpiece along the grinding direction. At time t the leading edge of the heat source (grinding zone) has traveled a distance $v_w t$ from the left end of the workpiece. The heat flux increases during cut in as the wheel depth of engagement with the workpiece in the downfeed direction increases from zero to the specified wheel depth of cut, and decreases during cut out as the engagement with the workpiece in the downfeed direction decreases from the specified wheel depth of cut to zero.

Grinding fluid is applied to the wedge formed by the wheel and work-piece top surface, at the leading edge side for down grinding with the wheel and workpiece velocities in the same direction in the grinding zone, and at the trailing edge side for up grinding with the velocities in opposite directions. A fraction of the applied fluid is carried through the grinding zone by the rotating wheel, while most of the remaining fluid falls back into the stream of fluid on the workpiece top surface [21, 22]. Most of the fluid passing through the grinding zone is ejected slightly upward by the high-speed rotation of the grinding wheel so that only a small amount actually falls on the workpiece top surface ahead of the grinding zone. Therefore, convective heat transfer on the top surface should be greatest on that side of the grinding zone where the grinding fluid is directly applied (trailing edge side for up grinding and leading edge side for down grinding), and much smaller on the opposite side. For simplicity, two different convective heat transfer coefficients are assumed for the regions ahead of and behind the grinding zone. Within the grinding zone, fluid being carried through by the porous wheel rotating at high speed may also provide convective cooling. Cooling by the fluid within the grinding zone can be considered to reduce the heat source intensity $\varepsilon q(x)$ by reducing the energy partition to the workpiece [14, 18].

An example illustrating the transition from a thermally short to a thermally long workpiece is presented in Figure 7-11 [14]. The maximum dimensionless workpiece temperature $\bar{\theta}_m$ uring the grinding pass is shown for thermal numbers $L = 1$ and $L = 5$ with dimensionless workpiece lengths ranging from $L_w = 0.5$ to $L_w = 4.5$. The results in Figure 7-11(a) are for a rectangular heat source and in Figure 7-11(b) for a triangular heat source. In both cases, the maximum temperature rises very rapidly during cut in as the grinding wheel engages the workpiece. The quasi-steady state value is subsequently reached at approximately $L_w = 1.5$ for $L = 5$ and $L_w = 4.5$ for $L = 1$. Whether a workpiece can be considered to be thermally long depends on the thermal number. At the end of the pass during cut out, there is also an abrupt additional temperature rise in each case due to the sudden unavailability of workpiece material to conduct heat away. Cooling of the workpiece end face by using an additional nozzle can significantly lower or eliminate this additional temperature rise [14].

As a practical matter, the maximum temperature during the entire grinding pass may be of particular concern. For a triangular heat source, the maximum dimensionless temperature can be approximated as [14]:

$$\bar{\theta}_m = 1.06e^{-0.06L_w}\left[\frac{0.37}{L^{0.26}} + 0.24e^{-0.62L}\right] + 0.10L^{-0.83}L_w^{0.66} \quad (7\text{-}25)$$

for $0.2 < L_w < 5$ and $0.4 < L < 5$. This result can be used to predict burnout in creep-feed grinding with thermally short workpieces.

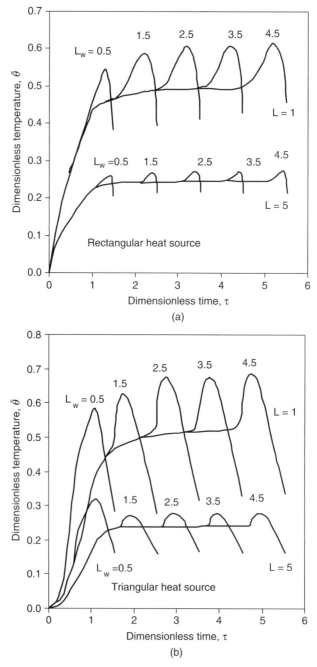

Figure 7-11 Maximum dimensionless temperature versus dimensionless time for various dimensionless workpiece lengths with (a) rectangular heat source and (b) triangular heat source.

7.7 THERMAL COMPARISON: REGULAR AND CREEP-FEED GRINDING

Thermal damage control may require imposing an upper limit on the grinding temperature. For creep-feed grinding, it is essential to avoid fluid burnout to avoid thermal damage to the workpiece. Eqs. (7-4) and (7-5) are the general forms of the critical power flux and grinding power at fluid burnout. The energy partition ε in these equations depends on the grinding condition, wheel types, and fluid application conditions.

In order to compare the thermal limit for creep-feed grinding with that of regular shallow-cut grinding, it is convenient to examine the relationship between the grinding power and the volumetric removal rate while maintaining a constant value of the parameter $R \equiv v_w a^{1/2}$. Since the volumetric removal rate is

$$Q_w = v_w ab \tag{7-26}$$

the workpiece velocity and wheel depth of cut are given by

$$v_w = \frac{bR^2}{Q_w} \tag{7-27}$$

and

$$a = \left(\frac{Q_w}{bR}\right)^2 \tag{7-28}$$

with a faster removal rate for a given value of R being associated with more creep-feed like conditions, i.e. slower workpiece velocity and bigger wheel depth of cut. The parameter R is also indicative of the undeformed chip thickness (Chapter 3). Furthermore, the surface finish in straight surface grinding without spark-out may also depend on this parameter (Chapter 10), so maintaining a fixed value of R might approximate grinding with a constant finish.

Returning to Eq. (7-5), the critical burnout power for creep-feed grinding can now be written in terms of R and Q_w as

$$P_c^* = \left(\frac{q^* d_e^{1/2}}{R}\right) Q_w \tag{7-29}$$

or for given q^*, b, d_e, and R, the critical power is simply proportional to the removal rate:

$$P_c^* = c Q_w \tag{7-30}$$

where

$$c \equiv \frac{q^* d_e^{1/2}}{R} \qquad (7\text{-}31)$$

From Eq. (6-14), the threshold power at the burning limit for conventional grinding of steels is given by

$$P^* = u_0 b v_w a + B^* b d_e^{1/4} a^{1/4} v_w^{1/2} \qquad (7\text{-}32)$$

or substituting for v_w and a from Eqs. (7-27) and (7-28):

$$P^* = u_0 Q_w + B^* b d_e^{1/4} R^{1/2} \qquad (7\text{-}33)$$

Since u_0 and B^* are constants, this relationship for given values of d_e, b, and R reduces to

$$P^* = u_0 Q_w + d \qquad (7\text{-}34)$$

where

$$d \equiv B^* b d_e^{1/4} R^{1/2} \qquad (7\text{-}35)$$

Together with these two relationships for the onset of thermal damage ((Eqs. (7-30) and (7-34)), an expression can also be derived for the grinding power consumption in terms of Q_w and R in place of the parameters v_w and a. Like the specific energy, the grinding power can be considered to consist of chip-formation, plowing, and sliding components:

$$P = P_{ch} + P_{pl} + P_{sl} \qquad (7\text{-}36)$$

as discussed in Chapter 5. The specific energy for chip formation u_{ch} can be considered to be constant, so that

$$P_{ch} = u_{ch} Q_w \qquad (7\text{-}37)$$

The tangential plowing force is considered to be constant, in which case the plowing power for a given wheel velocity

$$P_{pl} = F_{t,pl} v_s \qquad (7\text{-}38)$$

is also constant. The sliding power P_{sl} is proportional to the tangential sliding force component $F_{t,sl}$ which, in turn, is proportional to the wear-flat area A

and is given by the second term in Eq. (5-11):

$$P_{sl} = F_{t,sl}v_s = \mu\bar{p}b(d_e a)^{1/2}Av_s \tag{7-39}$$

Furthermore, assuming a linear relationship between the average contact pressure $\bar{p}$ and the curvature difference (see Figure 5-7) of the form

$$\bar{p} = p_0 + p_1\Delta \tag{7-40}$$

where p_0 and p_1 are constants, and substituting for Δ from Eq. (3.22), the sliding power can be written as:

$$P_{sl} = fQ_w + g \tag{7-41}$$

where

$$f \equiv \frac{\mu p_0 d_e^{1/2}Av_s}{R} \tag{7-42}$$

and

$$g \equiv \frac{4\mu p_1 bAR}{d_e^{1/2}} \tag{7-43}$$

The total grinding power is the sum of Eqs. (7-37), (7-38), and (7-41), which finally reduces to

$$P = (u_{ch} + f)Q_w + (P_{pl} + g) \tag{7-44}$$

which increases linearly with Q_w when the parameters f and g are fixed.

Eqs (7-30), (7-33), and (7-44) for the burnout, burning, and grinding powers (P_c^*, P^*, and P) are shown plotted versus removal rate in Figure 7-12. As mentioned above, the grinding parameters v_w and a are varied so as to maintain a fixed value of R while the remaining parameters remain constant. Of particular interest are the intersections at $Q_{w,1}$ and $Q_{w,2}$. The region $Q_w < Q_{w,1}$ corresponds to regular grinding without burning. Here, the grinding power exceeds the critical burnout power ($P > P_c^*$) so cooling is ineffective, but workpiece burn does not occur since the grinding power is less than the burning power without cooling ($P < P^*$). For $Q_w > Q_{w,2}$, creep-feed grinding conditions are obtained without workpiece burn, even though $P > P^*$. In this region, the burnout limit exceeds the burning power ($P_c^* > P^*$) owing to cooling, but the grinding power is less than the burnout power ($P < P_c^*$). Furthermore, the difference between P_c^* and P becomes progressively bigger at larger removal rates, which would suggest lower grinding temperatures. The removal rate $Q_{w,2}$ can be considered to define the boundary of the creep-feed grinding regime. In the intermediate region $Q_{w,1} < Q_w < Q_{w,2}$, thermal damage (workpiece burn) would prevail because the grinding power P exceeds both thermal damage limits P^* and P_c^*.

Some additional characteristics of creep-feed grinding can also be seen from Figure 7-12. If the slope for P versus Q_w exceeds that for P_c^* versus Q_w, it is apparent that creep-feed grinding cannot be performed without burnout. This condition can be expressed as

$$(u_{ch} + f) < c \qquad (7\text{-}45)$$

Noting that $u_{ch} \ll f$ for creep-feed grinding and substituting for c and f from Eqs. (7-31) and (7-42) leads to the minimum requirement that

$$\frac{\mu p_0 A v_s}{q^*} < 1 \qquad (7\text{-}46)$$

In order to 'enlarge' the creep-feed grinding regime and reduce thermal damage, it is generally desirable that the quantity on the left-hand side of

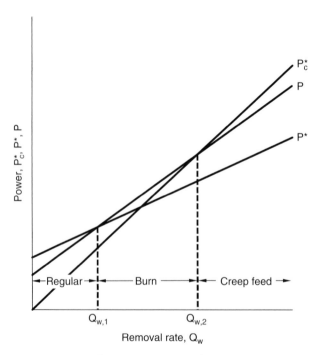

Figure 7-12 *Burnout power* (P_c^*), *burning power* (P^*), *and grinding power* (*P*) *plotted versus removal rate for a fixed value of the parameter* $R = v_w a^{1/2}$.

this inequality be as small as possible. This objective is met by keeping the wheel sharp (small A) and providing more effective cooling (large q^*). Because of the increased tendency for wheel dulling with creep-feed grinding (see Chapter 5), it becomes important to redress often. In many cases, especially when grinding high-strength aircraft alloys, continuous rotary dressing is used during grinding to maintain a constant degree of wheel sharpness [10, 23, 24]. More effective cooling requires delivery of more and/or cooler grinding fluid to the grinding zone. From Eq. (7-46) it would appear that low wheel velocities are beneficial, and this is sometimes true, but it should be noted that higher wheel velocities tend to enhance cooling and increase the burnout flux q^* by 'pumping' more fluid through the grinding zone and lowering the energy partition (e.g. Eq. (7-9))

Creep-feed grinding is characterized by very high specific energies, removal rates, and forces, so that very stiff and powerful machines are required. A pressurized grinding fluid system and a large tank are also necessary in order to provide the necessary cooling. Conventional grinding machines are usually not suitable for creep-feed grinding.

REFERENCES

1. Malkin, S. and Guo, C., 'Thermal Analysis of Grinding', *Annals of the CIRP*, 57/2, 2007, p.760.
2. Werner, P. G., Younis, M. A. and Schlingersiepen, R., 'Creep Feed - an Effective Method to Reduce Work Surface Temperatures in High Efficiency Grinding Processes', *Proceedings, Eighth North American Manufacturing Research Conference*, 1980, p. 312.
3. Dawson, P. R. and Malkin, S., 'Inclined Moving Heat Source Model for Calculating Metal Cutting Temperatures', *Trans. ASME, J. of Eng. for Ind.*, 106, 1984, p. 179.
4. Incropera, F. P. and Dewitt, D. P., *Introduction to Heat Transfer*, 4th edition, John Wiley and Sons, New York, 2002, p. 559.
5. Shafto, G. R., Howes, T. D. and Andrew, C., 'Thermal Aspects of Creep Feed Grinding', *Proceedings of the Sixteenth International Machine Tool Design and Research Conference*, 1975, p. 31.
6. Andrew, C., 'Coolant Application in Creep Feed Grinding', *Proceedings of the International Conference on Creep Feed Grinding*, Bristol, 1979, p. 67.
7. Powell, J. W. and Howes, T. D., 'A Study of the Heat Flux at which Burn Occurs in Creep Feed Grinding', *Proceedings of the Nineteenth International Machine Tool Design and Research Conference*, 1978, p. 629.
8. Pearce, T. R. A. and Howes, T. D., 'The Effect of Wheel Structure on Coolant Application in Creep Feed Grinding', *Fourth Polytechnics Symposium on Manufacturing Engineering, Birmingham*, 1984.
9. Salter, N. D., Pearce, T. R. A. and Howes, T. D., 'A Fundamental Investigation of Workpiece Burn in Creep Feed Grinding', *Milton C. Shaw Grinding Symposium*, PED-16, ASME, 1985, p. 199.

10. Andrew, C., Howes, T. D. and Pearce, T. R. A., *Creep Feed Grinding*, Holt, Rinehart and Winston, London, 1985, Chapters 4 and 5.
11. Yasui, H., 'On Limiting Grinding Condition for Fluid Supply Effect', *Proceedings of the 5th International Conference on Production Engineering*, Tokyo, 1984, p. 58.
12. Guo, C. and Malkin, S., 'Analytical and Experimental Investigation of Burnout in Creep-Feed Grinding,' *Annals of the CIRP*, 43/1, 1994, p. 283.
13. Furukawa, Y., Ohishi, S. and Shiozaki, S., 'Selection of Creep Feed Grinding Conditions in View of Workpiece Burning', *Annals of the CIRP*, 28/1, 1979, p. 213.
14. Guo, C. and Malkin, S., 'Analysis of Transient Temperature in Grinding,' *Trans. ASME, J. of Eng. for Ind.*, 117, 1993, p. 571.
15. Lavine, A.S., Malkin, S. and Jen, T.C., 'Thermal Aspects of Grinding with CBN Abrasives,' *Annals of the CIRP*, 38/1, 1989, p. 557.
16. Lavine, A., 'A Simple Model for Convective Cooling during Grinding Process,' *Trans. ASME, J. of Eng. for Ind.*, 110, 1988, p. 1.
17. Kim, N., Guo, C., and Malkin, S., 'Heat Flux and Energy Partition in Creep-Feed Grinding,' *Annals of the CIRP*, 46/1, 1997, p. 227.
18. Guo, C. and Malkin, S., 'Analysis of Energy Partition in Grinding,' *Trans. ASME, J. of Eng. for Ind.*, 117, 1995, p. 55.
19. Crisp, J. N., *Transient Thermal Aspects in Surface Grinding*, Ph.D. Thesis, Carnegie Mellon University, 1968.
20. Chiu, N., *Computer Simulation for Form Grinding Process*, PhD Thesis, University of Massachusetts, 1993.
21. Engineer, F., Guo, C. S. and Malkin, S., 'Experimental Measurement of Fluid Flow through the Grinding Zone', *Trans. ASME, J. of Eng. for Ind.*, 114, 1992. p. 61.
22. Guo, C. and Malkin, S., 'Analysis of Fluid Flow through the Grinding Zone', *Trans. ASME, J. of Eng. for Ind.*, 104, 1992, p. 427.
23. Howes, T. D., 'The Technique of Dressing during Grinding', *Proceedings of International Conference on Creep Feed Grinding*, Bristol, 1979, p. 184.
24. Pearce, T. R. A. and Howes, T. D., 'Applications of Continuous Dressing in Grinding Operations', *Proceedings of the Twenty-Third International Machine Tool Design and Research Conference*, 1982, p. 203.

BIBLIOGRAPHY

Andrew, C., Howes, T. D. and Pearce, T. R. A., *Creep Feed Grinding*, Holt, Rinehart and Winston, London, 1985.
Marinescu, I. D., Hitchiner, M., Uhlmann, E., Rowe, W. B. and, Inasaki, I., *Handbook of Machining with Grinding Wheels*, CRC Press, 2007, Chapter 16.

Chapter 8

Thermal Aspects: Grinding with CBN Abrasives[1]

8.1   INTRODUCTION

In the previous two chapters, grinding temperatures were analyzed by considering the grinding zone as a heat source which moves along the workpiece surface. Temperatures calculated for given operating parameters were generally found to be proportional to the rate of energy expended and to the fraction of that energy which is transported as heat to the workpiece at the grinding zone (energy partition). For regular shallow cut grinding with conventional abrasive wheels, it was seen in Chapter 6 that heat transfer to the workpiece is especially important, as the energy partition typically ranges from 60 to 90%. High temperatures are generated which may cause thermal damage to the workpiece. This is in sharp contrast to the situation for creep-feed grinding using aluminum oxide wheels as described in Chapter 7. In the absence of fluid burnout, the energy partition for creep feed grinding is typically only 3 to 6%. Thermal damage should not occur if the temperature remains below the burnout limit of the fluid.

The present chapter is concerned with the thermal aspects of grinding with cubic boron nitride (CBN) superabrasive wheels. As compared with conventional aluminum oxide wheels, thermal damage with CBN is generally found to be much less of a problem [2]. For grinding of steels with CBN wheels, workpiece burn is much less likely to occur, and residual stresses at the ground surface are usually found to be predominantly compressive. These observations are indicative of much lower temperatures with CBN than with aluminum oxide wheels. Lower temperatures with CBN were originally attributed to the somewhat smaller specific energies which are typically found in practice. However it was subsequently postulated that the

---

[1]*Parts of this chapter were adapted from Reference [1].*

main effect is due to the very high thermal conductivity of CBN, so that a much larger fraction of the grinding heat is transported to the grain rather than to the workpiece (lower energy partition) [3]. The thermal conductivity of CBN is approximately 35 times bigger than that of aluminum oxide. Cooling by the grinding fluid may also be an additional factor with CBN.

This chapter begins with a consideration of vitrified CBN wheels, and then proceeds to single layer electroplated CBN wheels. These two types of wheels account for the vast majority of CBN wheels used in industry. At one time, resin bonded CBN wheels were the most common, but their use has been largely supplanted by vitreous bonded products. In recent years, electroplated CBN wheels have been growing in popularity, especially for automotive and aerospace applications. The inherent structural porosity of vitrified CBN wheels would seem to provide favorable conditions for cooling by the fluid at the grinding zone. By contrast, this type of cooling would seem less likely to occur with electroplated CBN wheels due to their limited surface porosity, but it will be seen that this is not necessarily true. The chapter concludes with the thermal aspects of high efficiency deep grinding (HEDG), which typically uses electroplated wheels operating at high wheel speeds with very large depths of cut, comparable to or even bigger than for creep feed grinding, together with much faster work speeds. Despite the extreme removal rates which can be achieved, the HEDG process has found only limited application in industry up to now.

8.2   VITRIFIED CBN WHEELS

It was seen in Chapter 7 that very low energy partitions of only a few percent could be obtained for creep-feed grinding due to cooling by the fluid at the grinding zone. Two thermal models were developed to account for this behavior, one which matches the temperatures on the workpiece and the wheel/fluid composite at the grinding zone, and a second one which considers the thermal situation for a single active grain surrounded by grinding fluid. The single grain model was originally developed in order to assess how the high thermal conductivity of CBN abrasive grains might lower the energy partition [3], although the application of this model in the previous chapter was used to explain how cooling by the fluid in creep feed grinding could drastically lower the energy partition. The first energy partition measurements for grinding with CBN abrasive wheels were reported only some years later for regular shallow cut grinding [4]. In that investigation, temperature matching methods were used to obtain the energy partition for grinding of various steels using both vitrified and electroplated CBN wheels. The underlying idea was to ascertain the role of the thermal conductivity of CBN as implied by the single grain model. When conducting these experiments, no attempt was made to account for the possible role of the grinding fluid.

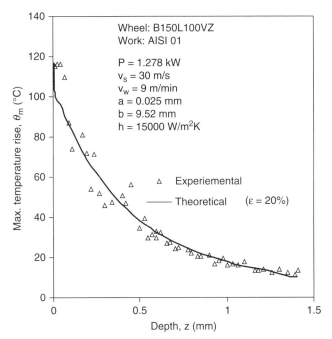

Figure 8-1 Temperature rise for grinding AISI 01 die steel with a vitrified CBN wheel

Up grinding experiments were performed on a simple surface grinder equipped with a typical low pressure flood application device at a low fluid flow rate (up to 2.4 liters/min). With this application method, the fluid does not directly reach the crevice between the wheel and the workpiece, but usually hits the rotating wheel and then falls on the workpiece (see Chapter 9).

    One set of results obtained for grinding of a die steel (AISI 01) with a vitrified CBN wheel is shown in Figure 8-1, where the maximum temperature rise is plotted versus the depth beneath the ground surface. This measured temperature was then matched to the theoretical temperature (Chapter 6), taking into account the dynamic response of the thermocouple, in order to obtain the energy partition. In this particular example, the effect of cooling ahead of the grinding zone where the fluid hits the workpiece top surface was also taken into consideration by specifying a cooling coefficient $h = 10,000$ W/m^2 K at this location. As compared with the usual assumption of an insulated surface, the main effect of this change in the boundary conditions is to increase the cooling rate of the workpiece material after it passes through the grinding zone. It has virtually no influence on the maximum surface temperature and energy partition at the grinding zone. The measured and theoretical temperatures in Figure 8-1 match quite well for an energy partition of 20%. For numerous other tests conducted under the same conditions on this die steel and on hardened AISI 52100 bearing steel, comparable energy partition values were obtained by applying temperature matching methods to both the maximum subsurface temperatures and to temperature responses during single grinding passes.

With flood cooling as in these tests, it is highly unlikely that the amount of fluid actually reaching the grinding zone would have been sufficient to fill up the pores and cool the workpiece [5, 6]. Furthermore the maximum surface temperature rise as seen in Figure 8-1 is close to 120°C which, when added to an ambient temperature of 20 – 25°C, brings the grinding temperature above the fluid burnout limit of 130°C. Because of the ineffective fluid application and/or burnout of the fluid, it would seem reasonable to neglect cooling by the fluid at the grinding zone when applying the single grain energy partition model, which can be done by setting $(k\rho c)_f = 0$ in Eqs. (7-21) and (7-22). The reduced energy partition in this case, relative to that for aluminum oxide wheels, would then be attributed to the high thermal conductivity of the CBN. Using the thermal properties for aluminum oxide $((k\rho c)_g = 0.14 \times 10^9$ J^2/m^4 K^2 s) in place of the thermal properties for CBN $((k\rho c)_g = 2.27 \times 10^9$ J^2/m^4 K^2 s) in the single grain model with no cooling by the fluid $((k\rho c)_f = 0)$ would lead to a much higher energy partition of about 60%, which is more typical of what is found for shallow cut grinding with aluminum oxide wheels.

Subsequent energy partition experiments were conducted with vitrified CBN wheels on a bearing steel (AISI 52100) and on a nodular cast iron cast iron at much higher removal rates on a much bigger machine [7, 8]. In this case, the fluid flow rate was much higher (38 liters/min) with the fluid stream emanating from the nozzle carefully directed at the crevice between the wheel and the workpiece. The maximum surface temperature at various removal rates tended to increase with wheel depth of cut as seen in Figure 8-2. In most

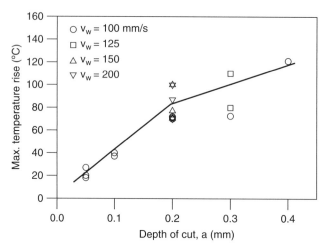

Figure 8-2 Maximum temperature rise for various grinding conditions: vitrified CBN wheel, AISI 52100 steel workpiece.

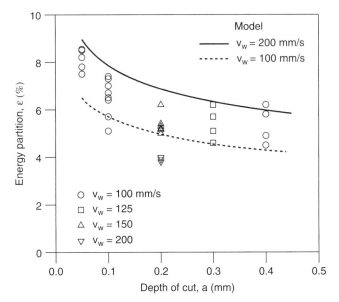

Figure 8-3 Energy partition for various grinding conditions: vitrified CBN wheel, AISI 52100 steel workpiece.

cases, the temperature remained below the burnout limit, so fluid burnout should not have occurred except perhaps in one or two cases. These experiments at higher removal rates were accompanied by lower specific energies of about 20 to 25 J/mm^3, as compared to about 35 J/mm^3 for the previous CBN experiments described above, which is one reason for the lower temperatures. By applying temperature matching and inverse heat transfer methods, the energy partition values were found to be extremely small, ranging from about 4% to 8%. These results were found to agree quite well with predictions from the single grain model (Eq. 7-21) as seen in Figure 8-3. In this case, cooling by the fluid was taken into account by setting $(k\rho c)_f = 2.72 \times 10^6 \, J^2/m^4 \, K^2 \, s$.

On the basis of these results, it can be concluded that low energy partition values with vitrified CBN wheels can be attributed to a number of factors. These include a low specific grinding energy, high thermal conductivity of the CBN grain which enhances heat removal from the grinding zone, and possible cooling by the fluid at the grinding zone.

8.3 ELECTROPLATED CBN WHEELS

The initial investigation of the energy partition with CBN wheels using flood cooling [4], as mentioned above, also included experiments with an electroplated wheel containing a single layer of 100 grit CBN abrasive.

The results obtained for grinding with this electroplated wheel were very similar to what was obtained with the vitrified wheel, both in terms of specific energy and energy partition. The specific energy was about 35 J/mm^3 and the energy partition approximately 20%. Therefore it would appear that the single grain model could also be applied in this case neglecting the influence of cooling by the fluid.

As with the vitrified wheels, additional experiments were subsequently conducted with electroplated CBN wheels on a much bigger grinding machine at much higher removal rates and fluid flow rates up to 113 liters/min [9]. One of the objectives in undertaking these experiments was to ascertain the prospects for cooling by the grinding fluid. Unlike vitrified CBN wheels which have a porous structure, electroplated CBN wheels have only a shallow surface porosity to a radial depth from the outermost grain tips to the nickel layer holding the single layer of abrasive grains on to the wheel hub. A further complication arises because the topography of these wheels progressively changes with continued use tending to increase the number of active grains and wear flat area while decreasing the depth of the porous abrasive layer as implied in Figure 8-4. According to the single grain energy partition model, dulled wear flat areas on the CBN grain tips should enhance heat conduction to the abrasive grains and thereby reduce the energy partition to the workpiece. This may not necessarily result in a lower grinding temperature, since wheel dulling should also cause bigger forces and higher power.

As with the vitrified wheels, the energy partition was found for grinding with an electroplated CBN wheel using temperature matching methods. Experiments were conducted during the life of the grinding wheel after various amounts of wheel wear. The inherent wear resistance of the CBN abrasive necessitated extensive grinding to wear the wheel down. During most of the wheel life, the wheel was worn down by grinding of hardened AISI 52100 steel, but a B1900 nickel base alloy workpiece was intermittently used in the later stages to accelerate the wheel wear.

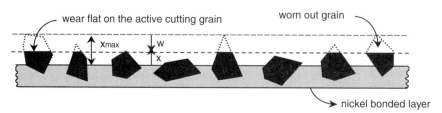

Figure 8-4 Illustration of an electroplated CBN wheel.

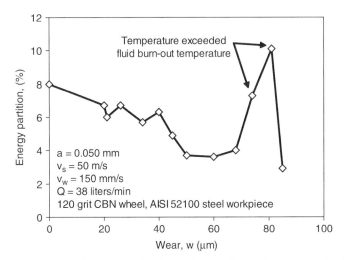

Figure 8-5 Energy partition versus wheel wear for an electroplated CBN wheel.

Experimental results for the energy partition in Figure 8-5 are shown plotted versus wheel wear. For these experiments, the applied fluid flow rate was 38 liters/min. The energy partition at the start of grinding with the new wheel began at about 8%, progressively decreased down to about 3.2% at 60 μm wear, but then increased to a maximum value of 10% at 80 μm of wear before suddenly dropping again to about 2.8% at 85 μm wear near the end of the wheel life. At this point numerous grains became dislodged from the wheel surface. Failure of the wheel subsequently occurred by stripping of the electroplated abrasive layer from the wheel hub. The decrease in the energy partition, at least up to about 60 μm wear, may be due to the increase in dulled wear flat area on the wheel surface. This is reflected in the corresponding increase in specific energy seen in Figure 8-6. Up to this point, the grinding zone temperature shown in Figure 8-7 remained well below the burnout limit. This was followed by a steep increase in the temperature above the burnout limit, which could be responsible for the corresponding rise in the energy partition. As compared with creep-feed grinding, exceeding the burnout limit with the plated CBN wheel resulted in a much less catastrophic change in the grinding behavior.

The single grain thermal model was then applied to analyze the energy partition results. The analysis was limited to those results below the burnout limit, thus allowing for cooling by the fluid at the grinding zone. Matching the thermal model to the experimentally measured energy partition was found to require a progressive increase in the wear flat area as seen

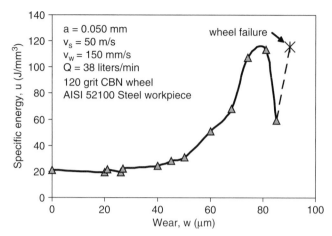

Figure 8-6 Specific energy versus radial wear for an electroplated CBN wheel.

in Figure 8-8. Unfortunately it was not possible to measure the actual wear flat area with the wheel mounted on the grinding machine. However a plot of the normal and tangential force components versus the estimated wear flat area, as seen in Figure 8-9, resulted in characteristic linear relationships similar to what were seen in Chapter 5.

The role of the grinding fluid as a coolant was further explored by investigating the effect of applied fluid flow rate on the energy partition. For these experiments, the radial wheel wear ranged from 20 to 35 μm. The

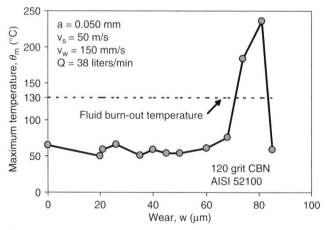

Figure 8-7 Maximum grinding zone temperature versus radial wear for an electroplated CBN wheel.

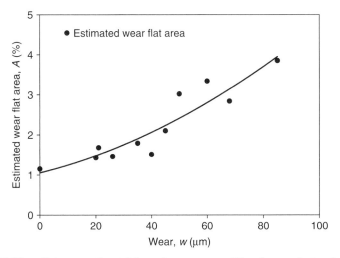

Figure 8-8 Wear flat area estimated from the energy partition for an electroplated CBN wheel.

results summarized in Figure 8-10 show nearly constant energy partition values of 4%–6% at flow rates above 9 liters/min. The maximum grinding temperature in this regime remained well below the burnout limit. At lower flow rates, the energy partition increased sharply and the burnout limit was exceeded. These results suggest that a critical applied flow rate of about

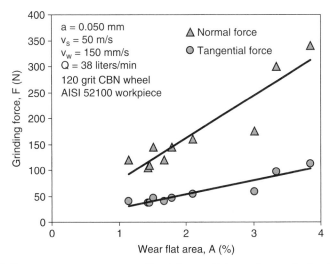

Figure 8-9 Force components versus estimated wear flat area for an electroplated CBN wheel.

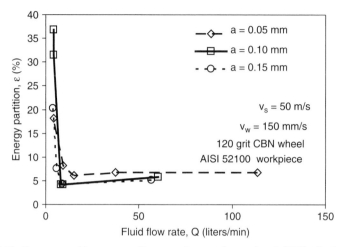

Figure 8-10 Energy partition versus flow rate for an electroplated CBN wheel.

9 liters/min might be the minimum needed to fill up the wheel surface porosity. A rough estimate of the corresponding flow rate of fluid actually passing through the grinding zone can be obtained as the product of the depth of the porous layer, its average porosity, the wheel velocity, and the grinding width. In the present case, the maximum porous depth with the new wheel was approximately 120 μm, its average porosity was estimated at about 60%, the wheel velocity was 50 m/s, and the grinding width was 25.4 mm. This leads to a flow rate through the grinding zone of about 5.3 liters/min to fill up the porous wheel surface, which is about 60% of the applied critical flow rate. Therefore it would appear that about 60% of the applied fluid actually passes through the grinding zone at the critical flow rate of 9 liters/min [10]. Further details regarding fluid flow through the grinding zone are presented in Chapter 9.

8.4   HIGH EFFICIENCY DEEP GRINDING (HEDG)

High Efficiency Deep Grinding (HEDG) utilizes mainly electroplated CBN wheels to achieve large depths of cut, comparable or bigger than for creep-feed grinding, and also relatively fast workpiece velocities, comparable to what is used in conventional shallow cut grinding. As such, HEDG processes are considered to provide the highest removal rates per unit width of any grinding process. Another important feature of HEDG grinding is the use of high wheel velocities, generally in excess of 100 m/s.

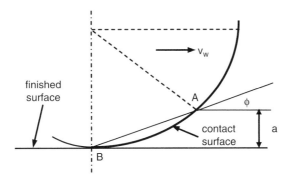

Figure 8-11 Illustration of contact at the grinding zone for HEDG.

Numerous studies of the thermal aspects of the HEDG process have been reported, especially in recent years [11-14]. While much of this work has highlighted the unique characteristics of the process, much uncertainly remains about how to estimate the grinding temperatures and predict the onset of thermal damage. However it has become apparent that a number of assumptions which were adopted when dealing with the thermal aspects of other grinding processes may not be valid for HEDG.

As with creep feed grinding, the large depths of cut might suggest the use of an inclined heat source for the grinding zone with HEDG as illustrated in Figure 8-11. The inclination angle ϕ of the inclined heat source at AB can be readily calculated in terms of the wheel diameter d_s and depth of cut a as

$$\phi = \cos^{-1}\left(\frac{d_s - 2a}{d_s}\right) \tag{8-1}$$

It was previously noted in Chapter 7 that inclination of the heat source was found to have only a minimal effect on creep-feed grinding temperatures (Chapter 7). The inclination angles for HEDG may be bigger than for creep feed grinding because of bigger wheel depths of cut and also smaller wheel diameters. However the main difference between the creep-feed situation and HEDG appears to be related not so much to the difference in inclination angle, but rather to the much faster workpiece velocities in HEDG. With faster workpiece velocities and, consequently, larger values of the thermal parameter L ($L = v_w l_c/4\alpha$), more heat remains in the path of the advancing grinding zone in the material being removed without enough time for it to be conducted downward into the remaining workpiece. This is analogous to the situation encountered in cut-off grinding (Chapter 6) where the infeed rate down into the workpiece proceeds sufficiently fast so that much of the grinding heat is removed together with the material being machined. This phenomenon can have important implications for the grinding mechanisms, since the material being ground is essentially preheated and could lower the energy for chip formation. It was shown in Chapter 5 that the chip formation in grinding is directly related to the melting energy per unit volume, which is the energy to adiabatically take the material from the initial ambient temperature up to the liquid state at the melting point (approximately 9.8 J/mm^3 for ferrous materials). With preheated material, the initial temperature would be higher, so the corresponding melting energy would be reduced accordingly. This could explain the observation that the specific energy decreases with faster removal rates, since increasing the removal rate by using a faster workpiece velocity and/or bigger depth of cut would increase the fraction of the heat remaining in the grinding path.

Another important factor which needs to be taken into account with HEDG is the heat source distribution. In the previous thermal analyses, both uniform and triangular heat sources were used, and both gave comparable results for the maximum grinding zone temperature. Because the localized removal rate is essentially proportional to the distance along the grinding zone, a triangular heat source is more realistic, and this is consistent with results obtained by applying inverse heat transfer to measurements of grinding temperatures. When considering the possibility of thermal damage with the inclined heat source, it should be of particular interest to consider the temperature occurring at the finished surface at point B in Figure 8-11. For a uniform heat source distribution, the maximum temperature along he grinding zone would tend to be skewed towards the trailing edge of the grinding zone. However for a triangular heat source, the maximum temperature would be skewed along the grinding zone away from the finished surface. In this case, the maximum temperature on the finished surface at B can be considerably lower than the maximum temperature within the grinding zone.

The moving heat source thermal analysis has been modified in order to take into account the effect of depth of cut and a triangular rather than rectangular heat source distribution [11]. On the basis of these results, the maximum temperature rise at the contact zone can be expressed, analogous to Eq. 6-2 as:

$$\theta_m = 1.595 \frac{C}{\sqrt{k\rho c}} \left(\frac{l_c}{v_w}\right)^{0.5} \varepsilon\, q \qquad (8\text{-}2)$$

where l_c is the grinding zone chard length (AB in Figure 8-11), q is the total average heat flux at the grinding zone, ε is the energy partition, and C is a dimensionless parameter ranging from 0 to 1 which takes into account the effect of both the thermal number L ($L = v_w l_c / 4\alpha$) and the inclination angle

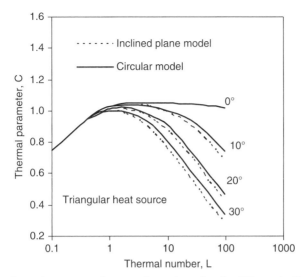

Figure 8-12 Thermal parameter C versus thermal number for different inclination angles.

ϕ. Values for C are summarized in Figure 8-12. As expected, a bigger thermal number and steeper inclination angle, as with HEDG, lead to smaller C. For an inclination angle $\phi \approx 0$ and thermal number $L > 5$, $C \approx 1$, which is the case of regular shallow grinding as discussed in Chapter 6.

For considering the possibility of thermal damage, the temperature of interest should be the maximum temperature at the finished surface at point B in Figure 8-11 rather the maximum temperature along the grinding zone. For the same inclined heat source analysis [11], the maximum temperature rise on the finished surface at B, $\theta_{f\,max}$, was also calculated as a fraction λ of the maximum grinding zone temperature θ_m:

$$\theta_{f\,max} = \lambda\theta_m \qquad (8\text{-}3)$$

Results for λ are summarized in Figure 8-13, which shows how an increase in the thermal number or inclination angle reduces the maximum temperature on the finished workpiece surface at B.

In order to calculate the temperature at the grinding zone and maximum temperature on the finished workpiece surface, it necessary to estimate the energy partition ε at the grinding zone. For this purpose, it would seem to be reasonable to apply the single grain model to take into account the heat transfer to the grain and to the fluid as discussed in Chapter 7. However a further reduction in the energy partition due to heat being removed with the chips should also be considered. While this effect was

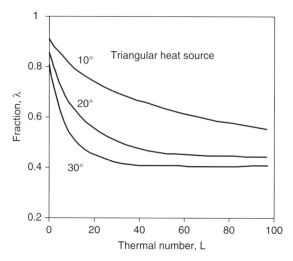

Figure 8-13 Maximum temperature at the finished surface as a fraction of the maximum temperature at the grinding zone contact surface.

neglected up to now when applying the single grain energy partition model, it could have a significant influence, especially if the specific grinding energy is low. For HEDG grinding of a low alloy steel, the specific grinding energy was found to decrease down to 10-15 J/mm³ at very high removal rates [12], and even a smaller specific energy of about 7 J/mm³ has been reported [14]. For grinding with aluminum oxide wheels, it was previously found (see Chapter 6) that about 45% of the chip formation energy ($u_{ch} \approx 13.8$ J/mm³) is removed with the chips.

To illustrate the implication of these results, consider the case of HEDG grinding of a steel with a CBN wheel of diameter $d_s = 150$ mm, wheel velocity $v_s = 150$ m/s, workpiece velocity $v_w = 250$ mm/s, and wheel depth of cut $a = 3$ mm corresponding to a specific removal rate of 750 mm³/mm·s. According to the single grain model, the energy partition for a CBN grain with an assumed wear flat area of 0.3% would be $\varepsilon = 15.5\%$ for grinding with a water-based soluble oil ($(k\rho c)_f = 2.72 \times 10^6$ J²/m⁴ K² s) and $\varepsilon = 26\%$ with a straight oil ($(k\rho c)_f = 0.246 \times 10^6$ J²/m⁴ K² s). These energy partition values, which take into account both conduction to the grains and cooling by the fluid, would apply if the grinding zone temperature is maintained below the fluid burnout temperature, which is about 130 °C for water-based fluids and 300 °C for straight oils. For a specific energy of 12 J/mm³, the maximum grinding zone temperature would reach about 676 °C for a water based fluid and 1125 °C for a straight oil, which means that burnout would occur in both cases. The energy partition would be 36.5% if cooling is not considered ($(k\rho c)_f = 0$), and the maximum

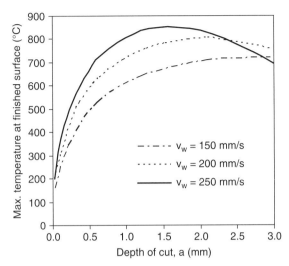

Figure 8-14 Maximum temperature at the finished surface versus wheel depth of cut for HEDG grinding example.

grinding zone temperature would reach 1575 °C. The inclination angle is about 10 degrees and the thermal number L is 92.6 for this case. The corresponding C factor is 0.79 (Figure 8-12) and λ is 0.56 (Figure 8-13). Therefore, the maximum temperature rise at the finished workpiece surface (Point B in Figure 8-11) would be only 44% of the maximum grinding zone temperature, or $\theta_{fmax} = 690$ °C.

Some results are presented in Figure 8-14 which show the effect of both wheel depth of cut and workpiece velocity on the maximum temperature at the finished surface. For these calculations the specific grinding energy was varied from 27 J/mm^3 at the lowest specific removal rate down to 12 J/mm^3 at the highest in accordance with experimental measurements [11]. It appears that beyond a certain point, the thermal situation with HEDG can become better with faster removal rates.

REFERENCES

1. Malkin, S. and Guo, C., 'Thermal Analysis of Grinding', *Annals of the CIRP*, 57/2, 2007, p. 760.
2. Malkin, S., 'Current Trends in CBN Grinding Technology', *Annals of the CIRP*, 43/2, 1985, p. 557.
3. Lavine, A. S., Malkin, S. and Jen, T., 'Thermal Aspects of Grinding with CBN', *Annals of the CIRP*, 38/1, 1998, p. 557.

4. Kohli, S. P., Guo, C., and Malkin, S., 'Energy Partition for Grinding with Aluminum Oxide and CBN Abrasive Wheels', *Trans. ASME, J. of Eng. for Ind.,* 117, 1995, p. 160.

5. Engineer, F., Guo, C. and Malkin, S., 'Experimental Investigation of Fluid Flow in Grinding', *Trans. ASME, J. of Eng. for Ind.*, 114, 1992, p. 61.

6. Guo, C. and Malkin, S., 'Analysis of Fluid Flow through the Grinding Zone', *Trans. ASME, J. of Eng. for Ind.,* 104, 1992, p. 427.

7. Varghese, V., Guo, C., Malkin, S., and Xiao, G., 'Energy Partition for Grinding of Nodular Cast Iron with Vitrified CBN Wheels', *Machining Science and Technology*, 4, 2000, p. 197.

8. Guo, C. Wu, Y, Varghese, V., Malkin, S., 'Temperature and Energy Partition for Grinding with Vitrified CBN Wheels', *Annals of the CIRP*, 48/1, 1999, p. 247.

9. Upadhyaya, R. P. and Malkin, S., 'Thermal Aspects of Grinding with Electroplated CBN Wheels', *Trans. ASME, J. of Manufact. Sci. and Eng.* 126, 2004, p. 107.

10. Krishnan, N., Guo, C., and Malkin, S., 'Fluid Flow through the Grinding Zone in Creep Feed Grinding', *Proceedings of the 1st International Machining and Grinding Conference*, SME, 1995, p. 905.

11. Jin, T. and Rowe, W. B., 'Temperatures in High Efficiency Deep Grinding (HEDG)', *Annals of the CIRP*, 50/1, 2001, p. 205.

12. Stephenson, D. J., Jin, T. and Corbett, J., 'High Efficiency Deep Grinding of a Low Alloy Steel with Plated CBN Wheels', *Annals of the CIRP*, 51/1, 2002, p. 241.

13. Jin, T., Stephenson, D. J., 'Analysis of Grinding Chip Temperature and Energy Partitioning in High-Efficiency Deep Grinding', *J. Engineering Manufacture*, 220, 2006, p. 615.

14. Tawakoli, T., *High Efficiency Deep Grinding*, Mechanical Engineering Publications Ltd, London, 1993.

Chapter 9

Fluid Flow in Grinding

9.1 INTRODUCTION

Most grinding operations are performed with the aid of a grinding fluid. Grinding fluids are generally considered to have two main roles: lubrication and cooling. Grinding fluids can also help to keep the wheel surface clean and provide corrosion protection for newly machined surfaces. Lubrication by grinding fluids reduces the friction and wear associated with the grinding process, thereby allowing for more efficient operation with less consumption of the abrasive, as will be seen in Chapter 11. Cooling by the fluid within the grinding zone is especially critical for creep-feed operations and also for many grinding operations with CBN wheels, as seen in Chapters 7 and 8. Bulk cooling of the workpiece by the applied fluid decreases the inaccuracies associated with thermal expansion and distortion of the workpiece.

Grinding fluids are usually applied from a nozzle as illustrated in Figure 9-1. The rotating grinding wheel serves as a pump to transport part of the applied fluid through the grinding zone. Furthermore, fluid entrained within the converging wedge formed between the wheel and the workpiece at the entrance to the grinding zone may generate hydrodynamic forces, especially when operating at high wheel speeds with non-porous wheels (e.g. resin and metal bonded superabrasive wheels). Hydrodynamic forces are generally not a problem with porous (vitrified) wheels, although the rough wheel surface interacting with the fluid may consume a significant amount of power especially when operating at high wheel speeds.

This chapter is concerned with the fluid mechanics aspects of grinding. It begins with experimental measurements and analytical modeling of the fluid flow thorough the grinding zone for flood application and for creep feed grinding where fluid is usually applied at higher rates and pressures. The chapter concludes with a consideration of hydrodynamic forces in grinding.

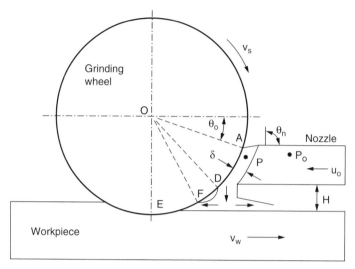

Figure 9-1 Illustration of flood fluid application in surface grinding.

9.2 FLUID FLOW THROUGH GRINDING ZONE: FLOOD APPLICATION

Flood application refers to low-pressure application of grinding fluid from a nozzle. This method is commonly used for shallow cut grinding. For those operations where the grinding area may not be completely enclosed, the flow rate is often kept small enough so as to limit splashing of the fluid. For straight surface grinding, some of the applied fluid usually hits the wheel and then falls on to the workpiece, and some of the remaining fluid may not even reach the wheel but fall directly on to the workpiece. This seemingly undesirable situation may not lead to poor grinding, since the grinding zone temperature, even with higher flow rates of the fluid carefully directed to the wedge between the wheel and the workpiece, would probably be well above the burnout limit, thereby limiting the potential for cooling at the grinding zone (see Chapters 7 and 8). Low flow rates with flood application are normally sufficient to provide lubrication and bulk cooling of the workpiece.

Because of the apparent complexity, the first attempts to investigate fluid flow in grinding were experimental rather than analytical [1]. A test rig was developed to measure the amount of 'useful' fluid actually passing through the grinding zone for shallow cut surface grinding as shown in Figure 9-2 [1]. Fluid is pumped from a tank though a flow meter (16) to the nozzle (13), and then to the grinding wheel (2) and the workpiece (5). That

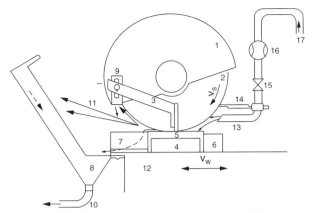

1) Wheel guard 2) Wheel 3) Fluid separating scraper 4) Magnetic parallel
5) Workpiece 6) Stop 7) Channel 8) Cone collector 9) Downstream
spray scraper 10) Fluid spray 12) Chuck 13) Nozzle 14) Air scaper
15) Valve 16) Flowmeter 17) Fluid from pump

Figure 9-2 Experimental apparatus for measuring fluid flow through the grinding zone.

portion of the fluid which passes through the grinding zone flows into a
cone collector (8). The remaining fluid is prevented from reaching the cone
collector by means of fluid separating scrapers (3) on both sides of the
wheel. A downstream spray scraper (9) and channel (7) contain the fluid
and direct it to the collector. A stop (6) and the channel walls at the ends of
the workpiece lock the workpiece in place. Fluid from the collector is fed
through a tube (10) to a canister for weighing.

Figure 9-3 shows some of the results obtained with six different grinding wheels at applied flow rates of ranging from 10 ml/s (0.6 liter/min) to
70 ml/s (4.2 liter/min) of water based soluble oil grinding fluid. The full
specifications for these wheels, indicated only by grade and structure number in Figure 9-3, are given in Table 9-1 together with the bulk porosities
reported by the wheel manufacture. Also included are values for the 'effective porosity', which will be considered at a later point in this chapter. All
these wheels contain the same 60 grit white alumina (38A) abrasive.
Wheels with the 'VCF2' vitreous bond are high porosity creep-feed wheels,
whereas wheels with the 'VBE' bond are less porous conventional wheels.
Experimental results plotted in Figure 9-3 show a proportional relationship
between the flow rate of fluid passing through the grinding zone (useful
flow rate) and the total applied flow rate from the nozzle, thereby indicating a nearly constant percentage of the applied fluid (percent utilization)
passing through the grinding zone for each wheel. The slopes of these
curves are steeper with more porous wheels, indicative of a higher percent

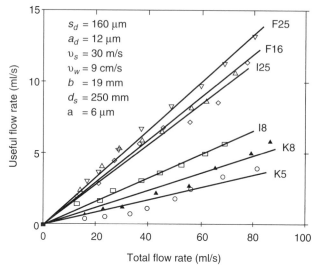

Figure 9-3 Effect of wheel porosity on fluid flow through the grinding zone.

utilization of the fluid. When the percent utilization is plotted versus bulk wheel porosity in Figure 9-4, a linear relationship is obtained with a distinct separation between the higher porosity creep-feed wheels (percent utilization from 14.2% to 16.8%) and the lower porosity conventional wheels (percent utilization from 4.5% to 8.4%).

In addition to the wheel porosity and applied flow rate, another important factor is the nozzle position as seen in Figure 9-5. The nozzle position in this case is characterized by a_1 and a_2 in Figure 9-6. For the distant nozzle position in Figure 9-5, $a_1 = 7$ mm and $a_2 = 5$ mm, and for the closer position, $a_1 = 2$ mm and $a_2 \rightarrow 0$. A much higher utilization of approximately 30% is obtained for the closer position, as compared with only about 4.5% for the distant position.

Table 9-1   Wheel porosity and effective porosity

Wheel Specification	Bulk Porosity V_p (%)	Effective Porosity V_e (%)	V_e/V_p
38A60K5VBE	40.8	34	0.83
38A60K8VBE	43.2	47	1.1
38A60I8VBE	46.0	58	1.3
38A60I25CF2	51.5	93	1.8
38A60F16CF2	52.8	94	1.8
38A60F25CF2	54.6	96	1.8

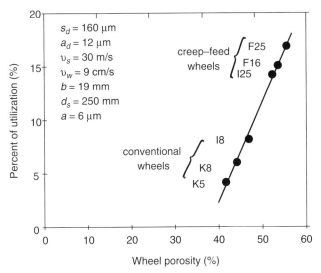

Figure 9-4 Effect of wheel porosity on percent utilization.

Numerous additional experiments were conducted to determine the effect of other parameters on the fluid flow through the grinding zone. With this test method, it was not possible to detect any significant effects of the wheel depth of cut, workpiece velocity, and dressing conditions [1].

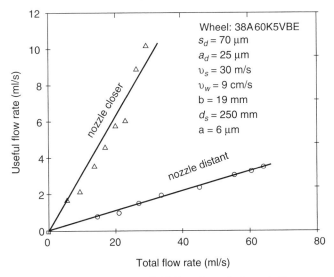

Figure 9-5 Effect of nozzle position on fluid flow through the grinding zone.

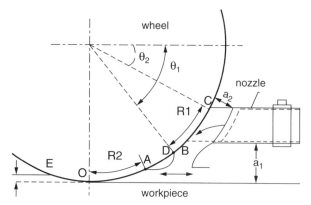

Figure 9-6 Characterization of the nozzle position.

At the high wheel velocities used in grinding, an air boundary layer forms on the wheel surface which can hinder access of the applied fluid from the nozzle to the grinding zone. One way to mitigate this effect is to provide a scraper on the wheel surface slightly upstream from where the applied fluid hits the wheel surface in order to break up this air barrier. Rather than measuring the fluid flow through the grinding zone, which is not feasible in production, the effectiveness of the air scraper was demonstrated in terms of the hydrodynamic pressure developed in the converging wedge at the entrance to the grinding zone [2]. (Hydrodynamic forces in grinding are considered in sections 9.5 and 9.6.) For grinding with a non-porous electroplated CBN wheel, the maximum hydrodynamic pressure was found to decrease to ambient when the wheel velocity reached a critical value if no air scraper was used. This lack of pressure was equated with no fluid reaching the grinding zone. With an air scraper, the critical wheel velocity at which the pressure decreased to ambient was more than 20% faster. Measurement of the hydrodynamic pressure in this way was also used to evaluate the effectiveness of eight commercial nozzles with rectangular and circular shapes [3].

9.3   FLUID FLOW THROUGH THE GRINDING ZONE: CREEP-FEED GRINDING

The results described above were for regular shallow cut grinding with flood application of the fluid. In creep-feed grinding, the fluid is applied at much higher pressures and flow rates. A subsequent investigation was conducted on a much bigger creep-feed machine with larger wheels (wheel diameter $d_s = 355$ mm) to investigate the useful flow rate in creep feed grinding [4]. A more robust test rig, similar in concept to the one shown in Figure 9-2, was developed in order to handle the much higher flow rates.

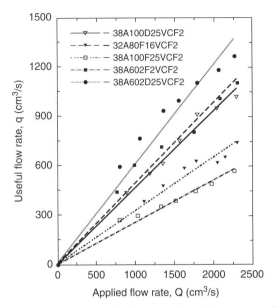

Figure 9-7 Useful flow rate versus applied flow rate for different creep-feed wheels.

Figure 9-7 and Figure 9-8 show some results obtained using the creep-feed wheels listed in Table 9-2. The wheel porosities indicated in Table 9-2 were measured using Archimedes principle by weighing the wheels while immersed in water. As with the previous experiments, the useful flow rate in Figure 9-7 is nearly proportional to the total applied flow rate. Significantly higher flow rates through the grinding zone were obtained with more porous wheels, as before, and also with a coarser grit size. Percentage utilization values ranged from about 25%, with the least porous 100-grit wheel, to 55% with the most porous 60-grit wheel. Increasing the wheel velocity proportionally raises the useful flow rate as seen in Figure 9-8, which is analogous to what occurs with a rotary pump.

Table 9-2 Creep feed wheel porosity

Wheel Specification	Porosity, V_p (%)
38A100D25VCF2	50.5
38A100F25VCF2	44.1
38A602D25VCF2*	53.7
38A602F25VCF2*	46.2
32A80F16VCF2	—

* *The number 2 after the grain size 60 indicates that these wheels contain an equal mixture of 60 and 80 grit abrasive grains rather than only one size.*

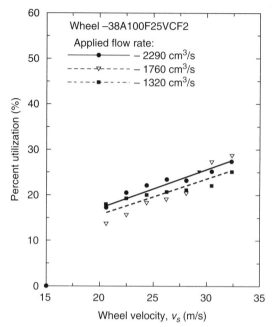

Figure 9-8 Percentage utilization versus wheel velocity.

As with flood application, a closer nozzle position was again found to increase the flow rate through the grinding zone as seen in Figure 9-9. With reference to Figure 9-10, all three nozzle positions were set with $\theta = 6$ degrees and $t = 1$ mm. The closest was Position 1 with $s = 38$ mm, followed by Position 2 with $s = 73$ mm, and Position 3 with $s = 117$ mm. Coarser wheel dressing also tended to slightly enhance the flow rate through the grinding zone, which may be due to a rougher and more porous wheel surface (see Chapter 4). However, as with flood application, wheel depth of cut and workpiece velocity were found to have no significant influence on the useful flow rate.

9.4   ANALYSIS OF USEFUL FLOW RATE THROUGH THE GRINDING ZONE

In the previous two sections, some experimental results were presented for the useful flow rate through the grinding zone for both shallow cut grinding with flood application and for creep feed grinding with much higher flow rates. In both cases, the fluid flow rate though the grinding zone was found to depend mainly on the wheel velocity, the applied flow rate,

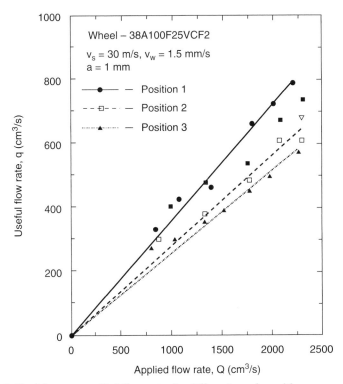

Figure 9-9 Useful versus applied flow rates for different nozzle positions.

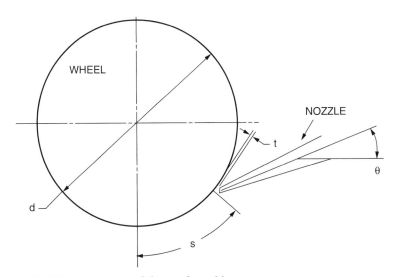

Figure 9-10 Characterization of the nozzle position.

the wheel porosity, and the nozzle position. The wheel depth of cut, workpiece velocity, and dressing conditions have little or no influence.

Various attempts have been made to analyze the fluid flow through the grinding zone. Three types of models have been developed. The first type assumes that the wheel is a smooth disk and there is a gap between the workpiece and the grinding wheel [5, 6]. But neglecting the wheel porosity or roughness does not adequately represent the actual grinding situation. The second one analyzes the fluid penetrating the grinding wheel using a shoe-type of nozzle which fits snugly on the grinding wheel periphery [7]. This model is valid only for creep-feed grinding with this particular type of nozzle. The third and more generic model, to be considered here, takes into account the porous wheel structure [8].

The analysis begins with the situation illustrated in Figure 9-1 where grinding fluid is applied from a horizontal nozzle at a velocity u_o. The fluid impinging on the rotating wheel at the fluid application zone AD will tend to infiltrate into the wheel pores. Whatever fluid flows into the wheel will be accelerated tangentially toward the peripheral wheel velocity by the permeability force exerted by the moving wheel matrix on the fluid. The high tangential velocity induced at the wheel surface will then cause a centrifugal force on the fluid which tends to force the fluid radially outward. As the fluid penetrates deeper into the wheel and the tangential velocity increases, the centrifugal force will also increase and the inward radial component of the velocity will decrease. The depth of fluid penetration will reach its maximum value when the inward radial velocity is reduced to zero, and thereafter will decrease very rapidly after the wheel matrix rotates past the fluid application zone because there will be no external pressure on the wheel surface to counteract the centrifugal force. The problem of calculating the amount of fluid passing through the grinding zone is equivalent to predicting the fluid depth of penetration h and the tangential velocity at plane OE, and the effective wheel porosity which represents the degree to which the wheel is impregnated with fluid.

When applied to typical flood application conditions, this analysis predicts that all the fluid applied from the nozzle at AD (Figure 9-1) should be ejected from the wheel before it can reach the plane OE. If fluid can enter the wheel only from the fluid application zone AD, this would imply that no fluid passes through the grinding zone. However, experimental measurements using flood application conditions (Figure 9-3 and Figure 9-5) indicate that some fluid passes through the grinding zone. When conducting these experiments, it was observed that part of the fluid which is applied to the wheel surface from the nozzle reaches the wheel and part of it falls directly on to the workpiece surface, as illustrated in Figure 9-1. That fluid which reaches and penetrates into the wheel is accelerated up to the wheel peripheral velocity and ejected from the wheel by the centrifugal force before it reaches the grinding zone, as predicted by the analysis. After exiting the wheel surface, this fluid also collects on the workpiece surface and, together with the other fluid which directly fell on the workpiece surface from the nozzle, forms a secondary fluid stream flowing toward the grinding zone. This secondary fluid stream impinges on the grinding wheel at EF in Figure 9-1, where it can now enter the pores in a manner similar to that for the original fluid application zone.

The fluid flow analysis as described above was modified to predict how much fluid from the secondary stream is carried into the grinding zone (beyond plane OE in Figure 9-1) [8]. Within the secondary fluid application zone, the fluid flows into a wedge bounded by the porous wheel surface and the workpiece surface. For simplicity, the flow in the secondary fluid application zone was modeled as a layer of fluid on the workpiece surface flowing toward the wedge. The analysis is insensitive to the thickness of this layer, although it was measured to be 1 to 3 mm. The wheel's effective porosity V_e rather than its bulk porosity V_p was used in the analysis since the fluid may not actually fill up the pores, and the actual porosity at the outer portion of the wheel to a depth of about one grain dimension is bigger than the bulk porosity. Theoretical flow rates obtained from this analysis were fitted to the experimental results in Figure 9-3. The straight lines shown in this figure were obtained by choosing appropriate V_e values, which are the implied effective porosities for each of the wheels also included in Table 9-1 together with the ratio of V_e to the bulk porosity V_p. All the creep feed wheels were found to have relatively much higher effective porosities than the conventional wheels, about 1.8 times their bulk porosities, as compared with 0.83 – 1.3 for the conventional wheels.

The fact that the effective porosity can substantially exceed the bulk wheel porosity suggests that the depth of fluid penetration at the grinding zone is much smaller than the grain dimension. From continuity considerations, the useful flow rate Q_u can be simply obtained as the product of the effective porosity V_e, the fluid penetration depth h, the grinding width b, and the wheel velocity v_s:

$$Q_u = V_e b h v_s \tag{9-1}$$

For the measured useful flow rates in Figure 9-3 and corresponding values of V_e, the penetration depths for these flood application experiments calculated from Eq. (9-1) were found to range from only 4 to 5 μm at the lowest flow rate and up to 20 to 25 μm at the highest. These penetration depths are much smaller than the grain dimension ($d_g = 220$ μm) for the 60 grit wheels used in these tests.

In principle the same fluid flow analysis described above could also be applied with some modifications to creep-feed grinding. The main problem in doing so would be to quantitatively define the fluid application zone, analogous to what is shown in Figure 9-1, and the corresponding pressure exerted by the fluid. For creep feed grinding, the fluid would be more carefully directed to the crevice between the wheel and the work-piece. The extent of contact between the fluid and the stream would depend not only on how well it is aimed at the crevice, but also on the thickness (spreading) of the fluid stream. For this reason, attempts have been made to develop nozzles for creep feed grinding which maintain parallel coherent flow with minimum spreading [9]. The problem of providing sufficient fluid becomes more difficult for creep-feed grinding of components with profiles across their width. This is especially important when fixtures, part shape, and other geometrical constraints limit the proximity between the nozzle and the grinding zone.

From the measurements of useful flow rate for creep feed grinding, such as presented in Figure 9-7, Eq. (9-1) can also be readily used to estimate the depth of fluid penetration into the grinding wheel. For this purpose, it is assumed that the effective porosity V_e for each creep feed wheel can be approximated by its bulk porosity V_p. The depths of penetration obtained in this way range from $h \approx 0.7$ mm at the lowest applied flow rate with the least porous finer grit wheel up to about $h \approx 4$ mm at the highest flow rate with the most porous coarsest grain wheel. These depths of penetration are much bigger than the grit dimensions, so assuming the bulk porosity to be equal to the effective porosity would be reasonable. It is important to note that such large fluid penetration depths are also much bigger than the thermal boundary layer thickness to be encountered [10], so there should be more than sufficient fluid to cool the workpiece at the grinding zone (see Chapter 8). Recognizing the importance of ensuring sufficient fluid penetration into the wheel for creep feed grinding, a special fluid application device was developed whereby fluid is applied by a pressurized shoe which fits snugly against the wheel to reduce fluid leakage and to force fluid into the pores [7]. The maximum depth of fluid penetration into the wheel matrix was measured by applying dyed fluid and then sectioning the wheel, but this maximum depth of penetration may not be indicative of the fluid penetration depth at the grinding zone because part of the fluid injected under pressure into the wheel surface is ejected before reaching the grinding zone.

9.5   MEASUREMENT OF HYDRODYNAMIC FORCES

Up to this point, we have considered how the applied fluid is carried through the grinding zone by the rotating porous grinding wheel. When grinding with non-porous wheels, such as resin or metal bonded diamond

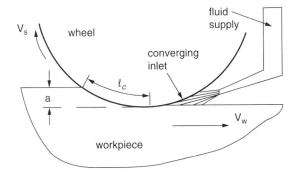

Figure 9-11 Illustration of the converging inlet region and the grinding zone of length l_c for up grinding.

and CBN wheels, a hydrodynamic force will be generated as the fluid flows into the converging inlet to the grinding zone between the rotating wheel and the workpiece as shown in Figure 9-11. The situation is analogous to what occurs with hydrodynamic fluid film bearings. The hydrodynamic force, which acts mainly in the vertical direction normal to the wheel and workpiece surfaces bounding the grinding fluid, can contribute to system deflections and adversely affect the dimensional accuracy of the component being ground.

An investigation of hydrodynamic forces began with a series of experiments [11]. The initial challenge was how to separate the hydrodynamic force from the total force. Straight surface grinding tests were performed using a resin bond 120 grit diamond wheel (ASD120-R75B56) of diameter $d_s = 305$ mm to machine a ceramic workpiece (sintered reaction bonded silicon nitride) 8.5 mm wide. (Although these tests were conducted for grinding of a ceramic, it is expected that these results concerning the hydrodynamic effects should also apply to grinding of metallic materials.) Grinding fluid consisting of 5% soluble oil in water was supplied at a flow rate of $Q = 2.2$ liters/second. For each test, three down grinding passes were taken. The first grinding pass was in a 'normal' grinding mode with a wheel velocity $v_s = 48$ m/s, depth of cut $a = 0.0254$ mm, and workpiece velocity $v_w = 20$ mm/s. The second pass was essentially a spark out pass using the same wheel and workpiece velocities, but with virtually no additional material removal due to the very high stiffness of the grinder. The third pass was then taken also without any additional downfeed, but with the wheel velocity reduced to a very low value of 0.5 m/s.

Measured vertical forces for all three passes are shown in Figure 9-12. The average vertical force obtained during Pass 1 was approximately 70 N.

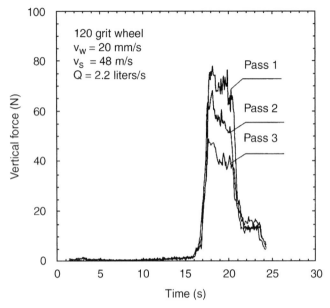

Figure 9-12 Vertical forces measured during successive grinding passes (120 grit wheel).

This force is assumed to have three components: abrasive-workpiece interaction for actual grinding; impingement of the applied fluid on the workpiece, fixture, and force dynamometer; and the hydrodynamic fluid force between the wheel and the workpiece. For Pass 2 with virtually no material removal, the abrasive-workpiece interaction force should become negligible, such that the total force can be attributed to the fluid impinging effect and the hydrodynamic effect. The force difference between Pass 1 and Pass 2 should be the force associated with actual grinding, which is 13 N in this case. For Pass 3, the hydrodynamic force should be virtually eliminated as the wheel was rotating extremely slowly, leaving only the force due to the fluid impinging effect. It was originally intended to run this final pass with zero wheel velocity in order to completely eliminate the hydrodynamic force, but the fluid pump on the CNC grinder would not operate with a wheel velocity less than 0.5 m/s. The force difference between Pass 2 and Pass 3 should provide a measure of the hydrodynamic force. For this test, the vertical force component due to impingement of the applied fluid is approximately 40 N, the force component for actual grinding is 13 N, and the hydrodynamic force is 17 N. The hydrodynamic component is bigger than the force for actual grinding.

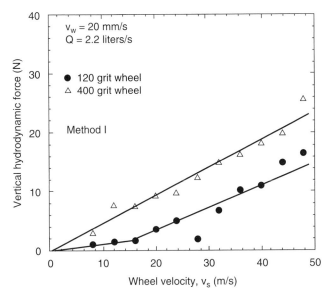

Figure 9-13 Vertical hydrodynamic force versus wheel velocity for coarser and finer grit wheel.

This same experimental procedure was also applied to a finer 400 grit wheel (DN400N100B¼) for a range of wheel velocities in Passes 1 and 2 from $v_s = 8$ m/s to $v_s = 48$ m/s. The results are summarized in Figure 9-13 for both wheels, where it can be seen that the vertical hydrodynamic force increases with wheel velocity, as expected. For the coarser 120 grit wheel, there appear to be two regimes. In the first regime up to $v_s = 16$ m/s, the force is approximately proportional to wheel velocity. In the second regime at faster velocities, the force increases linearly with wheel velocity at a steeper rate. One possible explanation for this behavior is that a full hydrodynamic film is generated above $v_s = 16$ m/s, whereas only a partial hydrodynamic film is developed at slower velocities. For the finer 400 grit wheel, the forces are bigger than with the 120 grit wheel and proportional to the wheel velocity with a slope approximately equal to that reached with the 120 grit wheel. This would suggest full hydrodynamic behavior with the finer grit wheel, which may be attributed to its 'smoother' surface. The peak-to-valley (total) roughness was about 40 μm for the coarser 120 grit wheel surface and 20 μm for the finer 400 grit wheel surface.

Another set of experiments was conducted to observe the influence of workpiece velocity on the hydrodynamic forces. Since the workpiece velocity is generally much slower than the wheel velocity, variations in workpiece velocity were expected to have little influence on the hydrody-namic forces. Experiments were conducted with the 400 grit wheel, but with a reduced flow rate of 0.5 liters per second at workpiece velocities of 2.1, 4.2 and 8.4 mm/s with wheel velocities ranging from 16 m/s to 48 m/s in the two passes. As expected, the workpiece velocity was found to have virtually no effect on the hydrodynamic forces.

The hydrodynamic force obtained as the force difference between the second and third passes, as described above, would apply to zero wheel depth of cut. For calculating the forces associated with actual grinding between the first and second passes, it was implicitly assumed that the hydrodynamic force is developed at the converging inlet to the grinding zone between the wheel and the workpiece as illustrated in Figure 9-11 for up grinding. In this case, the hydrodynamic force along the grinding zone length l_c would be negligible and, therefore, the hydrodynamic force would be independent of the wheel depth of cut a. A similar situation would apply to down grinding.

In order to check whether the hydrodynamic forces were indeed inde-pendent of the wheel depth of cut, a new test method (Method II) was developed which made it possible to observe the effect of different wheel velocities at a particular wheel depth of cut in a single experiment [11]. With the wheel running at $v_s = 48$ m/s, the specimen was ground in the up-grinding mode about half way along its length with the specified depth of cut, and then the workpiece was stopped. With the workpiece in this stopped position, the wheel velocity was first reduced to the minimum value of 0.5 m/s for 10 seconds, increased to 8 m/s for 10 seconds, and then increased in increments of 8 m/s for 10 second intervals up to $v_s = 48$ m/s. The last half of the cut was then completed with the wheel running at $v_s = 48$ m/s. Differences between forces measured at the various wheel veloci-ties and at the very slow wheel velocity of 0.5 m/s were taken as the hydro-dynamic forces for the corresponding wheel velocities. In this way, results analogous to those from the first method (Method I) but for a given wheel depth of cut could be obtained from a single experiment. An example of the vertical forces generated during one pass with this method is shown in Figure 9-14 with the 400 grit wheel for a depth of cut $a = 0.025$ mm. The first and last parts of the graph represent the force for grinding each half of the workpiece. The portions between these two represent the forces gener-ated at different wheel velocities with the workpiece stopped without any material removal.

Experimental results for the hydrodynamic forces obtained from this test method using the 400 grit wheel with four depths of cut are summa-rized in Figure 9-15. The hydrodynamic forces increase progressively with wheel velocity, but a simple proportional relationship was not obtained as with the previous method (Figure 9-13). Furthermore, it can be seen that the

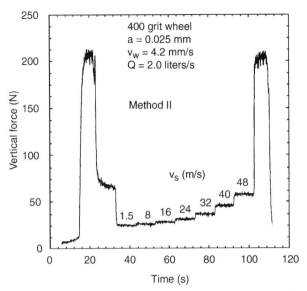

Figure 9-14 Vertical forces at various wheel velocities obtained with Method II.

vertical hydrodynamic force is insensitive to the depth of cut. This is consistent with the assumption that the hydrodynamic force is developed mainly at the converging inlet to the grinding zone. Additional experiments conducted at nozzle flow rates from 1 liter/second to 2 liters/second showed no influence of fluid flow rate.

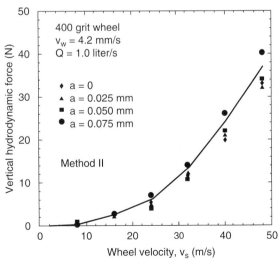

Figure 9-15 Vertical hydrodynamic force versus wheel velocity at various wheel depths of cut.

9.6 ANALYSIS OF HYDRODYNAMIC FORCES

An analysis was developed to quantitatively account for the hydrodynamic forces in grinding [12]. The hydrodynamic situation for this analysis is
illustrated in Figure 9-16 where fluid from the nozzle enters the converging
gap between the rotating wheel and the workpiece. The workpiece is approximated as being stationary because its velocity is usually negligible relative to
the peripheral wheel velocity. Since the gap thickness is much less than the
grinding width, the situation can be considered to be two-dimensional.

For an incompressible fluid of constant viscosity flowing between two
perfectly smooth rigid surfaces, the differential equation for the pressure
variation developed in a converging wedge (fluid film) can be written as [13]:

$$\frac{dp}{dx} = -12\mu\left(\frac{v_s}{2h^2} - \frac{Q}{h^3}\right) \tag{9-2}$$

where dp/dx is the pressure gradient in the x-direction, μ is the dynamic
viscosity of the grinding fluid, v_s is the wheel velocity, Q is the volumetric
fluid flow rate passing under the wheel, and h is the separation height
between the wheel and the workpiece which varies from a minimum value
h_o at $x = 0$ to a maximum value at the fluid inlet $(x = l)$. The negative sign
on the right hand side of Eq. (9-2) arises because of the definition of the coordinate system in Figure 9-16. The fluid flow rate Q is given as:

$$Q = \frac{v_s}{2}\frac{\displaystyle\int_0^l \frac{dx}{h^2}}{\displaystyle\int_0^l \frac{dx}{h^3}} + \frac{p(0) - p(l)}{12\mu \displaystyle\int_0^l \frac{dx}{h^3}} \tag{9-3}$$

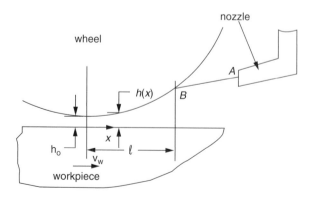

Figure 9-16 Illustration of inlet geometry.

where $p(0)$ at $x = 0$ is taken as atmospheric pressure and $p(l)$ is the pressure developed at the fluid inlet $(x = l)$. To facilitate the analysis, the circular wheel surface bounding the fluid film can be approximated as a parabola, in which case the separation height can be expressed as:

$$h = \frac{x^2}{d_s} + h_o \qquad (9\text{-}4)$$

where d_s is the grinding wheel diameter and h_o is the gap thickness corresponding to the minimum separation distance at $x = 0$.

In order to calculate the hydrodynamic force, it is necessary to obtain the pressure distribution from Equation (9-2) and integrate it from $x = 0$ to $x = l$. For this purpose, the boundary condition for the pressure at $x = l$ is approximated as the pressure head developed due to the velocity of the fluid from the nozzle:

$$p(l) = p(0) + \frac{\rho}{2}\left(\frac{Q_a}{A}\right)^2 \qquad (9\text{-}5)$$

where ρ is the density of the fluid, Q_a is the applied volumetric flow rate, and A is the cross-sectional area of the nozzle. Integrating Equation (9-2) from $x = 0$ and defining Q in terms of $p(l)$ leads to the pressure distribution:

$$p(x) = p(0) + 6\mu v_s \int_0^x \frac{dx}{h^2} - 12\mu Q \int_0^x \frac{dx}{h^3} \qquad (9\text{-}6)$$

An example of the pressure distribution is shown in Figure 9-17 for a gap thickness $h_o = 25$ μm and dynamic viscosity $\mu = 0.005$ N · s/m². A sharp pressure peak is generated near the minimum height separation towards the exit. The area under the pressure distribution curve is the force per unit width.

A critical parameter for computing the hydrodynamic force is the fluid gap length l which is the distance from $x = 0$ directly under the wheel center to the point where the fluid strikes the wheel. In Figure 9-16, point A is where the fluid exits the nozzle and point B is where it strikes the wheel. For the geometrical setup which was used, the fluid gap length l corresponding to the horizontal distance from $x = 0$ to point B was 70 mm. The orientation of the fluid flow exiting the nozzle and striking the wheel was 12 degrees from the horizontal.

In the experiments presented in the previous section, the hydrodynamic force was obtained as the measured difference between the force at the wheel velocity v_s minus that measured at the slowest velocity $v_o = 0.5$ m/s.

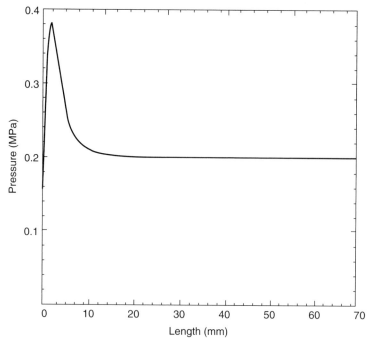

Figure 9-17 Hydrodynamic pressure distribution for $h_o = 25$ μm with wheel velocity $v_s = 48$ m/s.

Therefore for calculating the forces, the forces computed at the slowest wheel velocity $v_o = 0.5$ m/s were subtracted from the forces calculated at v_s in each case for comparison with the experimental results. For example with a wheel velocity $v_s = 48$ m/s, the force generated between the wheel and the workpiece is calculated by integrating the pressure distribution curve for a wheel velocity $v_s = 48$ m/s and subtracting the force calculated by integrating the pressure under the curve for a wheel velocity $v_o = 0.5$ m/s. Following this approach, the force per unit width at velocity v_s is obtained as:

$$F_{v_s} = p(0)l + 6\mu v_s \int_0^l \left(\int_0^x \frac{dx}{h^2} \right) dx - 12\mu Q \int_0^l \left(\int_0^x \frac{dx}{h^3} \right) dx \qquad (9\text{-}7)$$

and at velocity v_o as:

$$F_{v_o} = p(0)l + 6\mu v_o \int_0^l \left(\int_0^x \frac{dx}{h^2} \right) dx - 12\mu Q_o \int_0^l \left(\int_0^x \frac{dx}{h^3} \right) dx \qquad (9\text{-}8)$$

Subtracting Eq. (9-8) from Eq. (9-7), and multiplying by the grinding width *b,* leads to:

$$F_v = (F_{v_s} - F_{v_0})b = \mu(v_s - v_o)f(h_o) \qquad (9\text{-}9)$$

where

$$f(h_o) = b\left[12\left(\frac{Q - Q_o}{v_s - v_o}\right)\int_0^l\left(\int_0^x\frac{dx}{h^3}\right)dx - 6\int_0^l\left(\int_0^x\frac{dx}{h^2}\right)dx\right] \qquad (9\text{-}10)$$

and *Q* is given by Equation (9-3). It can be readily shown using Eq. (9-10) that $f(h_o)$ depends only on h_o for given values of v_s, v_o, μ and *l.* The hydrodynamic forces F_v computed for different gap thicknesses using a dynamic viscosity of $\mu = 0.005$ N · s/m^2 are summarized in Figure 9-18. (Selection of the appropriate viscosity μ is discussed below.)

In order to test this analytical model, experiments were conducted to measure the hydrodynamic forces developed with different set gap thicknesses, h_o, between the wheel and the workpiece (no actual grinding). For comparing the measurements with the computed forces, it is convenient to divide Eq. (9-9) by $(v_s\text{-}v_o)$, which leads to:

$$\frac{F_{v_s} - F_{v_o}}{(v_s - v_o)} = \mu f(h_o) \qquad (9\text{-}11)$$

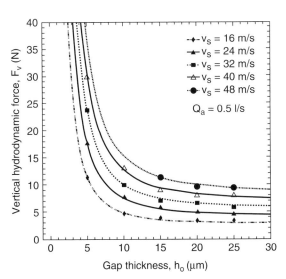

Figure 9-18 Calculated vertical hydrodynamic force versus gap thickness for $h_c = 0$.

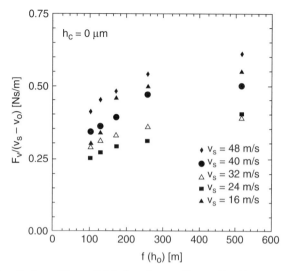

Figure 9-19 Results from Figure 9-18 plotted according to Eq. (9-11).

If the experimental results fit the hydrodynamic analysis as presented above, then a plot of $F_v/(v_s - v_o)$ (with $v_o = 0.5$ m/s) versus $f(h_o)$ should yield a straight line through the origin with a slope equal to μ. However, the experimental results plotted in this way in Figure 9-19 do not fit such a proportional relationship.

For the hydrodynamic analysis, the grinding wheel surface was assumed to be smooth, but this is certainly not true. As noted above, measured peak-to-valley roughnesses were about 20 μm for the 400 grit wheel surface and 40 μm for the coarser 120 grit wheel which was also used in previous experiments. From the data, it appears that the difference between the analytical and experimental results may be rationalized by adding a correction factor h_c to the gap thickness in order to account for the roughness of the rotating wheel surface, in which case Eq. (9-4) becomes:

$$h = \frac{x^2}{d_s} + h_o + h_c \qquad (9\text{-}12)$$

With the peak-to-valley wheel roughness as the correction factor h_c, the results obtained for the 400 grit wheel ($h_c = 20$ μm) in Figure 9-20 now fall reasonably close to the same straight line. In essence this would suggest that the hydrodynamic thickness is effectively increased by an

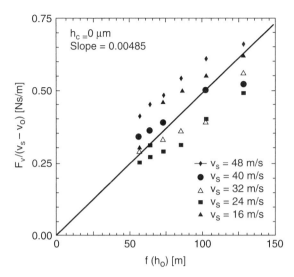

Figure 9-20 *Forces obtained with a correction factor of 20 μm plotted according to Eq. (9-11).*

amount h_c equal to the peak-to-valley wheel roughness of 20 μm. The slope of the straight line in Figure 9-20 obtained from least square fitting of the data (correlation coefficient $r = 0.90$) implies a dynamic viscosity of $\mu = 0.0048$ N · s/m². This is virtually identical to the dynamic viscosity of $\mu = 0.005$ N · s/m² for the 5% solution of soluble oil in water, which was estimated from the viscosities of the oil and the water using a simple law of mixtures.

For the hydrodynamic force measurements presented in the previous section, and for actual grinding, the set gap thickness is zero ($h_o = 0$). However approximating the remaining gap thickness by the peak-to-valley wheel roughness seems to provide better agreement between the measured and calculated hydrodynamic forces. The hydrodynamic force measurements from Figure 9-13 and Figure 9-15 are presented again in Figure 9-21 and Figure 9-22, together with the theoretical results calculated according to the analytical model with $h_o = 0$ and $h_c = 40$ μm for the 120 grit wheel and $h_c = 20$ μm for the 400 grit wheel. The hydrodynamic forces calculated in this way now seem to agree quite well with the measured forces, especially when considering the complexity of the process. This implies that the hydrodynamic forces can be estimated using this analysis for a given fluid viscosity, wheel velocity, and peak-to-valley roughness of the wheel surface.

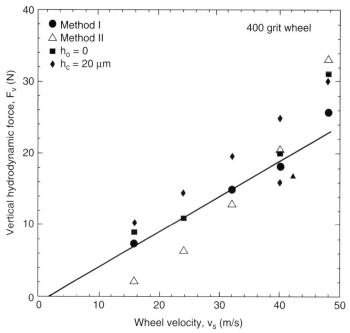

Figure 9-21 Comparison of experimental and calculated hydrodynamic forces for
$h_c = 20 \ \mu m$ with 400 grit wheel.

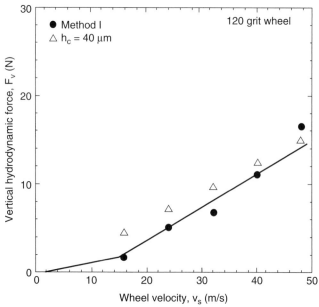

Figure 9-22 Comparison of experimental and calculated hydrodynamic forces for
$h_c = 40 \ \mu m$ with 120 grit wheel.

REFERENCES

1. Engineer, F., Guo, C. and Malkin, S., "Experimental Measurement of Fluid Flow through the Grinding Zone", *Trans. ASME, J. of Eng. for Ind.,* 114, 1992, p. 61.
2. Campbell, J. D., "Optimized Coolant Application," *Proceedings of the 1st International Machining and Grinding Conference*, SME, 1995, p. 893.
3. Campbell, J. D., "Tools to Measure and Improve Coolant Application Effectiveness," *Proceedings of the 2nd International Machining and Grinding Conference*, SME, 1997, p. 657.
4. Krishnan, N., Guo, C. and Malkin, S., 1995, "Fluid Flow Through the Grinding Zone in Creep Feed Grinding," *Proceedings of the 1st International Machining and Grinding Conference*, SME, 1995, p. 905.
5. Schumack, M. R., Chung, J. B., Schultz, W. W. and Kannatey-Asibu, E., "Analysis of Fluid Flow Under a Grinding Wheel," *Trans. ASME, J. of Eng. for Ind.,* 113, 1991, p. 190.
6. Akiyama, T., Shibata, J. and Yonetsu, S., "Behavior of Grinding Fluid in the Gap of the Contact Area between a Grinding Wheel and a Workpiece — A Study on Delivery of the Grinding Fluid", *Proceeding of the 5th International Conference on Production Engineering,* Tokyo, 1984, p. 52.
7. Powell, J. W., *The Application of Grinding Fluid in Creep Feed Grinding*, PhD Dissertation, University of Bristol, 1979.
8. Guo, C. and Malkin, S., "Analysis of Fluid Flow through the Grinding Zone", *Trans. ASME, J. of Eng. for Ind.,* 104, 1992, p. 427.
9. Webster, J. A., Cui, C. and Mindek, R. B., Jr., "Grinding Fluid Application System Design", *Annals of the CIRP*, 44/1, 1995, p. 333.
10. Lavine, A. S., 1988, "A Simple Model for Convective Cooling during the Grinding Process", *Trans. ASME, J. of Eng. for Ind.*, 110, 1988, p. 1.
11. Ganesan, M., Guo, C. and Malkin, S., "Measurement of Hydrodynamic Forces in Grinding", *Transactions of NAMRI/SME*, 22, 1995, p. 103.
12. Ganesan, M., Ronen, A., Guo, C. and Malkin, S., "Analysis of Hydrodynamic Forces in Grinding," *Transactions of NAMRI/SME*, 23, 1996, p. 105.
13. Schlichting, H., *Boundary Layer Theory*, McGraw Hill, New York, 1955.

Chapter 10

Surface Roughness

10.1   INTRODUCTION

Grinding processes are often selected for final finishing of components because of their ability to satisfy stringent requirements of surface roughness and tolerance. Surface roughness and tolerance are closely interrelated, as it is generally necessary to specify a smoother finish in order to maintain a finer tolerance in production. For many practical design applications, it is the tolerance requirement which imposes a limit on the maximum allowable roughness, although the proper operation of many devices also necessitates smooth surfaces.

The reliability of mechanical components, especially for high strength applications, is often critically dependent upon the quality of the surface produced by machining. Surface quality may be considered to consist of two aspects: surface integrity and surface topography [1]. Surface integrity is associated with mechanical and metallurgical alterations to the surface layer induced by machining. For grinding, the most important aspects of surface integrity are associated with thermal damage caused by excessive grinding temperatures, as we saw in Chapter 6. Surface topography refers to the geometry of machined surfaces, which is usually characterized by surface roughness, although there are other parameters which may also be of interest.

The present chapter is mainly concerned with surface roughness in grinding. We begin by observing the distinctive morphological features of ground surfaces. Quantitative characterization of surface topography is then briefly reviewed, and the possible basis for a direct interrelationship between roughness and tolerance is assessed. Various models are then considered for describing the generation of the 'ideal' ground surface topography to theoretically predict the workpiece roughness in terms of the wheel topography and the operating parameters. This provides a certain degree of insight into how various factors might affect the ground surface topography. But theoretical surface roughness relationships are found to be of limited practical use, and it is generally necessary to use empirical relations to assess the relative influence of the operating parameters.

10.2   GROUND SURFACE MORPHOLOGY

The fine-scale morphology of the surfaces generated by grinding consists mostly of overlapping scratches produced by the interaction of abrasive cutting points with the workpiece. An example of a typical ground surface is shown in the scanning electron microscope (SEM) photograph in Figure 10-1. For this example of straight plunge grinding, as with other types of grinding, the grit motion relative to the workpiece is readily identified from the directionality of the scratches and grooves. Sideways displacement of material from some scratches by plowing is also evident.

Scratches on the finished surfaces correspond to only the bottom portions of the cutting paths of the outermost cutting points on the wheel surface (Chapter 3). Here the undeformed chip thickness is considerably smaller than the maximum grit penetration at the top of the cutting path and is more likely to be less than the critical minimum depth for chip formation, so plowing is more likely to occur (Chapter 5). The degree of sideflow

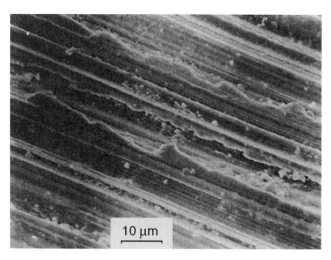

Figure 10-1 SEM micrograph of a medium carbon steel workpiece surface after straight surface grinding with a vitrified aluminum oxide wheel (32A60I8VBE) and soluble oil.

plowing depends upon the particular workpiece material being ground. Metals which are more adhesive, such as titanium, nickel-base alloys, and austenitic stainless steels, tend to exhibit more sideways flow. Conversely, grinding fluids which are more effective in lessening grit-workpiece adhesion by lubrication reduce plowing [2].

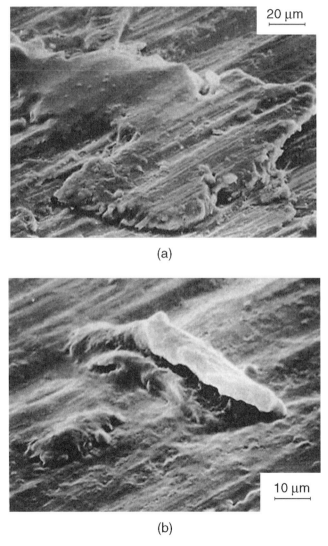

(a)

(b)

Figure 10-2 SEM micrographs of a titanium alloy (Ti-6Al-4V) surface after straight surface grinding with a vitrified aluminum oxide wheel (32A60I8VBE) and soluble oil.

The ground surface morphology is further complicated by numerous other phenomena. Back transfer of workpiece metal often occurs, especially with adhesive metals, whereby metal particles adhering to the abrasive grits are redeposited on the workpiece [2,3]. An example of this behavior is seen in Figure 10-2(a) for grinding of a titanium alloy. Interruption of the cutting action by fracture of the abrasive grit may leave a crater on the workpiece, as in Figure 10-2(b), possibly with an abrasive fragment embedded in the surface. When grinding steels, craters are more frequently observed at the start of grinding after wheel dressing when the rate of wheel wear by grit fracture is more rapid, and also with coarser dressing conditions [3]. Some difficult-to-grind metals, including titanium and austenitic stainless steels, seem to exhibit more extensive cratering and grit embedding. This type of surface damage provides a source of localized stress concentrations, which can be expected to have an adverse effect on in-service strength and fatigue properties.

The topography of surfaces produced by grinding can be recorded by surface profiles taken along and across the grinding direction (lay), as shown in Figure 10-3 for straight surface grinding of a mild steel [4]. Adjacent peaks and valleys within the profile along the grinding direction are much more widely spaced apart than across the grinding direction. Spectral (autocorrelation) analysis of these profiles revealed a predominant wavelength (correlation length) of 0.25 mm along the lay as compared with only 0.034 mm across the lay.

10.3   SURFACE TEXTURE AND TOLERANCE

The characteristic patterns of peaks and valleys on the finished workpiece are known as surface texture. As a basis for quantitatively describing the texture of ground surfaces, and in order to avoid confusion, the characterization of surface texture will be briefly reviewed. For this purpose, we will refer to the American standard for characterization of surface texture [5], which is also in substantial agreement with British (BS), German (DIN), and international (ISO) standards.

The concept of surface texture is illustrated in Figure 10-4 [5]. Most machined surfaces exhibit a predominant lay coinciding with the direction of cutting-tool motion relative to the surface, as seen in the previous section. The surface profile for the cross-section in Figure 10-4 is considered to have components of roughness and waviness. Roughness is associated with closely spaced perturbations which are superposed on waviness components more widely spaced apart. Surface flaws, such as cracks, adhered metal, and craters, also contribute to the surface texture.

Surface texture is commonly measured by a stylus instrument which traces the profile of the surface (e.g. Figure 10-3). For quantitative assessment

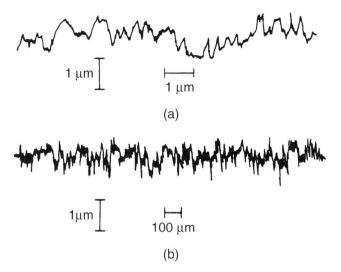

1 μm

1 μm

(a)

1μm

100 μm

(b)

Figure 10-3 Profiles of a medium carbon steel surface after grinding with a 60-grit vitrified aluminum oxide wheel: (a) along the lay, and (b) across the lay [4].

of surface roughness, a sampling length along the profile (Figure 10-4) is selected, long enough to include a representative number of roughness per-turbations but shorter than the waviness spacing. Therefore, the sampling length is the basis for discriminating between roughness and waviness. The sampling length is also referred to as the cut-off length, as some stylus instruments electronically filter (cut off) longer wavelengths for analyzing surface roughness. A typical sampling length for roughness measurements on ground surfaces is 0.8 mm. Each roughness measurement with a typical stylus instrument is taken over a number (usually five) of successive sam-pling lengths representing the total traversing (assessment) length.

A number of roughness parameters are defined from the surface pro-file, but for our purposes we will start with only two: arithmetic average roughness and peak-to-valley roughness. The arithmetic average roughness R_a is the mean value of the average deviation of the surface profile from the centerline (mean line) in each sampling length. The peak-to-valley rough-ness R_t, also referred to as the total roughness, is defined as the difference in elevation between the highest peak and lowest valley in the traversing length.

It is apparent that R_t should be substantially bigger than R_a. For a per-fect sinusoidal profile, it can be readily shown that $R_t = \pi R_a$. For grinding, differences between R_t and R_a are considerably bigger, which is a conse-quence of the broad distribution of peak heights and valley depths within the surface profile. The profile height distribution of ground surfaces

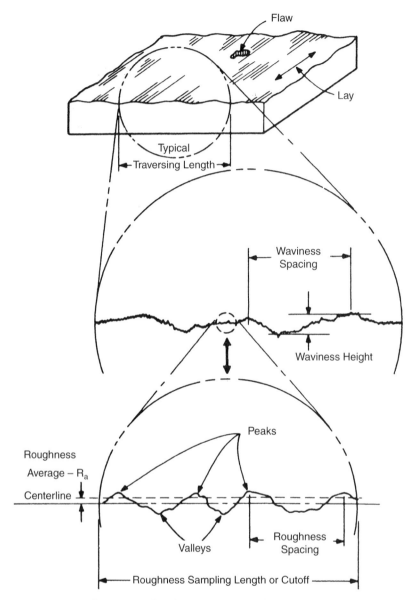

Figure 10-4 Illustration of surface texture according to ANSI Standard [5].

appears to be nearly Gaussian [6], and the peak and valley height distributions might also be approximately Gaussian. For ground surfaces, the R_t roughness is typically 7-14 times R_a [7]. However somewhat smaller ratios of 4-7 are obtained if the extreme values for the highest peak and lowest valley defining the peak-to-valley roughness are smoothed out either by ignoring unusually high peaks and low valleys in the profile [8,9] or by taking the elevation differential between the average of the five highest peaks and five lowest valleys in the traversing length (10-point roughness) [7].

The heights associated with the waviness component of the profile can be obtained for perturbations identified as having characteristic spacings longer than the cut-off length. Waviness is caused mainly by grinding vibrations. There are generally two types of vibrations: forced and self-excited [10-14]. Forced vibrations arise from external vibration sources, such as an unbalanced wheel or other rotating elements, and the frequency coincides with that of the vibration source or some harmonic thereof. The generation of waviness by forced vibration of an unbalanced wheel is illustrated in Figure 10-5 for straight surface grinding. This is likely to cause a visible pattern of successive straight lines or bands across the workpiece width spaced apart along the grinding direction by the wavelength:

$$\lambda = \frac{v_w}{f} \tag{10-1}$$

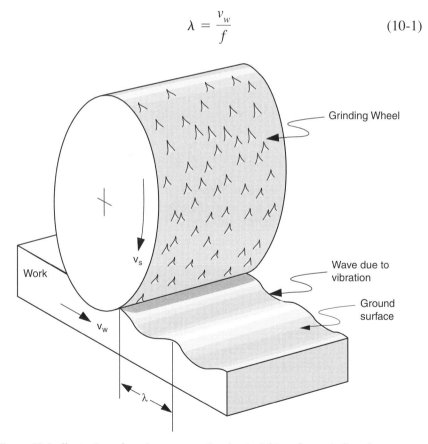

Figure 10-5 *llustration of waviness generation in straight surface grinding due to forced vibration. Adapted from Reference [3].*

where v_w is the workpiece velocity and f is vibration frequency which in this case corresponds to the wheel rotational frequency. Such waviness may be difficult to measure by profilometry.

Self-excited vibrations are caused by a regenerative effect whereby a fundamental instability of the machine is dynamically excited during successive workpiece revolutions (see Chapter 12). Regenerative chatter builds up progressively by the growth of waves (lobes) around the wheel and workpiece peripheries. The vibration frequency is usually much higher than with forced vibrations, so the characteristic wavelength on the workpiece, which is also given by Eq. (10-1), is proportionally shorter. In many practical cases, lobing of the workpiece is significantly attenuated by a mechanical filtering effect because the wavelength is less than the arc length of contact at the grinding zone. Inhomogeneities in the wheel structure increase the tendency for self-excited vibrations and a mottled workpiece appearance [15].

Forced and self-excited vibrations may also occur during dressing, producing irregularities in the wheel shape and workpiece waviness [12,16,17]. With single-point dressing, forced vibrations due to wheel unbalance are especially problematic when the wheel unbalance during dressing is different from that during grinding [16]. Self-excited vibrations may occur during rotary diamond dressing resulting in wheel lobing, and this was one of the main difficulties encountered with its initial introduction as an alternative to fixed-point dressing. During subsequent grinding, vibrations at the wheel lobing frequency cause workpiece waviness.

Surface topography is often the main factor limiting the tolerance which can be obtained in production. In general, the tolerance represents the acceptable deviation from the nominal intended dimension or geometrical form. Dimensions on machined components are measured from one surface to another or, in the case of diametrical measurements, between two opposite locations on the same cylindrical surface. The roughness of the finished surface can be thought of as a measure of the uncertainty in exactly specifying the location of the surface, as illustrated in Figure 10-6 for a cylindrical component, which means that the dimensional uncertainty depends upon the combined surface roughnesses at the measuring points. Therefore, it is generally necessary to have smoother surfaces in order to maintain tighter tolerance control. The surface roughness requirement is often a consequence of the dimensional tolerance requirement, and both factors are similarly affected by the grinding conditions. There are also a host of other factors which contribute to poor tolerance, including machine deflection (Chapter 12), thermal expansion and distortion of the machine and workpiece [18-25], and wheel wear (Chapter 11). In centerless grinding, unstable geometric and kinematic conditions cause workpiece lobing, and the situation is further aggravated by machine-tool vibrations [26-31].

Typical arithmetic average surface roughnesses for production grinding operations range from about 0.15 μm to 1.5 μm, although finishes

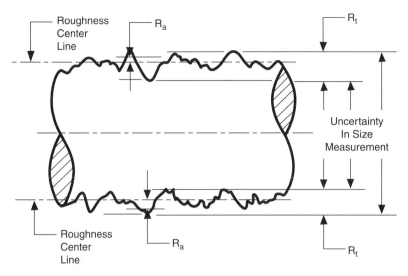

***Figure 10-6** Illustration of the interrelationship between roughness and tolerance. Adapted from Reference [5].*

outside this range are not uncommon. Corresponding dimensional tolerances specified in production are normally 10 to 50 times the arithmetic average roughness. This ratio will depend upon such factors as the machine-tool condition, allowable rejection rate, assembly requirements, and component size. One way to effectively maintain tolerances at the lower end ($\approx 10R_a$) is by selective grouping of nominally identical machined parts according to size, so that the tolerance within each group is much smaller than that of the entire lot. Selective grouping is commonly applied to rolling element bearing components.

   Many of the same factors which affect dimensional tolerances similarly affect form tolerances, since form is also specified in terms of linear as well as angular dimensions. However, the most significant form errors in grinding are usually caused by wheel wear, especially for profile (form) grinding of cross-sectional shapes having sharp radii or deep grooves. Better form control usually requires a slower wearing wheel, but this can be expected to cause bigger forces (Chapter 11). Form errors are also associated with elastic deflections during grinding (Chapter 12).

10.4   IDEAL SURFACE ROUGHNESS

   As with other machining processes, it is possible to theoretically predict an 'ideal' surface roughness for grinding by modeling how the abrasive cutting points on the rotating wheel kinematically interact with the workpiece.

For this purpose, the surface topography is assumed to be generated by clean cutting, whereby the cutting edges remove all material that they encounter in their paths, leaving behind the resulting cutting grooves. For machining processes with well defined cutting tool geometry, including turning and milling, the analysis is straightforward. The 'ideal' roughness is usually found to be less than the actual roughness due to such factors as material sideflow, built-up-edge phenomena, and vibrations. For grinding, the 'ideal' roughness is much more difficult to model owing to the random undefined topography of the wheel surface and cutting points (Chapter 3), so it is necessary to introduce some simplifying assumptions.

As a first step, let us consider what happens with an idealized wheel having cutting points equally spaced a distance L apart around the wheel periphery and radially protruding to the same height. This same uniform wheel topography was considered for the initial estimation of undeformed chip thickness in Chapter 3. The 'ideal' longitudinal surface profile generated by straight surface grinding with such a wheel, as illustrated in Figure 10-7, consists of successive identical scallops each with a radius of curvature corresponding to that of the cutting path. For practical grinding situations with the workpiece velocity v_w much less than the wheel velocity v_s, it was seen in Chapter 3 that the radius of the trochoidal path for straight surface grinding can be approximated by the wheel radius ($d_s/2$). Analogous to plain horizontal milling which would give the same profile geometry, the spacing between successive scallops along the workpiece is the feed per cutting point s_c which is given by:

$$s_c = \frac{v_w L}{v_s} \tag{10-2}$$

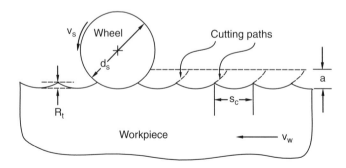

Figure 10-7 Illustration of an ideal longitudinal surface profile generated by a uniform wheel topography.

For the profile in Figure 10-7, the peak-to-valley roughness is

$$R_t = \frac{s_c^2}{4d_s} \tag{10-3}$$

which combined with Eq. (10-2) leads to:

$$R_t = \frac{1}{4}\left(\frac{v_w L}{v_s d_s^{1/2}}\right)^2 \tag{10-4}$$

The arithmetic average roughness for this profile is

$$R_a = \frac{1}{9\sqrt{3}}\left(\frac{v_w L}{v_s d_s^{1/2}}\right)^2 \tag{10-5}$$

which corresponds to $R_t = 3.9R_a$.

According to Eqs. (10-4) and (10-5), the roughnesses parameters R_t and R_a depend mainly on the velocity ratio v_w / v_s and the cutting-point spacing L, and to a lesser degree on the wheel diameter d_s. It is interesting to note that the wheel depth of cut a has no effect provided that $a > R_t$, which is also apparent from the cutting-path trajectories in Figure 10-7. For typical grinding conditions, Eqs. (10-4) and (10-5) give unrealistically low roughness values. Taking, for example, $v_w / v_s = 0.01$, $d_s = 200$ mm, and $L = 1$ mm, the roughnesses are $R_t \approx 1.25 \times 10^{-4}$ µm and $R_a = 3.25 \times 10^{-5}$ µm. The smallest roughnesses obtained in production grinding are about 1000 times bigger!

One factor contributing to the large discrepancy between the 'ideal' and actual roughnesses is the radial cutting-point distribution on the wheel surface, as was discussed in Chapter 3 [9, 32]. From the longitudinal 'ideal' surface profile illustrated in Figure 10-8 for cutting points protruding to differing heights, it is apparent that the peak-to-valley roughness is

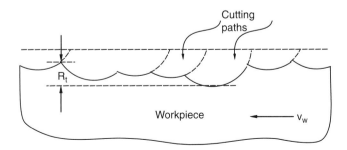

Figure 10-8 Illustration of an ideal longitudinal surface profile generated by a non-uniform wheel topography.

increased over the uniform wheel surface at least by the maximum height differential between those cutting points whose trajectories generate the 'ideal' profile. Therefore, the peak-to-valley roughness would seem to include a kinematic component depending on the grinding parameters plus an additional wheel-roughness component, although these two contributions should not be completely independent of each other. Other expressions for 'ideal' roughness will be seen to also include kinematic and wheel-roughness terms.

Only the outermost of the active cutting points on the wheel surface actually generate the ground profile, insofar as their cutting paths remove surfaces produced by preceding cutting points which protrude less according to the kinematic condition expressed by Eq. (3-81). With smaller v_w/v_s velocity ratios, particularly in creep-feed grinding, a further reduction in the number of cutting points contributing to the 'ideal' workpiece profile also occurs owing to overlapping of scratches produced by the most protruding cutting points with successive wheel revolutions [3,33,34]. In this case, the ground surface appears to consist of more nearly continuous grooves rather than discrete scratches.

An 'ideal' longitudinal workpiece surface profile, like that in Figure 10-8, can be generated by computer simulation of the relative wheel and workpiece motion for a measured circumferential wheel profile. As with the simplified case of uniformly protruding cutting points equally spaced apart, a bigger velocity ratio should increase the workpiece roughness, while the wheel depth of cut should have no effect provided that $a > R_t$. In one case [35], the influence of wheel topography and workpiece velocity on the simulated 'ideal' roughness was reported to be similar to that actually found, but the simulated 'ideal' roughness was still much smaller than the actual roughness by one or two orders of magnitude. This discrepancy may be associated with plowing effects and the lack of clean cutting action. It has been suggested that an additional contribution to the longitudinal roughness from this effect can be estimated by assuming that the material in the cutting paths to be removed at a depth of cut below the critical value (see Chapter 5) remains in place on the workpiece surface [36]. However the actual situation is much more complex, involving sideflow plowing into ridges which also adds to the workpiece roughness [37, 38]. On the other hand, elastic deflection of the grits and rubbing of wear flats over the workpiece may improve the surface finish.

Up to this point we have considered the longitudinal workpiece profile along the grinding direction, but it is the transverse roughness across the grinding direction which is usually measured. The surface roughness across the lay is usually somewhat bigger, owing to the relative magnitudes of the sampling (cut-off) length and characteristic wavelength in each direction. The typical sampling length of 0.8 mm may be comparable to, or only a few times longer than, the characteristic wavelength of the longitudinal roughness component. For the particular example in Figure 10-3, the characteristic wavelength was found to be 7-8 times bigger along the grinding direction than across it, and the roughness was $R_a = 0.26$ μm along the lay and $R_a = 0.30$ μm across the lay.

It is conceptually more difficult to physically model or simulate the 'ideal' surface profile across the lay than along it. One simple model assumes a transverse profile like that of the longitudinal one in Figure 10-7, but with the radius of curvature of the scallops corresponding to the tip radius of the abrasive grits equally spaced across the wheel and each protruding to the same height [9]. A subsequent analysis also takes into account the radial distribution of the active grain tips, which are assumed to correspond to the active cutting points, leading to an expression for the peak-to-valley roughness [9]:

$$R_t = a_1 \left(\frac{v_w}{v_s (\rho d_s)^{1/2}} \right)^{1/2} + b_1 \tag{10-6}$$

where a_1 and b_1 are parameters defined from the measured radial distribution of active cutting points (grain tips) and ρ is the grain-tip radius which is assumed to correspond to half the abrasive grain dimension (spherical grits). Smaller values of a_1 and b_1 correspond to a greater accumulation of cutting points per unit area with radial depth into the wheel, which in turn leads to a smoother ideal roughness. As in Eqs. (10-4) and (10-5), the kinematic contribution to the 'ideal' roughness in Eq. (10-6) depends on the combination of grinding parameters $v_w / v_s d_s^{1/2}$ and is independent of depth of cut. The best possible peak-to-valley roughness would be equal to b_1. In view of the results in Chapter 3, the assumption of spherical cutting points and grits is highly questionable, although the predicted roughness is of the right order of magnitude.

In a number of other studies [32,36,39-49], the 'ideal' transverse surface profile has been generated by computer simulation for measured transverse wheel profiles or statistical models thereof. In the former case, an actual workpiece surface profile is generated, while in the latter a statistical model of the profile is obtained. Each subsequent transverse wheel profile passing a particular location on the workpiece is considered to remove material in its path left behind by previous profiles, as illustrated in Figure 10-9, thereby making the surface progressively smoother. For computational simplicity, the outer points on successive wheel profiles are assumed to protrude to the same height. In one particular computer simulation, the smoothing effect followed an approximate relationship of the form:

$$R_a = \frac{R_o}{i} + R_\infty \tag{10-7}$$

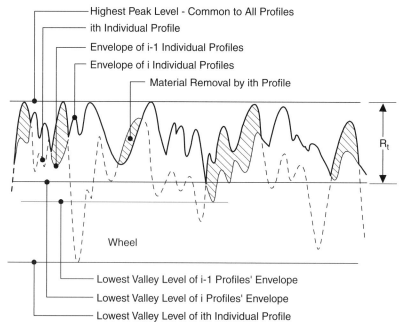

Highest Peak Level - Common to All Profiles
ith Individual Profile
Envelope of i-1 Individual Profiles
Envelope of i Individual Profiles
Material Removal by ith Profile

R_t

Wheel

Lowest Valley Level of i-1 Profiles' Envelope
Lowest Valley Level of i Profiles' Envelope
Lowest Valley Level of ith Individual Profile

Figure 10-9 Ideal workpiece envelope profile across the grinding direction generated by successive transverse wheel profiles. Adapted from Reference [44].

where i is the number of wheel profiles, and R_o and R_∞ are empirical constants. For a given wheel topography, the number of profiles contributing to the workpiece profile would depend on the grinding parameters. Assuming that there is an effective circumferential length of wheel surface L_e contributing to the transverse workpiece profile at any location, the number of profiles can be expressed as

$$i = \frac{v_s}{v_w} \frac{L_e}{L}$$ (10-8)

where L is the spacing between successive active cutting points or profiles. It was suggested that L_e might correspond to the wheel workpiece arc length of contact for a depth of cut a [36,41,42], but it is apparent from kinematic considerations (e.g. Figures 10-7 and 10-8) that the depth of cut should not affect the 'ideal' roughness. The arc length of contact for a wheel depth of cut equal to R_t would appear to be a more reasonable choice [39], in which case

$$i = \frac{v_s}{v_w} \frac{(R_t d_s)^{1/2}}{L}$$ (10-9)

Assuming a fixed ratio between R_t and R_a (i.e. $R_t = mR_a$), the arithmetic average surface roughness obtained by combining Eqs. (10-7) and (10-9) within a practical range of roughness values can be approximated by

$$R_a \approx \left(\frac{R_o}{m^{1/2}}\right)\left(\frac{v_w L}{v_s d_s^{1/2}}\right)^{0.8} + R_\infty \qquad (10\text{-}10)$$

Neglecting any influence of the grinding parameters on L, the kinematic contribution to the roughness in Eq. (10-10) again depends on $v_w/v_s d_s^{1/2}$, with greater sensitivity than in Eq. (10-6) and less than in Eqs. (10-4) and (10-5).

In the limiting case, as the number of profiles becomes very large ($i \to \infty$), the roughness given by Eq. (10-7) or (10-10) approaches R_∞. This situation would correspond to complete spark-out at the end of the grinding cycle (see Chapter 12) where the rotating wheel repeatedly passes over the workpiece without any additional downfeed increment. The actual roughness measured after spark-out was found to be reasonably close to the 'ideal' value R_∞ for the particular case which was simulated [44]. In this sense, the improvement in surface finish by spark-out is attributable to more wheel profiles passing the same location on the workpiece. The parameter R_∞ is the best possible finish which can be obtained with this wheel topography. In order to obtain a smoother finish, it is necessary to have a smoother wheel surface, which would require finer dressing.

Predictions of 'ideal' surface roughness (Eqs. (10-4)–(10-6), and (10-10)) were derived for straight surface grinding without crossfeed. It can be readily shown that these relationships would be more generally applicable to also include external and cylindrical plunge grinding if the wheel diameter d_s is replaced by the equivalent diameter d_e (Eq. (3-8)). Transverse roughness analyses can also be readily extended to include traverse grinding with crossfeed for the ideal case of a non-wearing wheel having the same topography across its width. With crossfeed, additional smoothing of the workpiece can be considered to occur by successive passes of adjacent wheel sections over the workpiece, as seen in Figure 3-7. As compared with plunge grinding, the number of profiles i in Eq. (10-9) would be multiplied by a factor called the 'overlap ratio', which is equal to the ratio of the wheel width b_s to crossfeed s_t in Figure 3-7. Introducing this effect into Eq. (10-10), the ideal surface roughness for grinding with crossfeed would be

$$R_a = \left(\frac{R_o}{m^{1/2}}\right)\left(\frac{v_w L}{v_s d_e^{1/2}} \frac{s_t}{b_s}\right)^{0.8} + R_\infty \qquad (10\text{-}11)$$

Aside from the combination of grinding parameters $v_w/v_s d_e^{1/2}$, and also the overlap ratio for grinding with crossfeed, additional parameters needed for calculating the 'ideal' roughness are related to wheel topography.

However, this information is not readily available. In order to overcome this problem, attempts have been made to describe the grinding wheel topography in terms of the dressing parameters, and subsequently predict the 'ideal' surface roughness in terms of both the grinding and the dressing parameters [41,42,50-52]. For this purpose, a single-point dressing tool is assumed to cut a clean thread with a pitch equal to the dressing lead into a solid (non-porous) wheel, even though the wheel is certainly not solid and, as seen in Chapter 4, material is removed during dressing by fracturing the abrasive grits and dislodging them from the binder in a way that cannot produce a wheel topography conforming to a pure threaded shape. However, many points on the wheel surface should coincide with the locus of the dressing diamond path, and this can account for a component in the transverse workpiece profile having a characteristic wavelength equal to the dressing lead [45,53].

For plunge grinding without crossfeed, the 'ideal' peak-to-valley roughness along the grinding direction for a wheel with a single dressed helix is predicted as [50,51]:

$$R_t = \frac{\pi^2}{4} \frac{d_s}{d_e} \left(\frac{v_w}{v_s} \right)^2 \tag{10-12}$$

which is independent of the dressing conditions. With the possible exception of some internal and creep-feed situations, this result would not apply because the predicted workpiece roughness exceeds the wheel roughness, which is the maximum limiting workpiece roughness. A statistical model of the transverse wheel surface profile takes into account the influence of subsequent spark-out dressing passes each having the same lead but randomly phased relative to each other [41,42]. A complex relationship obtained for the transverse workpiece profile and roughness predicts that smoothing of the wheel surface by increasing the number of spark-out dressing passes across the wheel should have a more significant influence on reducing the surface roughness than an increase in the number of successive wheel profiles (see Figure 10-9). For grinding with crossfeed, an ideal workpiece profile has also been generated by superposing subsequent passes by adjacent sections of the wheel each having the same single helix (saw-tooth) profile [52]. Complex expressions for the 'ideal' roughness indicate smoother surfaces with finer dressing, a slower workpiece velocity, and a bigger overlap ratio.

10.5   EMPIRICAL ROUGHNESS BEHAVIOR

Theoretical analyses of 'ideal' surface roughness provide physical insight into how ground surfaces are generated and what may be the controlling factors. However, the relationships presented in the previous section are found to be of limited practical use for predicting how the grinding and dressing parameters affect the actual surface roughness, and for this purpose it is generally necessary to rely upon empirical relationships.

One starting point for empirically relating surface roughness to the grinding parameters has been to assume a direct correlation with the undeformed chip thickness [54,55]. In Chapter 3 it was seen that the undeformed chip thickness for a given wheel topography and equivalent diameter depends on both the velocity ratio v_w/v_s and the wheel depth of cut a combined together as $(v_w/v_s)a^{1/2}$. From kinematic considerations of surface generation in the previous section, there is no apparent reason why wheel depth of cut should have any effect on surface roughness, although it could influence wheel wear thereby altering the wheel topography and the corresponding workpiece surface roughness. For straight surface grinding without spark-out, the measured surface roughness is generally found to increase with v_w/v_s, but a larger depth of cut is usually also found to have an influence but to a somewhat lesser degree [8,34,56,57]. For the purpose of comparing conventional and creep-feed grinding in Chapter 7 (Figure 7-12), we assumed that the surface roughness depends upon $(v_w/v_s)a^{1/2}$, which is equivalent to assuming dependence upon the undeformed chip thickness. In any case, the relatively greater sensitivity of surface roughness to workpiece velocity rather than to wheel depth of cut is an especially important factor in creep-feed grinding, because it enables smooth finishes to be maintained at higher removal rates.

For cylindrical plunge grinding, experimental surface roughness data tend to follow a relationship of the form [58,59]:

$$R_a = R_1\left(\frac{Q'_w}{v_s}\right)^x = R_1\left(\frac{v_w a}{v_s}\right)^x \qquad (10\text{-}13)$$

where Q'_w is the volumetric removal rate per unit width ($Q'_w = v_w a$), and R_1 and x are experimentally determined constants. The exponent x is typically in the range $0.15 < x < 0.6$. The quantity within the parentheses in Eq. (10-13) is the equivalent chip thickness h_{eq} which was introduced in Chapter 5.

Unlike straight surface grinding, the velocity ratio and wheel depth of cut in Eq. (10-13) have the same relative influence on surface roughness. However, it should be noted that the surface roughness for cylindrical grinding is actually obtained after some degree of spark-out, so that the final depth of cut to which the workpiece is subjected is very much less than that input to the machine. With external cylindrical grinding, for example, it is necessary to retract the wheel instantaneously to avoid any spark-out, in which case the workpiece, instead of being round, would show a spiral surface feature with a radial step height equal to the wheel depth of cut (Figure 3-1(b)). After stopping the infeed control prior to retracting the wheel, spark-out occurs while material continues to be removed at a diminishing rate as the elastic deflection of the grinding system is recovered. An analysis of this phenomenon and the effect on out-of-roundness is presented in Chapter 12. During spark-out, the helical feature tends to disappear and the workpiece becomes more nearly round as the wheel depth of cut decreases, and this is accompanied by an improvement in surface finish. Complete spark-out, which is considered to have been reached prior to wheel retraction if the removal rate diminishes to zero and the grinding sparks cease, reduces the surface roughness in cylindrical grinding by about half [58,59]. Comparable improvements by spark-out might be expected for straight surface grinding.

Up to this point, we have considered only the grinding parameters. Perhaps of even greater significance are the dressing parameters. Finer dressing generally produces a more even and smoother wheel surface, so the ground workpiece roughness is also smoother. For rotary diamond dressing, it was pointed out in Chapter 4 that the dressing severity depends on the angle δ at which the diamonds on the rotary dresser surface initially cut into the wheel, which is readily expressed in terms of the dresser infeed and velocity (Eq. (4-9)). A larger value of δ causes more grit fracture, less plastic deformation of the abrasive, and a rougher wheel surface. The surface roughness has been found to be proportional to $\delta^{1/3}$ [60], as seen in Figure 10-10. Combining this effect with Eq. (10-13) suggests a surface roughness dependence

$$R_a = R_2 \delta^{1/3} \left(\frac{Q'_w}{v_s} \right)^x \qquad (10\text{-}14)$$

where R_2 is a new empirical constant. An additional complication is introduced by spark-out during rotary diamond dressing, analogous to grinding spark-out, whereby the dressing action continues after the rotary dresser infeed is stopped. During dressing spark-out the wheel becomes smoother, and a corresponding reduction in workpiece surface roughness is obtained for longer dressing spark-out times [60].

A similar situation also prevails with single-point diamond dressing. A finer dressing lead and/or smaller dressing depth produce a smoother wheel surface (Chapter 4), and consequently a smoother workpiece. The influence of dressing lead s_d and dressing depth a_d on surface roughness [61], combined with the surface roughness relationship of Eq. (10-13), suggests that

$$R_a = R_3 s_d^{1/2} a_d^{1/4} \left(\frac{Q'_w}{v_s} \right)^x \qquad (10\text{-}15)$$

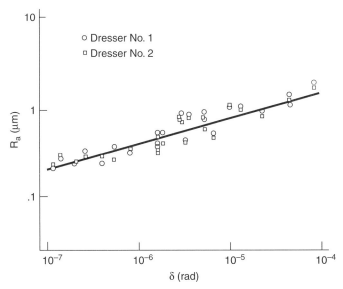

Figure 10-10 Surface roughness versus rotary dressing parameter δ: straight surface grinding, AISI 1090 steel workpiece, 32A46I8VBE wheel, v_s = 32 m/s, v_w = 13 m/min, a = 25 μm [60].

where R_3 is an empirical constant. The dressing lead has a much bigger influence than the dressing depth. Measured roughness values for internal grinding of bearing cups over a wide range of removal rates and dressing conditions, which were correlated with a semi-empirical relationship for undeformed chip thickness [62], can also be shown to follow a relationship like that of Eq. (10-15) but with somewhat smaller exponents for the dressing parameters. Other results for conventional and creepfeed surface grinding suggest slightly less sensitivity of surface roughness to dressing lead than in Eq. (10-15) [34,57], whereas other results for external cylindrical grinding suggest slightly greater sensitivity [63]. While the influence of dressing parameters on surface roughness in Eq. (10-15) was obtained for single-point diamond dressers, it is also found to work reasonably well for multipoint diamond dressing tools. As with rotary dressing, additional spark-out passes by a fixed dressing tool at the end of the dressing operation smooths the wheel and reduces workpiece roughness.

While a better surface finish can be produced by resorting to finer dressing conditions, this will cause the wheel to be duller, thereby raising the grinding power and specific energy and increasing the risk of thermal damage (Chapter 6). This behavior is illustrated in Figure 10-11 which shows a trade-off between specific energy and surface roughness as the

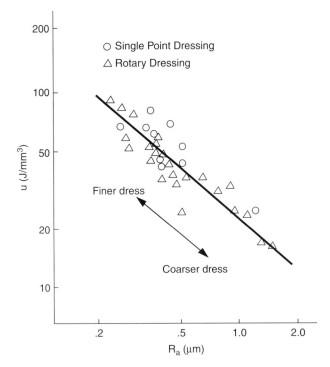

Figure 10-11 Tradeoff between specific grinding energy and workpiece surface rough-
ness due to change in wheel dressing conditions with both single-point
and rotary diamond dressing. Grinding conditions are as in Figure 10-10
[61].

dressing severity is varied for otherwise identical grinding conditions [61]. In this case, changing only the dressing severity caused as much as a five-fold variation in surface roughness and specific energy. It is of interest to note that the results for both single-point and rotary dressing fall on the same trade-off curve.

The effect of grinding and dressing parameters on workpiece surface roughness as described in the foregoing would normally apply to short grinding cycles with frequent wheel redressing. In such cases, the wheel may be dressed only once per part. With longer grinding cycles and less frequent dressing, wear of the grinding wheel alters its topography and so the surface roughness also changes with time. Such behavior can be seen in Figure 10-12, which shows the peak-to-valley wheel surface roughness (measured on an imprint of the wheel surface) versus the accumulated volume of metal removed per unit grinding width for various dressing leads [64]. Finer dressing provides an initially smoother wheel surface as expected. But with continued use after dressing, coarser dressed wheels become smoother and finer-dressed wheels become rougher, tending in each case towards the same 'steady state' condition as the effect of dressing progressively disappears. The 'steady state' wheel roughness is found to be bigger at faster removal rates.

With continued use, the workpiece surface roughness follows a similar trend to that of the wheel roughness [65]. The transition of the surface roughness R_a from its initial value $R_{a,o}$ after dressing towards the 'steady state' value $R_{a,\infty}$ can be described by an exponential relationship of the form [65-68].

$$\frac{R_a - R_{a,\infty}}{R_{a,o} - R_{a,\infty}} = \exp\left(\frac{V'_w}{V'_o}\right) \tag{10-16}$$

where V'_w is the accumulated volumetric removal per unit width and V'_o is a constant which characterizes how fast the roughness changes. This behavior is illustrated in Figure 10-13 for internal cylindrical grinding of a hardened bearing steel [65]. For a given wheel-workpiece combination, V'_o appears to be relatively insensitive to the removal rate and dressing conditions whereas $R_{a,\infty}$ has a power function dependence as seen in Eqs. (10-13) – (10-15). The data in Figure 10-12 would correspond to $V'_o \approx 300$ mm^2, whereas surface roughness data for cylindrical grinding of a hardened tool steel [63] and the roughness data in Figure 10-13 have a smaller value of $V'_o \approx 200$ mm^2. In general, the wheel needs to be dressed fine enough so as to satisfy the required surface roughness which is usually much smaller

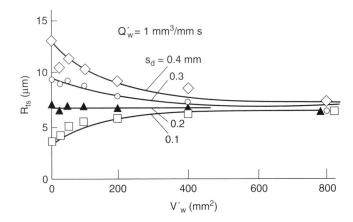

Figure 10-12 Wheel surface roughness versus accumulated metal removal for various dressing leads: cylindrical plunge grinding, vitrified alumina wheel, $v_s = 29$ m/s, $v_w = 19$ m/min, a = 3 µm [64].

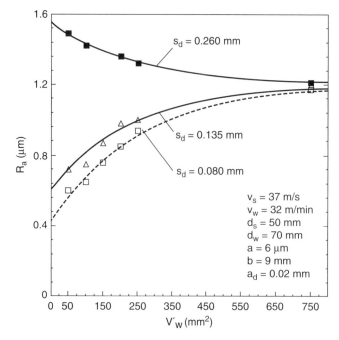

Figure 10-13 Workpiece surface roughness versus accumulated metal removal for various dressing leads: internal cylindrical plunge grinding, 32A80M6VBE wheel, AISI 52100 steel workpiece [65].

than the steady state roughness. The amount of material which can be removed per wheel dressing, or the number of parts per dress, depends on the initial roughness and the rate of roughness degradation.

One factor which is usually assumed to significantly affect the surface roughness is the abrasive grit size. But for conventional abrasive wheels dressed prior to use, the grit size is usually found to have only a minor influence on the initial workpiece roughness after dressing. In one particular investigation, a refinement in abrasive size from 36 to 120 grit, representing a dimensional grit size reduction of more than five to one, lowered the roughness by less than 10% [56]. Insensitivity to grit size is not surprising when it is realized that active cutting points are generated on abrasive grits by localized deformation and fracture during dressing, and that the undeformed chip thickness is typically much smaller than the grit dimension. While there are fewer grains per unit area on coarser-grit wheels (Figure 4-8), each active grain has, on the average, more cutting points. Thus we saw in Figure 4-16 that the radial distribution of active cutting points is not very sensitive to grit size, and so the surface roughness is also relatively unaffected. However, with continued grinding after the initial effects of dressing are removed, smoother workpiece surfaces are obtained with finer-grit wheels. An analysis of surface roughness data for external cylindrical grinding of a plain carbon steel [63] implies a dependence of steady roughness on average grit dimension d_g (see Figure 2-2) as

$$R_{a,\infty} \propto d_g^y \tag{10-17}$$

where $0.3 < y < 0.7$ for 36- to 100-grit wheels (450 to 107 μm average dimension). Larger values of the exponent and hence a greater sensitivity to grit size apply to smoother finishes obtained at slower removal rates.

The above results apply to plunge grinding without crossfeed. For traverse grinding with crossfeed, the workpiece becomes progressively smoother by successive passes of adjacent wheel sections over the workpiece. The radial depth of cut taken by each subsequent section depends upon the rate of wheel wear (see Figure 3-7). Available data in the literature suggest that the combined effects of crossfeed s_t and wheel width b_s on surface roughness can be expressed in terms of the overlap ratio (b_s/s_t) as [59,69,70]

$$R_a \propto \left(\frac{b_s}{s_t}\right)^{-z} \tag{10-18}$$

where $0.5 < z < 1$. Either a wider wheel or a finer crossfeed should improve the finish in crossfeed grinding.

Grinding with crossfeed might be expected to provide better finishes than plunge grinding owing to the repeated grinding action by adjacent sections. However, non-uniform wear across the wheel width may lead to uncertainties in the final dimensions obtained, and the situation may be further aggravated by traversing back and forth across the workpiece, which tends to 'crown' the wheel. For production grinding operations, it may be preferable to use a wider wheel and plunge grind, where possible, and this will also reduce the grinding time.

The foregoing discussion applies to conventional abrasive wheels which are periodically redressed. With CBN wheels, the situation is similar in some ways and much different in others. After initial wheel preparation (truing and dressing), the initial surface roughness may be as good as that obtained with a finely dressed conventional wheel [71-73]. However, the wheel surface, although smooth, is likely to be very dull, thereby causing extremely large forces (Figure 5-13). With continued grinding, the wheel tends to self-sharpen and the forces become smaller, but the workpiece becomes rougher [71,72] as seen in Figures 5-13 and 5-16. A comparable wheel condition and grinding performance may be obtained directly, while avoiding the high initial forces, by 'sharpening' the wheel prior to grinding using methods described in Chapter 4. For a completely sharpened wheel having almost no flattened areas remaining on the grain tips (Figure 4-11), the workpiece surface roughness reaches a limiting value which is insensitive to the grinding parameters and cannot be improved to any significant degree by spark-out. In this condition, the wheel roughness is the overriding factor controlling the workpiece finish, and the kinematic contribution becomes insignificant by comparison.

With CBN wheels, a better finish may be obtained by resorting to a finer grit size, insofar as this will reduce the roughness of the wheel in its sharpened condition. Finer grit sizes are generally required with CBN wheels than with conventional abrasive wheels in order to obtain comparable roughnesses.

REFERENCES

1. *Machining Data Handbook*, Volume 2, 3rd edn, Machinability Data Center, Cincinnati, Ohio, 1980, Section 18.
2. Tripathi, K. C. and Rowe, G. W., 'Grinding Fluid, Wheel Wear and Surface Generation', *Proceedings of the Seventeenth International Machine Tool Design and Research Conference*, 1976, p. 181.
3. Bhateja, C. P., 'The Intrinsic Characteristics of Ground Surfaces', *Proceedings, Abrasive Engineering Society International Conference*, 1975.
4. Thomas, T. R., 'Correlation Analysis of the Structure of a Ground Surface', *Proceedings of the Thirteenth International Machine Tool Design and Research Conference*, 1972, p. 303.
5. 'Surface Texture', American National Standard ANSI B46.1, 1978.
6. Williamson, J. B. P., Pullen, J. and Hunt, R. T., 'The Shape of Solid Surfaces', *Surface Mechanics*, ASME, 1969, p. 24.
7. Reason, R., 'Surface Topography', *Tribology Handbook*, M. J. Neale, Ed., Butterworths, London, 1973, Section F2.
8. Farmer, D. A., Brecker, J. N. and Shaw, M. C., 'Study of the Finish Produced in Surface Grinding. Part 1 – Experimental', *Proc. Instn Mech. Engrs*, 182, pt. 3K, 1966-1967, p. 171.
9. Nakayama, K. and Shaw, M. C., 'Study of the Finished Produced in Surface Grinding. Part 2 – Analytical', *Proc. Instn Mech. Engrs, 182, pt. 3K*, 1966-1967, p. 179.
10. Inasaki, I., Karpuschewski, B. and Lee, H. S., 'Grinding Chatter – Origin and Suppression', *Annals of the CIRP*, 50/2, 2001, p. 515.
11. Snoeys, R. and Brown, D., 'Dominating Parameters in Grinding Wheel and Workpiece Regenerative Chatter', *Proceedings of the Tenth International Machine Tool Design and Research Conference*, 1969, p. 325.
12. Hahn, R. S. and Lindsay, R. P., 'Principles of Grinding. Part V, Grinding Chatter', *Machinery*, New York, 1981.
13. Tönshoff, H. K., Kuhfuss, B. and Tellsbuscher, E., 'Improvement of Work Quality by Vibration Reduction', SME Paper No. MR84-535, 1984.

14. Tönshoff, H. K. and Kuhfuss, B., 'Influence of Mechanical Vibrations on Workpiece Shape in External Plunge Grinding', *Proceedings, Twelfth North American Manufacturing Research Conference*, SME, 1984.

15. Hahn, R. S. and Price, R. L., 'A Nondestructive Method of Measuring Local Hardness Variations in Grinding Wheels', *Annals of the CIRP*, 16, 1967, p. 19.

16. Kaliszer, H. and Trmal, G., 'Generation of Surface Topography on a Ground Surface', *Proceedings of the Fourteenth International Machine Tool Design and Research Conference*, 1973, p. 677.

17. Kaliszer, H. and Trmal, G., 'Factors Influencing the Topography of Ground Surfaces', *Proceedings, International Conference on Production Engineering*, Pt. 1, JSPE, Tokyo, 1974, p. 689.

18. Yokogawa, K., 'Influence on Thermal Deformation of Cylindrical Grinding Machine on Grinding Accuracy, *Bull., Japan Soc. of Prec. Engg.*, 2, No. 2, 1966, p. 60.

19. Masuda, M. and Shiczaki, S., 'Out of Straightness of Workpiece, Considering] the Effect of Thermal Work Expansion on Depth of Cut in Grinding', *Annals of the CIRP*, 23/1, 1974, p. 91.

20. Nakano, Y. and Hashimoto, K., 'A Surface Grinding Technique for the Automatic Elimination of Profile Effects due to Thermal Deformation of Workpieces', *Bull., Japan Soc. of Prec. Engg.*, 9, No. 5, 1976, p. 139.

21. Thé, J. H. L. and Scrutton, R. F., 'Thermal Expansions and Grinding Forces Accompanying Plunge-cut Surface Grinding', *Int. J. Mach. Tool Des. Res.*, 13, 1973, p. 287.

22. Kops, L. and Hucke, L. M., 'Thermal Simulation of the Grinding Process', *Proceedings of the North American Metalworking Research Conference*, 3, 1973, p. 95.

23. Kops, L. and Hucke, L. M., 'Simulated Thermal Deformation in Surface Grinding', *Proceedings, International Conference on Production Engineering*, Pt. 1, JSPE, Tokyo, 1974, p. 683.

24. Yokoyama, K. and Ichimiya, R., 'Thermal Deformation of Workpiece in Surface Grinding', *Bull., Japan Soc. of Prec. Engg.*, 11, No. 4, 1977, p. 195.

25. Dakiwada, T. and Kanauchi, T., 'Thermal Deformation and Thermal Stress in Workpiece under Surface Grinding', *Bull., Japan Soc. of Prec. Engg.*, 15, No. 3, 1981, p. 195.

26. Dall, A. H., 'Rounding Effect in Centerless Grinding', *Mechanical Engineering*, ASME, 58, 1946, p. 325.

27. Gurney, Y. P., 'An Analysis of Centerless Grinding', *Trans. ASME, J. of Eng. for Ind.*, 86, 1964, p. 163.

28. Furukawa, Y., Miyashita, M. and Shiozaki, S., 'Vibration Analysis and Work-rounding Mechanism in Centerless Grinding', *Int. J. Mach. Tool Des. Res.*, 11, 1971, p. 145.

29. Rowe, W. B. and Barash, M. M., 'Computer Method for Investigating the Inherent Inaccuracy of Centerless Grinding', *Int. J. Mach. Tool Des. Res.*, 4, 1964, p. 91.

30. Rowe, W. B., 'Research into the Mechanics of Centerless Grinding', *Precision Engineering*, 2, 1979, p. 75.

31. Bhateja, C. P., 'Current State of the Art of Workpiece Roundness Control in Precision Centerless Grinding', *Annals of the CIRP*, 33/1, 1984, p. 199.

32. Orioka, T., 'Probabilistic Treatment of the Grinding Geometry', *Bull. Japan Soc. Grinding Eng.*, 1, 1961, p. 27.

33. Schleich, H., 'Tief- und Pendelschleifen-Vergleich der Oberflachenstruktur', *Industrie-Anzeiger*, 102, No. 50, 1980, p. 32.

34. Andrew, C., Howes, T. D. and Pearce, T. R. A., *Creep Feed Grinding*, Holt, Reinhart, and Winston, London, 1985, Chapter 7.

35. Steffens, K., 'Closed Loop Simulation of Grinding', *Annals of the CIRP*, 32/1, 1983, p. 255.

36. Pandit, S. M. and Sathyanarayanan, G., 'Data Dependent Systems Approach to Surface Generation in Grinding', *Trans. ASME, J. of Eng. for Ind.*, 106, 1984, p. 205.

37. Tsukizo, T., Hisakado, T. and Hasegawa, M., 'On the Generating Mechanism of Surface Roughness (Forming Process of Single Asperity)', *Bulletin of the JSME*, 16, No. 22, 1973, p. 363.

38. Hasegawa, M., Tukizoe, T. and Hisakado, T., 'On the Generating Mechanism of Surface Roughness "Forming Process of Single Asperity by Two Grooves' Interference'", *Bulletin of the JSME*, 17, No. 105, 1974, p. 367.

39. Peklenik, J., 'Contribution to the Correlation Theory for the Grinding Process', *Trans. ASME, J. of Eng. for Ind.*, 86, 1964, p. 85.

40. Yoshikawa, H. and Sata, T., 'Simulated Grinding Process by Monte Carlo Method', *Annals of the CIRP*, 16, 1968, p. 297.

41. Hasegawa, M., 'Statistical Analysis for the Generating Mechanism of Ground Surface Roughness', *Wear*, 29, 1974, p. 31.

42. Hasegawa, M., 'Order Statistical Approach to Ground Surface Generation', *Trans. ASME, J. of Eng. for Ind.*, 103, 1981, p. 22.

43. Bhateja, C. P., Pattinson, E. J. and Chisholm, A. W. J., 'The Influence of Dressing on the Performance of Grinding Wheels', *Annals of the CIRP*, 21/1, 1972, p. 81.

44. Bhateja, C. P., 'An Enveloping Profile Approach for the Generation of Ground Surface Texture', *Annals of the CIRP*, 25/1, 1977, p. 333.

45. Vickerstaff, T. J., 'The Influence of Wheel Dressing on the Surface Generated in the Grinding Process', *Int. J. Mach. Tool Des. Res.*, 16, 1976, p. 145.

46. Law, S. S., Wu, S. M. and Joglekar, A. M., 'On Building Models for the Grinding Process', *Trans. ASME, J. of Eng. for Ind.*, 95, 1973, p. 983.

47. Law, S. S. and Wu, S. M., 'Simulation Study of the Grinding Process', *Trans. ASME, J. of Eng. for Ind.*, 95, 1973, p. 972.

48. Pandit, S. M., Suratkar, P. T. and Wu, S. M., 'Mathematical Model of a Ground Surface Profile with the Grinding Process as a Feedback System', *Wear*, 39, 1976, p. 205.

49. Pandit, S. M. and Sathyanarayan, G., 'A Model for Surface Grinding Based on Abrasive Geometry and Elasticity', *Trans. ASME, J. of Eng. for Ind.*, 104, 1982, p. 349.

50. Verkerk, J., 'Kinematical Approach to the Effect of Wheel Dressing Conditions on the Grinding Process', *Annals of the CIRP*, 25/1, 1976, p. 209.

51. Verkerk, J. and Pekelharing, A. J., 'The Influence of the Dressing Operation on Productivity in Precision Grinding', *Annals of the CIRP*, 28/2, 1979, p. 487.

52. Fletcher, N. P., 'A Simple Model for Predicting the Possible Surface Roughness of a Cylindrically Traverse Ground Workpiece when Using Wheels Dressed with Single Point Diamond Tools', *Proceedings of the Twenty-first International Machine Tool Design and Research Conference*, 1980, p. 329.

53. Pahlitzsch, G. and Schneidemann, H., 'Neue Erkenntnisse beim Abrichtenvon Schleifdscheiben', *wt-Z. lnd. Fertig.*, 61,1971, p. 622.

54. Rubenstein, C., 'The Factors Influencing the Surface Finish Produced by Grinding', *Proceedings of the Fourteenth International Machine Tool Design and Research Conference*, 1973, p. 625.

55. Yang, C. T. and Shaw, M. C., 'The Grinding of Titanium Alloys', *Trans. ASME*, 77, 1955, p. 645.

56. Kannappan, S. and Malkin, S., 'Effects of Grain Size and Operating Parameters on the Mechanics of Grinding', *Trans. ASME, J. of Eng. for Ind.*, 94,1972, p. 833.

57. Brandin, H., 'Vergleichende Untersuchung zwischen Pendel- und Tiefschleifen', *TZ für Praktische Metallbearbeitung*, 71, No. 1, 1977, p. 6.

58. Snoeys, R., Peters, J. and Decneut, A., 'The Significance of Chip Thickness in Grinding', *Annals of the CIRP*, 23/2, 1974, p. 227.

59. Kedrov, S. M., 'Investigation of Surface Finish in Cylindrical Grinding Operations', *Machines and Tooling*, 51, 1980, p. 40.

60. Murray, T. and Malkin, S., 'Effects of Rotary Dressing on Grinding Wheel Performance', *Trans. ASME, J. of Eng. for Ind.*, 100, 1978, p. 297.

61. Malkin, S. and Murray, T., 'Comparison of Single Point and Rotary Dressing of Grinding Wheels', *Proceedings, Fifth North American Metalworking Research Conference*, 1977, p. 278.

62. Lindsay, R. P., 'On the Surface Finish-Metal Removal Relationship in Precision Grinding', *Trans. ASME, J. of Eng. for Ind.*, 95, 1973, p. 815.

63. Makino, H., 'Roughness of Finished Surface in Grinding Operation of Hardened Steels', *Bull., Japan Soc. of Prec. Engg.*, 1, No. 4, 1965.

64. Saljé, E. and Weinert, K., 'Die Veranderung der Wirkrauftiefe der Schleifscheibe beim Schleifen', *Z. fur wirtschalftliche Fertigung*, 70, 1975, p. 63.

65. Xiao, G. and Malkin, S., 'On-Line Optimization for Internal Plunge Grinding' *Annals of the CIRP*, 45/1, 1996, p. 287.

66. Saljé, F. and Heidenfelder, H., 'Comparison of CBN and Conventional Grinding Processes', *Proceedings, Thirteenth North American Manufacturing Research Conference*, SME, 1985.

67. Jacobs, U. and Saljé, E., 'Vergleich zwischen Korund- und CBN-Schleifscheiven', *Z. fur Metallbearbeitung*, 76, 1982, p. 47.

68. Malkin, S., 'Practical Approaches to Grinding Optimization', *Milton C. Shaw Grinding Symposium*, PED-16, ASME, 1985, p. 289.

69. Banerjee, J. K. and Hillier, M. J., 'Some Observations on the Effects of Wheel Wear-land on Workpiece Surface Finish during Flat Surface Grinding with Cross-Feed', *First World Conference on Industrial Tribology*, New Delhi, 1972, Paper G5.

70. Vickerstaff, T. J., 'Wheel Wear and Surface Roughness in Cross Feed Surface Grinding', *Int. J. Mach. Tool Des. Res.*, 13, 1973, p. 183.

71. Malkin, S., 'Current Trends in CBN Grinding Technology', *Annals of the CIRP*, 34/2, 1985, p. 557.

72. Pecherer, E. and Malkin, S., 'Grinding of Steels with Cubic Boron Nitride (CBN)', *Annals of the CIRP*, 33/1, 1984, p. 211.

73. Tönshoff, H. K. and Grabner, T., 'Cylindrical and Profile Grinding with Boron Nitride Wheels', *Proceedings of the Fifth International Conference on Production Engineering*, JSPE, Tokyo, 1984, p. 326.

74. Pinson, A. G., 'CBN Overcomes the Finish Barrier', SME Paper No. MR84-549, 1984.

Chapter 11

Wheel Wear and Lubrication

11.1 INTRODUCTION

Grinding of materials is accompanied to a greater or lesser degree by wear of the grinding wheel. Historically, wheel performance has been judged mainly in terms of its wear resistance, but wheel wear is only one of many factors which should be taken into account.

The practical significance of wheel wear depends upon the particular type of grinding. For precision grinding using conventional abrasives, the termination of the grinding wheel 'tool life' and the need for redressing is likely to be determined either by excessive forces and wheel dulling due to attritious wear at the grain tips, or by loss of finish, form, or size due to 'bulk' wheel wear. More of the wheel may be consumed by periodic dressing than by wear, but the wheel itself is usually not a significant cost factor. By contrast, wheel consumption has more significant economic consequences for grinding with superabrasive wheels and for heavy-duty grinding with conventional wheels, owing to high abrasive costs with the former and high rates of wheel wear with the latter.

Grinding wheel wear is a complex process. The overall wheel wear is the culmination of numerous individual wear events from encounters of abrasive grains with the workpiece. Research studies have identified and quantified the prevailing wear mechanisms and the influence of various factors. Lubrication by the grinding fluid is one of the most significant factors affecting wheel wear, mainly through its influence on the chemical reactions which occur at the grinding zone. Wheel-wear behavior can be empirically correlated to some extent with the operating parameters which, together with an understanding of the mechanisms of wheel wear, provides a rational basis for deciding what steps might be taken to alleviate limitations on the grinding process imposed by wheel wear.

11.2   QUANTIFYING WHEEL WEAR

The wear of a grinding wheel is usually expressed as a volumetric loss of material. For plunge grinding, the volume of radial wheel wear is simply

$$V_s = \pi d_s \Delta r_s b \qquad (11\text{-}1)$$

where Δr_s is the measured decrease in wheel radius, d_s is the mean of the wheel diameters before and after wear has occurred, and b is the grinding width. In most practical cases, Δr_s is only a very small fraction of the wheel diameter.

An illustration of typical wheel-wear behavior is shown in Figure 11-1 as a plot of volumetric wheel wear versus accumulated metal removed V_w [1]. Since the removal rate is essentially constant, the accumulated metal

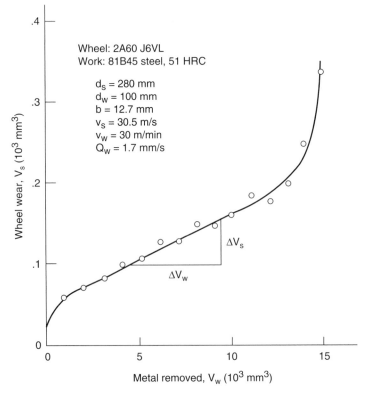

Figure 11-1 *Volumetric wheel wear versus accumulated metal removed for an external cylindrical plunge grinding operation.*

removed on the horizontal axis is proportional to time. The wear behavior seen here is similar to that observed with other wear processes. High initial wear is followed by a steady-state regime with a nearly constant wear rate. A third regime of accelerating wear, as seen in Figure 11-1, usually indicates a 'catastrophic' situation and the need to redress the wheel. Accelerating wear, if and when it occurs, may be associated with workpiece burn or chatter.

A performance index commonly used to characterize wheel-wear resistance is the 'grinding ratio', also referred to as the G-ratio or G, which is the volume of material removed per unit volume of wheel wear. This may be computed for the entire test as

$$G = \frac{V_w}{V_s} \qquad (11\text{-}2)$$

which corresponds to $G \approx 50$ in Figure 11-1, or only for the steady-state wear regime as

$$G = \frac{\Delta V_w}{\Delta V_s} \qquad (11\text{-}3)$$

which corresponds to $G \approx 100$.

G-ratios cover an extremely wide range of values. On vanadium-rich high-speed steels, G-ratios less than unity may be obtained [2], in which case the work appears to be abrading the wheel rather than vice versa. At the other extreme, G-ratios of above 60,000 have been reported for internal grinding of bearing races using CBN wheels with a straight oil as the grinding fluid [3].

An important consequence of wheel wear, as indicated by the grinding ratio, is that on feed-controlled grinders, the actual stock removal rate is less than the infeed rate on the machine since part of the infeed motion corresponds to following the retreating wheel surface as it wears. With cylindrical plunge grinding, for example, it can be readily shown from volumetric continuity requirements (neglecting machine deflections) that the actual radial size reduction rate $\dot{r}_w$ of the workpiece is given by

$$\dot{r}_w = \frac{v_f}{1 + \dfrac{d_w}{d_s G}} \qquad (11\text{-}4)$$

where v_f is the radial infeed velocity set on the machine (Figures 3-1(b) and 3-1(c)), and d_w and d_s are the workpiece and wheel diameters, respectively.

The discrepancy between $\dot{r}_w$ and v_f becomes significant with small G-ratios and relatively small wheels (e.g. internal grinding).

For grinding to a required shape, it is usually not the overall wear across the face of the wheel, but localized wear at corners and sharp protrusions in the profile which is likely to necessitate wheel redressing. Corner wear is especially problematic with straight plunge grinding of the type illustrated in Figure 11-2, such as for grinding of crankshafts [4-7]. In addition to uniform radial wear Δr_b across the wheel face, initially sharp corners become progressively rounded [4-8]. A worn corner may be approximated as being nearly round with radius Δr_k, but a more accurate indication of material loss at a corner is the associated reduction in profile area ΔA_k (see Figure 11-2). The uniform wear area ΔA_b plus the sum of the localized wear area ΔA_k at each corner give the total cross-sectional wear area ΔA_s. The corresponding volumetric wear is

$$\Delta V_s = \pi \bar{d}_s \Delta A_s \tag{11-5}$$

where $\bar{d}_s$ is the mean wheel diameter measured to the centroid of the wear area ΔA_s. It is generally more practical to describe corner wear in terms of the radius Δr_k rather than ΔA_k, since it is easier to measure and more directly related to the need for redressing and loss of form. To regenerate a nominally sharp corner by redressing, the minimum wheel depth to be removed is equal to Δr_k.

A situation similar to that for corner wear also applies to other cross-sectional shapes. An example of a V-shaped profile is illustrated in Figure 11-3. A more generalized approach has been proposed for quantifying wheel wear with any arbitrary wheel shape [8].

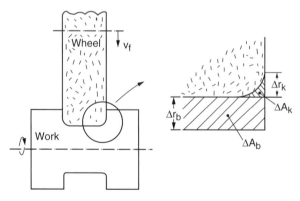

Figure 11-2 Illustration of uniform wear and localized corner wear in cylindrical plunge grinding of a workpiece wider than the wheel.

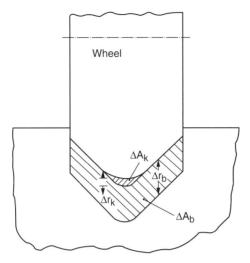

Figure 11-3 Illustration of uniform wear and localized corner wear with a V-shaped profile.

Non-uniform wear across the wheel face also occurs in any operation with traverse motion. The simplest example is straight traverse grinding with crossfeed, as was seen in Figure 3-7. At the start of grinding, when the wheel face is flat, virtually all material removal occurs across the leading width s_t corresponding to the crossfeed per workpiece revolution. But as the first section wears down, both across its width and by rounding at the leading corner, part of the material to be removed to a depth a is left behind for removal by the next section of width s_t, and its wear leaves behind material to be removed by the subsequent section, etc. In this way, a series of steps is generated across the wheel, equal in number to the overlap ratio and having a cumulative height not exceeding the wheel depth of cut [9]. This may give the appearance of an inclined worn profile across part of the wheel width over which most of the actual material removal occurs [10,11].

11.3 WHEEL-WEAR MECHANISMS

It is generally recognized that there are three main mechanisms of wheel wear as illustrated in Figure 11-4 [12-14]: attritious wear, grain fracture, and bond fracture. Attritious wear involves dulling of abrasive grains and the growth of wear flats by rubbing against the workpiece. Grain fracture refers to removal of abrasive fragments by fracture within the grain, and bond fracture to dislodging of the abrasive grain from the binder. Both of these types of fracture wear may lead to self-sharpening, which reduces

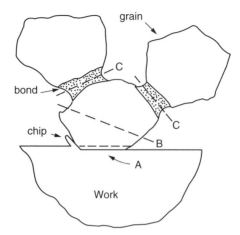

Figure 11-4 Illustration of wheel-wear mechanisms: A – attritious wear, B – grain fracture, and C– bond fracture [13,14].

the dulled wear flat area caused by attritious wear. Another type of wear is binder erosion, which is likely to reduce the bond strength and promote grain dislodgement, especially with resin- and metal-bonded wheels.

Volumetric wear measurements tell us little, if anything, about the mechanisms of wheel wear. In some research investigations, wheel wear has been characterized instead in terms of the weight of the wear particles removed from the wheel and their size distribution [12,13,15-17]. For this purpose, it is necessary to collect all the grinding debris and then separate the wheel-wear particles from the metallic swarf. The size distribution of the wear particles is related to the prevailing wheel-wear mechanisms.

Some wear results obtained in this way are shown in Figure 11-5 for grinding a low carbon steel with a series of five vitrified aluminum oxide wheels varying only in grade [15]. Here again as in Figure 11-1 we see a high initial rate of wear followed by a steady-state regime of nearly constant wear rate. The greatest amount of wear occurred with the softest G-grade wheel, but the least wear in this case was not with the hardest K-grade wheel but with the intermediate I-grade wheel.

Size distributions of the wear particles recovered in the steady-state regime are shown in Figure 11-6 for each of the five wheels. Also included for comparison is the size distribution of the original abrasive grain used in making these wheels. These results are presented on a cumulative weight basis from larger to smaller particles (smaller to larger sieve numbers), so a higher elevation of the curve corresponds to less fragmentation. It is apparent that abrasives in softer wheels undergo less fragmentation during the wear process, which is similar to that previously observed for particles removed by dressing (Figure 4-6).

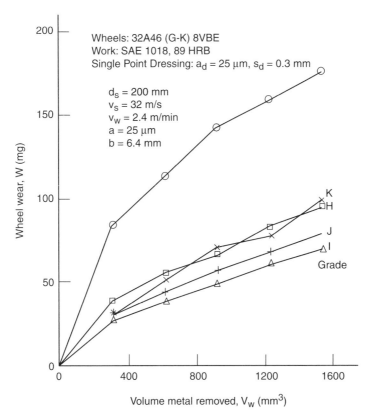

Figure 11-5 Weight of wheel wear versus volume of metal removal for wheels of different grades [15].

The relative contribution of each of the three wear mechanisms — attritious wear, grain fracture, and bond fracture — can be estimated from the wear particle size distribution together with measurements of wear-flat area and volumetric wear. We begin with attritious wear. As seen in Chapter 5, attritious wear leads to the growth of flat areas on the grain tips. An upper limit on the weight of abrasive worn away by attrition can be approximated by the product of the volumetric wear, the wear-flat-area fraction, and the density of the alumina abrasive [15]. Results from this calculation show that attritious wear amounts at most to only a few per cent of the total weight loss.

The amount of bond fracture is estimated from a statistical analysis of the wear particle-size distribution. For this purpose, we use the same method as in Chapter 4 for analyzing the dressing particles. The main assumptions of this analysis are that particles removed by bond fracture are the biggest ones, and that there is only one bond fracture per grain. The

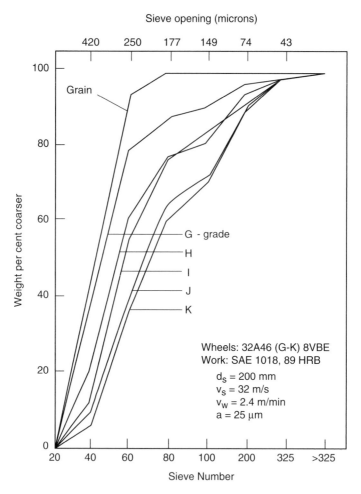

Figure 11-6 Size distribution of wheel-wear particles. Corresponds to results in Figure 11-5 [15].

bond fracture components calculated from the wear particle-size distributions in Figure 11-6 are shown in Figure 11-7, together with additional results using these same wheels and finer dressing conditions. With increasing binder content there is a corresponding reduction of bond fracture, from about 90% with the G-grade down to about 50% with the K-grade wheel.

Since attritious wear is negligible, this would suggest that most of the wear not generated by bond fracture is due to grain fracture. This neglects any wear contribution from the binder, which should be relatively insignificant at least with typical porous vitrified wheels containing much more

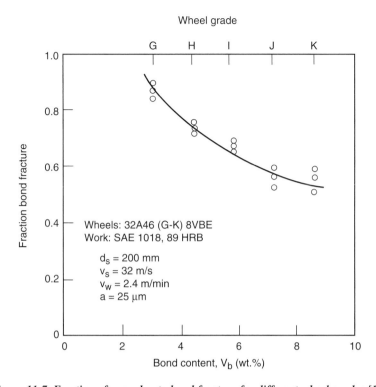

Figure 11-7 Fraction of wear due to bond fracture for different wheel grades [15].

grain than binder (see Chapter 2). With increasing wheel hardness, there is relatively more grain fracture and less bond fracture, because of the greater bond strength allied to the greater probability of the abrasive grain undergoing fracture prior to being finally dislodged. Nevertheless, although the amount of observed attritious wear is negligible, it is often the most important form of wear since it controls the grinding forces and hence also the rate of bond fracture, as will be seen in the following section.

Aside from wheel grade, the relative amounts of grain and bond fracture should also depend on the friability and size of the abrasive grains and the particular bond material. Although these factors have not been systematically investigated, it seems appropriate at this point to say something about grain friability. More friable grains are more susceptible to grain fracture prior to their being finally dislodged from the bond, which is an important factor in restricting the size of the wear-flat areas initially dressed on to the grains and their subsequent growth by attrition. In the absence of self-sharpening in this way, the abrasive grains would become extremely dull, thereby causing large grinding forces and thermal damage to the workpiece.

This is why tough abrasives are generally not suitable for precision grinding operations [18]. On the other hand, heavy-duty grinding operations with powerful machines usually operate more efficiently using tougher abrasives, owing to reduced grain breakdown and lower abrasive consumption.

11.4   ANALYSIS OF WHEEL WEAR

A grinding wheel consists of an agglomeration of hard abrasive grains held together by a weaker bond material. In practice, the overall wear of a grinding wheel can proceed only as fast as bond fractures occur. Prior to bond fracture, pieces of the abrasive are lost by grain fracture and also to a much lesser extent by attritious wear, but the overall wear rate is governed by the frequency of bond fracture [15].

We now proceed to develop a quantitative description of grinding wheel wear in terms of the factors which might affect the frequency of bond fracture [15]. Each time an active abrasive grit encounters the workpiece, there is a certain probability p_b that it will be dislodged by bond fracture. The total weight of abrasive worn away W is equal to the product of the probability of bond fracture, the average weight of a single whole abrasive grain w, and the total number of encounters N between active grains and the workpiece

$$W = p_b w N \tag{11-6}$$

Again, this relationship accounts for the total wear and not only that lost by bond fracture.

With vitrified wheels, the binder is a glassy brittle material, so the likelihood of bond fracture should depend on the magnitude of the tensile stress induced at the bond bridges by the grinding force. A schematic illustration in Figure 11-8 shows an isolated active grain subject to a tangential force component f_t and a normal component f_n. Although the stress causing bond fracture, say across the plane AB, cannot be readily calculated, it seems likely that the tangential force component induces a predominantly tensile stress and the normal force a compressive stress. Assuming that the stresses are proportional to the force components, the tensile stress in the bond bridge might be simply written as

$$\sigma_t = c_1 f_t - c_2 f_n \tag{11-7}$$

where c_1 and c_2 are constants which would depend on the geometry of the grain and the bond.

With harder wheels containing more binder, the bond bridges between adjacent grains are bigger, which means that the stress should be

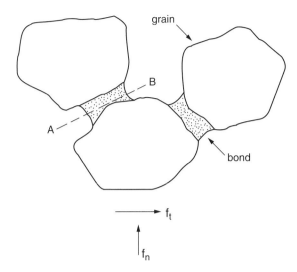

Figure 11-8 Illustration of an active grain subject to tangential and normal force components [15].

lower. Assuming a proportional relationship between the cross-sectional bond area at AB and the weight fraction V_b of binder, the tensile stress in Eq. (11-7) can be rewritten as

$$\sigma_t = K\left(\frac{f_t - \beta f_n}{V_b}\right) \tag{11-8}$$

where K and β are new constants. The quantity within the parentheses is called the 'bond stress factor'.

From the foregoing discussion we should expect the probability of bond fracture to depend upon the bond stress factor. A direct correlation has indeed been found as seen in Figure 11-9 where a semi-logarithmic plot of p_b versus the bond stress factor (with $\beta = 0.2$) gives a reasonable straight-line fit for the wheel-wear results in Figure 11-5 together with others referred to previously on the same mild steel workpiece material and on a bearing steel with finer dressing using the same wheels. The values of f_t and f_n used in calculating the bond stress factor in Figure 11-9 were obtained by dividing the measured force components F_t and F_n by the number of active grains instantaneously in contact with the workpiece (measured number of active grains per unit area times the geometric wheel-workpiece contact area). The probability p_b was calculated in each case, using Eq. (11-6), by dividing the wear increment W by the product of the average weight of a

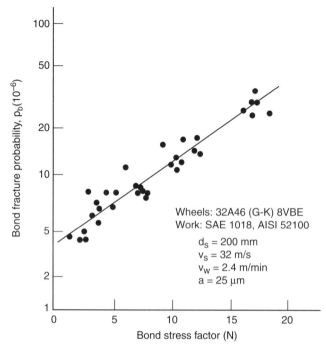

Figure 11-9 Probability of bond fracture versus bond stress factor. Includes wear results in Figure 11-5 together with others for finer dressing on the same workpiece material and on a hardened bearing steel [15].

whole abrasive grain w and the number of encounters N (number of instantaneous active grains times the number of wheel revolutions). A correction can be introduced to take into account the weight of binder in addition to the weight of abrasive in the wear debris, but this would be significant only with much harder vitrified wheels higher in bond content. For each active grain encounter with the workpiece, the probability of bond fracture in Figure 11-9 ranges from only about 4×10^{-6} with the smallest bond stress factor to 3×10^{-5} with the biggest. In this sense, bond fracture is an extremely rare event.

This wheel-wear analysis might be further developed in a more generalized form for quantitatively predicting the influence of grinding forces on wheel wear with different wheel-workpiece combinations. However, it seems unlikely that a comprehensive model of this type could take into account all the important factors. Despite its limitations, the present analysis does present a physical explanation of the wheel-wear process. It can be appreciated that attritious wear, although negligible, is probably the most important type of wear, insofar as it controls the grinding forces and thus governs the probability of bond fracture and the overall rate of wheel wear.

(Additional aspects of attritious wear are considered in the following sections.) Here we also see the significance of wheel grade. Harder wheels provide stronger bonds and less of a tendency for bond fracture. Furthermore, with less bond fracture during wheel dressing and subsequent wear, there are more active grits per unit area, as was seen in Chapter 4. This is why harder wheels have more total wear-flat area [19], thereby also resulting in bigger grinding forces and a greater tendency for thermal damage.

While the results presented above apply mainly to vitrified aluminum oxide wheels, there is considerable evidence that wear of CBN wheels also occurs in a similar manner by attrition, by grain fracture, and by grain pullout analogous to bond fracture [20-24]. It is difficult to quantify the contributions of these various types of wear in vitrified CBN wheels since the wear rates are usually so low. However some success has been achieved with electroplated CBN wheels by periodic microscopic observations and measurements of the wheel surface during grinding [20,21]. As with most grinding operations, the wear of electroplated CBN wheels is characterized by a high wear rate regime at the start of grinding with a new wheel followed by a long steady state wear regime at a much lower rate. A third catastrophic wear regime occurs and the wheel reaches the end of its useful life when the metal binder begins to strip off the wheel hub. The high initial wheel wear in the first regime was found to be mainly due to pullout of the most protruding weakly held grains, and tended to increase with coarser grained wheels. Grain pullout accounted for 60 – 80% of the total transient wear in this regime, with the remaining wear mostly caused by grain fracture. Wheel wear in the steady state regime was dominated by grain fracture. Attritious wear accounted for a negligible portion of the overall wear but, as with aluminum oxide wheels, the dulled wear flat area associated with this type of wear is directly related to increased forces and power.

Many attempts to analyze wheel wear have followed an empirically oriented approach of exploring possible correlations between overall wheel wear and grinding severity, without any explicit consideration of the detailed wear mechanisms. For example, it was postulated that the radial wheel wear per unit distance of sliding on the workpiece might depend on the average radial infeed velocity $\bar{v}_r$ at which the workpiece can be considered to be infeeding normal to the wheel surface as illustrated in Figure 11-10 [25]. The sliding distance is equal to the arc length of contact l_c times the total number of wheel revolutions N during the grinding interval. The average radial infeed velocity $\bar{v}_r$, which can be approximated by the radial component of the workpiece velocity v_w at the mid-point of the grinding zone in Figure 11-10 is readily obtained as

$$\bar{v}_r = v_w \left(\frac{a}{d_s} \right)^{1/2} \tag{11-9}$$

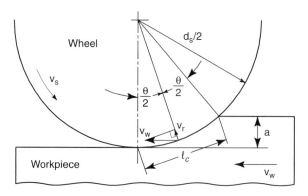

Figure 11-10 *Illustration of average radial infeed in straight plunge grinding. Adapted from Reference [25].*

or in terms of the removal rate per unit width ($Q'_w = v_w a$) as

$$\bar{v}_r = \frac{Q'_w}{(ad_s)^{1/2}} = \left(\frac{Q'_w v_w}{d_s} \right)^{1/2} \qquad (11\text{-}10)$$

A direct relationship was previously shown between the radial infeed velocity and workpiece infeed angle (Chapter 3) and it is also apparent from Eq. (11-9) that the radial infeed velocity for a given wheel velocity is directly related to the magnitude of the undeformed chip thickness. Instead of the radial infeed velocity, the maximum value at the top of the arc length of contact, which is $\sqrt{2}$ times bigger, has also been used for characterizing grinding severity [26].

In Figure 11-11, experimental results for external cylindrical plunge grinding covering a wide range of conditions show a direct relationship between the radial wheel wear per unit sliding distance ($\Delta r_s / l_c N_s$) and $\bar{v}_r$ [25]. Such good correlation is very encouraging, but significant departures from the curve in Figure 11-11 were also found whose cause was traced to inadvertent changes in the dressing condition by progressive wear of the dressing tool. One way the dressing process is likely to affect wheel wear is by its influence on the grinding forces. Another way is by its influence on how much wear occurs during the initial wear regime (Figure 11-1), which may be considered to result from the removal of abrasive material weakened by dressing-induced damage [27].

Another parameter used for empirically correlating grinding behavior is the equivalent chip thickness h_{eq} (Eq. (5-39)). In Chapters 5 and 10, we mentioned power function correlations of grinding forces and surface roughness with this parameter. The G-ratio has also been found to have a

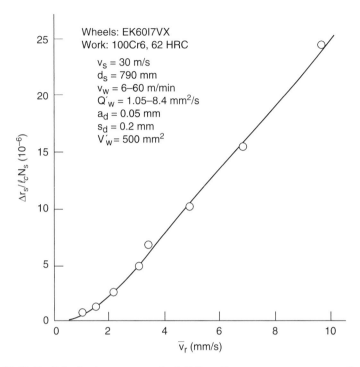

Figure 11-11 *Radial wheel wear per unit of sliding distance versus average radial infeed velocity [25].*

similar dependence of the form [28]

$$G = G_1 h_{eq}^{-g} = G_1 \left(\frac{Q'_w}{v_s} \right)^{-g} \qquad (11\text{-}11)$$

where G_1 and g are constants. Typical values of the exponent for precision grinding are $g \approx 0.1-0.5$ [28]. The wheel-wear results in Figure 11-11 can also be shown to be reasonably consistent with Eq. (11-11). However, the usefulness of this relationship is somewhat limited for predicting wheel wear since the constants G_1 and g need to be experimentally evaluated for each particular wheel-workpiece combination, dressing condition, grinding fluid, etc.

A similar relationship also applies to heavy-duty grinding (snagging and cut-off) at constant peripheral wheel velocity between the G-ratio and stock removal rate Q_w [29-32]:

$$G = G_2 Q_w^{-g} \qquad (11\text{-}12)$$

where G_2 is a constant. Usually the G-ratio decreases with faster removal rates ($g > 0$) or is relatively insensitive to removal rate ($g \approx 0$). But in some cases, especially at lower rates of stock removal, the G-ratio has been found to increase with faster removal rates ($g < 0$).

The analyses for wheel wear up to this point apply to grinding without profiles. The situation is more complex with profile grinding owing to the non-uniform working conditions across the active wheel surface, and localized wear leading to profile errors is likely to be more important than the overall wear. However, there appears to be a direct relationship between overall wear and localized wear [5,6], so better profile holding can be expected with lower rates of wheel wear (higher G-ratios). With this in mind, it is of interest to compare regular grinding with creep-feed grinding. (For such comparisons, regular straight grinding is sometimes referred to as pendulum grinding, apparently because of the need to take many more passes by going back and forth.) From Eq. (11-10), it is apparent that, for a given removal rate, the average radial infeed velocity becomes smaller with more creep-feed-like conditions (larger a and smaller v_w), which should tend to reduce the radial wheel wear. Longer wheel-workpiece contact lengths with creep-feed grinding should also result in lower forces per grain, which implies a lower overall rate of wheel wear. This explains why it is almost invariably found that creep-feed grinding results in less profile wear and better form holding [33, 34].

But the longer contact length with creep-feed grinding, while reducing the grinding severity by distributing the grinding action over a larger area, has a disadvantage of imposing a condition of more sliding. For straight plunge grinding, the sliding length per unit volume of metal removal, L_s', can be readily obtained as

$$L_s' = \frac{a^{1/2} v_s}{\pi\, Q_w d_s^{1/2}} \tag{11-13}$$

so that increasing the depth of cut while maintaining the same removal rate (more creep-feed-like conditions) and wheel velocity results in relatively more sliding. This should lead to more attritious wear which, together with a reduction in the tendency for self-sharpening by fracture wear, means larger wear-flat areas and much bigger forces. One method to overcome this problem has been to provide continuous dressing of the wheel during grinding, which controls the wheel sharpness and also maintains the profile [35,36]. A particularly successful application of continuous rotary diamond dressing is creep-feed profile grinding of jet engine turbine blades and vanes.

11.5   ATTRITIOUS WEAR AND GRINDING CHEMISTRY

The suitability of abrasive grain materials for grinding particular workpiece metals depends to a very large degree on their attritious wear resistance. Where possible, the abrasive should be considerably harder than the material being ground, but hardness by itself is often not the prevailing factor. If it were, diamond would be the most wear-resistant abrasive for grinding any metal, and silicon carbide would rank higher than aluminum oxide. In practice, neither diamond nor silicon carbide is the best choice for grinding most ferrous alloys.

Attritious wear and the dulling of abrasive grits are both chemical and mechanical [37-43]. Chemical effects are likely to be more significant when the abrasive is somewhat harder than the workpiece and any of its included phases. As an abrasive grain interacts with the workpiece at the elevated temperatures reached in the grinding zone, numerous chemical reactions may occur involving the abrasive, workpiece metal, binder, atmosphere, and grinding fluid in various combinations [37]. Here we will consider only a few of the known reactions which seem to be particularly relevant. Some additional aspects of grinding chemistry will be considered in the following section on lubrication and grinding fluids.

As mentioned above, diamond, despite its extreme hardness, is not suitable for grinding most ferrous alloys. This anomalous behavior can be attributed to excessive attritious wear mainly by the reversion of diamond to graphite [44]. Degradation of diamond appears to be more rapid in the presence of iron and other ferrous metals unsaturated in carbon, owing to their affinity for carbon. This could explain why diamond is successfully used for grinding some cast irons high in carbon. Cubic boron nitride (CBN), although somewhat softer than diamond, is more chemically stable to higher temperatures and wears much less on most ferrous metals.

For grinding ferrous metals with aluminum oxide wheels, the important chemical reactions usually involve the oxidation of iron and the reaction of the oxide with the abrasive to form the spinel $FeAl_2O_4$ [37]:

$$2Fe + O_2 + 2Al_2O_3 \rightarrow 2FeAl_2O_4$$

This spinel is an intermediate compound between the oxidized workpiece metal and aluminum oxide, and its formation has been linked to stronger bonding between iron and aluminum oxide [45]. A higher attritious wear rate of aluminum oxide on steel in humid air than in dry air has been attributed to the catalytic effect of water on the oxidation of iron [37]. The affinity of alumina for metal oxides may also be a factor in grinding other metals, although the particular metal oxide of interest is not necessarily that of the main constituent of the metal alloy being ground. In grinding a cobalt-base superalloy, chrome oxide has been identified on the metal surface, which should strongly adhere to aluminum oxide owing to the mutual solid solubility of these two oxides [42]. It has been suggested that chrome oxide on stainless steels might have a similar effect.

Despite the apparent role of oxygen and water vapor in promoting adhesion and attrition, their elimination by grinding in a vacuum or inert atmosphere has a drastic effect on the process [46-48]. Although chemical bonding between the abrasive and metal may be reduced, nascent metal workpiece particles with freshly formed uncontaminated surfaces tend to physically adhere to each other and to the wheel surface, thereby loading and clogging the wheel. Surface oxidation and corrosion in a normal grinding atmosphere reduce the adhesion of metal particles to each other and their re-adhesion to the workpiece. This same phenomenon also seems to account for some of the difficulties encountered when grinding high-temperature oxidation-resistant metals, including some stainless steels, nickel-base alloys, and titanium [49].

Silicon carbide abrasives are harder than friable aluminum oxide (see Figure 2-4), but they are usually inferior for grinding most ferrous materials. This has been explained by the tendency for silicon carbide to react with and adhere to iron at elevated temperatures [39]. The main chemical reaction appears to be the dissociation of silicon carbide [43, 50, 51], and this reaction could also promote attritious wear when grinding titanium and other non-ferrous metals. Dissociation of silicon carbide at elevated grinding temperatures could be driven by the affinity of silicon or carbon for the workpiece. Therefore, silicon carbide tends to work better than aluminum oxide on some ferrous metals with excess carbon, but not on carbon-hungry ferrous metals unsaturated in carbon [51], which is analogous to the situation mentioned above for diamond. It has also been suggested that the superiority of silicon carbide on some cast irons is due to the presence of small amounts of SiC as a normal constituent in the iron, which would have a much more drastic effect on the wear of the softer aluminum oxide [37].

Aside from chemical activity, purely mechanical factors contribute significantly to attritious wear. In grinding some carbon and alloy steels, the G-ratio is usually found to be reduced somewhat when grinding the material in its fully hardened condition as compared with its annealed state [52, 53], which would suggest a mechanical effect. But hardness, by itself, is not necessarily indicative of grindability and attritious wear for grinding of materials, including those whose hardest phases are softer than the aluminum oxide abrasive.

A rather different situation arises when grinding high-speed tool steels. Dispersed carbide phases in these materials are hard enough mechanically to cut or shatter the aluminum oxide and cause very low

G-ratios, which accounts for the particular success of the much harder CBN, compared with aluminum oxide, in grinding high-speed steels. The hardest carbides in high-speed steels are of tungsten, molybdenum, and vanadium. Carbides of tungsten and molybdenum are comparable in hardness to friable aluminum oxide grains, and vanadium carbide is even somewhat harder. The volume fraction, C^*, of these carbides in tool steels relative to that of tungsten carbide by itself can be approximated in terms of the weight percentages of tungsten (W), molybdenum (Mo), and vanadium (V) as [52]:

$$C^* = W + 1.9Mo + 6.3V \tag{11-14}$$

The effect of this parameter on the G-ratio is summarized in Figure 11-12 with data taken from numerous sources, where G^* is the G-ratio for different tool steels expressed as a percentage relative to that of an M-2 steel [52]. Although the results fall within a rather wide band, it is clear that a higher carbide content makes grinding much more difficult. A similar correlation has also been established between the G-ratio and the vanadium content for grinding of high-speed steels with aluminum oxide wheels [2].

　　Aside from the particular carbide content, grinding wheel wear is also affected by the carbide morphology resulting from how the high-speed

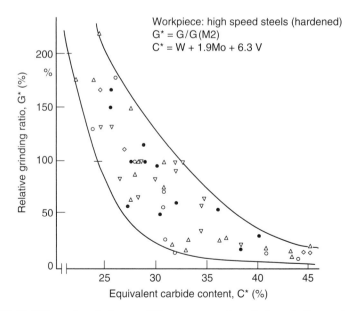

Figure 11-12 Effect of equivalent carbide content on the relative grinding ratio for hardened high-speed steels [52].

steel was processed. Of particular practical interest in this regard are high-speed steels produced by powder metallurgy techniques, which results in very fine uniformly dispersed carbides. Small, hard particles tend to be less abrasive than larger ones, which can account for lower wheel-wear rates and higher G-ratios obtained with high-vanadium tool steels produced by powder metallurgy methods [54, 55].

11.6 GRINDING FLUIDS AND LUBRICATION

Many grinding operations are performed with the aid of a grinding fluid. The grinding fluid is generally considered to have two main roles: cooling and lubrication. Grinding fluids are commonly referred to as coolants, but their role as lubricants is often more important.

Most grinding fluids can be categorized either as straight (or neat) oils or as soluble oils. Straight oils for grinding are mineral-oil-based fluids with additions of fatty materials for lubrication and wettability, and usually sulfur and/or chlorine for added wear reduction. Soluble oils are water-based fluids containing oil emulsions and numerous other ingredients which may include fatty materials, soaps, sulfur, and chloride for lubrication, surfactants for wetting and detergency and to prevent foaming, rust inhibitors, water conditioners, and germicides.

Straight oils are generally found to be better lubricants than soluble oils, as evidenced by higher G-ratios, lower grinding forces, and better surface quality [3, 51, 56-61]. Of course, the relative performance of the grinding fluid depends upon its particular formulation and application. The presence of water can have an adverse effect on the strength of the abrasive grain and the binder [58, 62, 63], thereby promoting fracture wear with water-based fluids, but the superior performance of straight oils appears to be related mainly to their ability to reduce attritious wear [56]. This is seen, for example, in Figure 11-13, which shows the influence of the grinding fluid (air (dry), a soluble oil at two concentrations in water, and a straight oil) on the wear-flat area in grinding a hardened bearing steel with a vitrified aluminum oxide wheel. Compared with dry grinding in air, the grinding fluids reduced attritious wear of the abrasive grains. The least wear-flat area by far was obtained with the straight oil, and this led to much lower sliding forces as well. Lubrication by the straight oil was also most effective in reducing the chip-formation and plowing energy components, but this effect was much less significant than wear-flat-area reduction in lowering the grinding forces.

In view of the importance of chemical reactions on attritious wear, it seems likely that the influence of the grinding fluid as a lubricant may be related to how it affects the grinding chemistry. The formation of lubricating

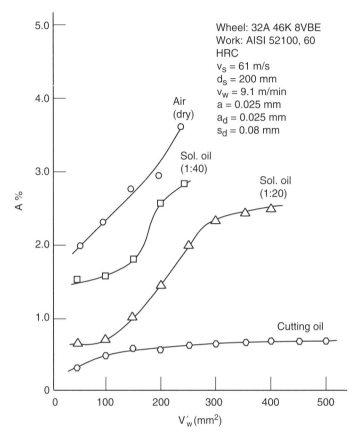

Figure 11-13 Wear-flat area versus accumulated metal removed per unit width with various fluids [55].

films, either by chemical or by physical action, can reduce workpiece-metal adhesion and also inhibit those chemical reactions which promote attritious wear [37]. For grinding of steels with aluminum oxide abrasives, lubricating films obtained with an active straight oil may provide a barrier to inhibit spinel formation, which is consistent with SEM observations of the wheel surface showing much less adhered metal and smaller wear flats with a straight oil than with a soluble oil [64]. For grinding of steels with silicon carbide, the straight oil may reduce attritious wear in a similar way by inhibiting the dissociation of silicon carbide and the diffusion of carbon to the workpiece [51]. On the other hand, the superior lubrication effectiveness of straight oils over soluble oils with resin-bonded CBN wheels has been attributed to reduced binder erosion [3], although the tendency for CBN to react with water may also be a factor favoring straight oils.

Commercial grinding fluids typically contain sulfur and/or chlorine as lubrication additives. Analogous to what has been postulated regarding the role of these elements in boundary lubrication of metals, it is generally believed that sulfur and chlorine react with the metal to form sulfide and chloride lubricating films, although the actual reactions seem to be much more complex. Beneficial lubricating effects of sulfur are also realized when it is incorporated as an addition to the workpiece (e.g. sulfurized free-machining steels) or by a sulfur treatment of the wheel. Sulfur additions to stainless and high-speed steels can increase the G-ratio by an order of magnitude [65].

In Chapters 6-8, we considered the role of grinding fluids as coolants, and here we have seen the importance of grinding fluids as lubricants. Cooling is most efficient with water-based fluids and lubrication with straight oils. With the notable exception of creep-feed grinding, cooling by grinding fluids appears to be generally ineffective in lowering the peak temperature within the grinding zone (Chapter 7). With improved lubrication and reduced wheel dulling, the grinding forces are reduced, thereby lowering the grinding zone temperature and the tendency for thermal damage. These considerations would seem to weigh heavily in favor of straight oils as opposed to soluble oils.

In actual practice, straight oils are used somewhat less than water-based fluids, which appears to be contrary to how well they perform. One advantage of water-based fluids as coolants is their superior ability to control the bulk temperature of the workpiece, which can reduce part-to-part size variations associated with thermal deformation of the workpiece, but this seems to be a secondary factor favoring the use of soluble oils. Unless lubrication is critical for form and finish, straight oils tend to be avoided mainly because of pollution and safety considerations. Oil-based fluids create mist and fumes in the atmosphere, and they may also present a fire hazard. Special environmental and safety precautions and equipment are necessary.

An alternative to straight oils are synthetic fluids, which are water-based chemical solutions with additions, containing little or no mineral oil, for lubrication, cleaning and corrosion control, and bacteria control. Synthetic fluids appear to be generally inferior to straight oils as lubricants. However, one particular area where synthetic fluids may be better lubricants than soluble oils and straight oils is for grinding titanium alloys [57, 59, 66, 67]. For this purpose, alkaline phosphate solutions, which can be buffered to nearly neutral pH to avoid skin irritation and chemical attack on the paint of the machine [68], are found to be particularly effective in reducing wheel wear. Their lubrication effectiveness has been attributed to strong physical adsorption of phosphate ions on titanium. In grinding titanium with silicon carbide wheels, a sharp drop in wheel wear when using a grinding fluid containing sodium phosphate was linked to a much lower silicon content at the finished workpiece surface, as measured by X-ray diffraction [69]. The adsorbed phosphate ions would seem to inhibit diffusion of silicon, and possibly also carbon, to the titanium workpiece, thereby retarding the dissociation of the abrasive. Phosphate solutions have not been adopted, at least to any significant extent, for grinding titanium.

For a grinding fluid to be effective as a lubricant and coolant, it must be delivered in sufficient quantity to the grinding zone as seen in Chapter 9. There are two main methods of grinding fluid delivery: low pressure (flood) and high pressure (jet) by means of nozzles. Most conventional grinders are equipped with simpler low-pressure systems, whereas creep-feed and high-speed grinders are more likely to have high-pressure systems. For creep-feed grinding at high rates of stock removal, large quantities of heat must be removed by pumping large volumes of fluid through the grinding zone (Chapter 7). With high-speed grinding, the fluid velocity with a low-pressure system may be insufficient to penetrate the boundary layer of air surrounding the wheel, thereby preventing the fluid from reaching the grinding zone [60, 61, 70, 71]. Aside from using high pressures, one simple solution to this problem is to break up the air film by positioning a scraper plate close to the wheel surface at a location just ahead of where the fluid hits the wheel. With high-pressure systems, multiple nozzles are often used to enhance fluid flow to the grinding zone, and this may have the added benefit of reducing the sensitivity of the system to nozzle location and orientation. Another approach is to use a 'shoe', rather than nozzles, which is closely fitted to the wheel so as to force the fluid into the wheel surface [61, 72].

In addition to their roles as coolants and lubricants, grinding fluids applied at high pressure also mechanically clean the wheel by removing adhered metal [73-76]. Jet infusion of fluids is particularly effective when grinding high-strength oxidation resistant alloys which tend to load the wheel, and for creep-feed grinding where there is a greater need to keep the wheel sharp as well as providing high flow rates for cooling. Separate nozzles for wheel cleaning may be directed towards areas on the wheel surface away from the grinding zone. Fluid supply pressures typically range from about 5 to 20 atmospheres, and much higher pressures of 50 atmospheres or more work much better [75, 76]. Aside from the additional pumping capacity, a practical limitation on jet pressure is nozzle wear by contaminant particles suspended in the fluid. Pressures of 50 atmospheres are feasible only with fluid filtration to remove particles bigger than 5 μm [76].

During grinding, most of the applied grinding fluid is recovered, pumped back to a tank, filtered to remove debris, and then reused. In most cases, the fluid is frequently replenished or treated to make up for losses or degradation, or completely replaced in which case the old fluid is either recycled or disposed. For soluble oils, costly disposal treatment usually consists of several steps [77]: breaking the emulsion, separation of oils and fats, and secondary treatment of the water and oil phases (precipitated oil and saturated sludge). Recycling of straight oils is simpler than for soluble oils because no detoxification and water conditioning are needed. The recycling consists mainly of filtering and replacing additives. Instead of recycling, the used oil may be used as a combustible fuel in energy production [77].

Grinding fluids present environmental and safety hazards, so there is considerable interest in replacing them with more benign fluids, reducing their use, or eliminating them altogether. One novel approach has been to combine the beneficial effects of lubrication by oil and the cooling properties of water in a high concentration solution of non-hazardous vegetable oil in water [78]. For CBN grinding, the best grinding behavior with 45% concentration of vegetable oil was comparable to what was obtained with straight oil. Cryogenic liquefied gases have also been considered as possible alternative fluids. Liquid nitrogen, the least expensive of these fluids, was reported to successfully lower grinding forces and reduce thermal damage [79, 80], although a subsequent study indicates much poorer lubrication than with soluble oils [81]. Another approach has been to apply Minimum Quantity Lubrication (MQL), whereby a miniscule amount of a non-hazardous ester oil is applied at a controlled rate using a precision dispenser. MQL has been found to provide comparable or better lubricating performance than soluble oil [82, 83]. However one notable drawback with MQL is insufficient bulk cooling of the workpiece which may lead to dimensional errors especially with long grinding cycles. Going one step further, attempts have also been made to completely eliminate the use of fluids altogether by treating the grinding wheel with molybdenum disulfide [84] or graphite [85] solid lubricants.

11.7   EVALUATING WHEEL PERFORMANCE

An important practical problem is how to evaluate the performance of grinding wheels. Historically, the most widely used parameter to judge whether one wheel is better than another has been the G-ratio. As pointed out at the beginning of this chapter, high G-ratios are generally desirable, but a more wear-resistant wheel may give higher forces and energy, thereby increasing the likelihood of thermal damage to the workpiece.

A more meaningful test for evaluating wheel performance, which is used by some wheel manufacturers, is to measure both the G-ratio and the grinding power under fixed grinding conditions [86, 87]. The general approach is to test a series of wheels covering a range of grades rather than a single wheel, as this makes it possible to distinguish between inherent wheel quality and wheel hardness effects. An underlying theoretical basis for this approach can be seen by deriving a relationship between the specific grinding energy and G-ratio as follows.

For the grinding model in Chapter 5, it was seen that the total specific energy includes chip-formation, plowing, and sliding components:

$$u = u_{ch} + u_{pl} + u_{sl} \qquad (11\text{-}15)$$

At a fixed removal rate, both the chip-formation and the plowing components remain constant. However, the specific sliding energy can be expressed in terms of the volumetric wheel-wear rate Q_s and attritious wear parameter q_s of the abrasive as

$$u_{sl} = \frac{P_o}{Q_{w,n}} \left(\frac{q_s}{Q_s} \right) \qquad (11\text{-}16)$$

where P_o and $Q_{w,n}$ are structural wheel characteristics which are assumed to be independent of grade. For a volumetric removal rate Q_w, the G-ratio can be simply written as

$$G = \frac{Q_w}{Q_s} \qquad (11\text{-}17)$$

Combining Eqs. (11-15), (11-16) and (11-15) leads to the final result

$$u = (u_{ch} + u_{pl}) + \left(\frac{P_o Q_w}{Q_{w,n}} \right) q_s G \qquad (11\text{-}18)$$

Therefore a plot of specific energy versus G-ratio with different wheel grades should yield a straight line whose slope is proportional to q_s.

An illustration of the application of this testing method is shown in Figure 11-14 for evaluating two series of wheels having identical grain content but different vitrified bonds referred to as VX and VY. In general, harder-grade wheels result in higher specific energies, but the corresponding G-ratios only tend to follow the expected linear behavior with the softer wheels. At about K-grade, in each case there is a reversal in the curve, which is caused by workpiece burn and accelerating wear (Figure 11-1) during a portion of the grinding interval. Of the two bond materials, clearly the VY is superior, as its curve lies to the right of the VX, which means higher G-ratios at the same grinding energy level and removal rate. Furthermore, the use of the J-grade wheel might be the best choice in order

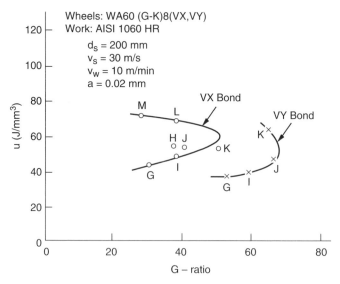

Figure 11-14 *Evaluating the performance of vitrified grinding wheels with two different bond materials. The letters on each graph refer to the wheel grades.*

to obtain the least wear while avoiding burning with the particular grinding conditions used in this test. According to Eq. (11-18), the straight-line extensions of both curves in Figure 11-14 to zero G-ratio should lead to the specific energy for a perfectly sharp wheel ($u_{ch} + u_{pl}$).

This same test method also shows the effect of other wheel parameters, including grain material and grit size, on grinding wheel performance [86,87]. The results are usually summarized with grinding power rather than specific energy plotted against G-ratio, but this has no effect on the final result for testing at a fixed removal rate. Of course, the conclusions to be drawn from comparative testing apply only to the particular type of grinding and test material.

Another use for this type of testing is for quality control in wheel manufacture. For example, the H-grade wheel with the VX bond in Figure 11-14 seems to fall out of line, as its energy requirement is close to that of the harder K-grade wheel. Its G-ratio also seems to be inconsistent with the other results, although this effect could not be reliably confirmed since G-ratio data are subject to more scatter. Wheel-grade shifts were also found with sonic testing (Figure 2-8). With vitrified wheels, shifts of two grades or more were not uncommon several years ago, but the situation has greatly improved owing to better production control. Wheel-to-wheel consistency and uniformity within the wheel may be more important than the inherent performance of the grain-bond system.

REFERENCES

1. Backer, W. R. and Krabacher, E. J., 'New Techniques in Metal-cutting Research', *Trans. ASME*, 78, 1956, p. 1497.
2. Tarasov, L. P., 'Grindability of Tool Steels', *Trans. ASME*, 43, 1951, p. 1144.
3. Tönshoff, H. K. and Graber, T. 'Cylindrical and Profile Grinding with Boron Nitride Wheels', *Proceedings of the 5th International Conference on Production Engineering*, JSPE, Tokyo, 1984, p. 326.
4. Radhakrishnan, V. and Achyutha, B. T., 'In-process Monitoring of Corner Wear in Cylindrical Plunge Grinding', *Milton C. Shaw Grinding Symposium*, PED-16, ASME, New York, 1985, p. 341.
5. Werner, G., 'Analytical Description of Wheel Wear', *Proceedings of the International Conference on Production Engineering*, Pt. 2, JSPE, Tokyo, 1974, p. 64.
6. Werner, G., 'Relation between Grinding Work and Wheelwear in Plunge Grinding', SME Paper No. MR 75-610, 1975.
7. Oliveira, J. F. G., Silva, E. J., Gomes, J. J. F., Klocke, F. and Friedrich, D., 'Analysis of Grinding Strategies Applied to Crankshaft Manufacturing', *Annals of the CIRP*, 54/1, 2005, p. 269.
8. Saljé, E., 'Wear Characteristics for the Description of the Grinding Process', CIRP Report, 1982.
9. Pekelharing, A. J., Verkerk, J. and Van Beukering, F. C., 'A Model to Describe Wheel Wear in Grinding', *Proceedings of the International Grinding Conference*, Pittsburgh, 1972, p. 412.
10. Vickerstaff, T. J., 'Wheel Wear and Surface Roughness in Cross Feed Surface Grinding', *Int. J. Mach. Tool Des. Res.*, 13, 1973, p. 183.
11. Banerjee, J. K. and Hillier, M. J., 'Wheel Wear Pattern in Surface Grinding', *The Tool and Manufacturing Engineer*, Feb. 1969, p. 59.
12. Yoshikawa, H., and Sata,T., 'Study on Wear of Grinding Wheels', *Trans. ASME, J. of Eng. for Ind.*, 85, 1963, p. 39.
13. Yoshikawa, H., 'Fracture Wear of Grinding Wheels', *Production Engineering Research Conference*, ASME, 1963, p. 209.
14. Malkin, S. and Cook, N. H., 'The Wear of Grinding Wheels. Part 1 – Attritious Wear', *Trans. ASME, J. of Eng. for Ind.*, 93, 1971, p. 1120.
15. Malkin, S. and Cook, N. H., 'The Wear of Grinding Wheels. Part 2 – Fracture Wear', *Trans. ASME, J. of Eng. for Ind.*, 93, 1971, p. 1129.
16. Stetiu, G. and Lal, G. K., 'Wear of Grinding Wheels', *Wear*, 30, 1974, p. 229.
17. Pande, S. J. and Lal, G. K., 'Wheel Wear in Dry Surface Grinding', *Int. J. Mach. Tool Des. Res.*, 16, 1976, p. 179.
18. Pahlitzsch, G. and Thormählen, K. H., 'Grinding with Alloyed and Unalloyed Corundum', *Proceedings of the International Grinding Conference*, Pittsburgh, 1972, p. 127.
19. Kannappan, S. and Malkin, S., 'Effects of Grain Size and Operating Parameters on the Mechanics of Grinding', *Trans. ASME, J. of Eng. for Ind.*, 94, 1972, p. 833.
20. Shi, Z., and Malkin, S., 'An Investigation of Grinding with Electroplated CBN Wheels', *Annals of the CIRP*, 52/1, 2003, p. 267.
21. Shi, Z, and Malkin, S., 'Wear of Electroplated CBN Grinding Wheels', *Trans. ASME Journal of Manufacturing Science and Engineering*, 128, 2006, p. 110.

22. Foster, M., Ramanan, N., 'Wear Mechanism of an Electroplated CBN Grinding Wheel during Grinding of a Nickel Base Alloy with Aqueous-Based Coolant', *Proceedings of the 2nd International Machining and Grinding Conference*, 1997, p. 25.

23. Stokes, R. J. and Valentine, T. J., 'Wear Mechanisms of ABN Abrasive', *Industrial Diamond Review*, 44, 1984, p. 34.

24. Kumar, K. V, 'Technical Advancements in CBN Grinding Products and Applications', *Proceedings of the 2nd International Machining and Grinding Conference*, 1997, p. 93.

25. Verkerk, J., 'Characterization of Wheel Wear in Plunge Grinding', *Annals of the CIRP*, 26/1, 1977, p. 129.

26. Andrew, C., Howes, T. D. and Pearce, T. R. A., *Creep Feed Grinding*, Holt, Rinehart, and Winston, London, 1985, Chapter 4.

27. Tsuwa, H. and Yasui, H., 'Microstructure of Dressed Abrasive Cutting Edges', *Proceedings of the International Grinding Conference*, Pittsburgh, 1972, p. 142.

28. Snoeys, R., Peters, J. and Decneut, A., 'The Significance of Chip Thickness in Grinding', *Annals of the CIRP*, 23/2, 1974, p. 227.

29. Brecker, J. N., Sauer, W. J. and Shaw, M. C., 'Conditioning of Steel', *Proceedings of the International Grinding Conference*, Pittsburgh, 1972, p. 562.

30. Farmer, D. A. and Shaw, M. C., 'Economics of the Abrasive Cutoff Operation', *Trans. ASME, J. of Eng. for Ind.*, 89, 1967, p. 514.

31. Davis, H. F. and Patch, J., 'Choosing Least-cost Abrasives for Snag Grinding', *Foundry*, Feb. 1969, p. 52.

32. Shaw, M. C., 'Cost Reduction in Stock Removal Grinding', *Annals of the CIRP*, 24/2, 1973, p. 539.

33. Andrew, C., Howes, T. D. and Pearce, T. R. A., *Creep Feed Grinding*, Holt, Rinehart, and Winston, London, 1985, Chapter 7.

34. Kita, Y., Damlos, H. H. and Saljé, E., 'Wheel Wear in Profile Grinding', *Proceedings of the 5th International Conference on Production Engineering*, JSPE, Tokyo, 1984, p. 612.

35. Andrew, C., Howes, T. D. and Pearce, T. R. A:, *Creep Feed Grinding*, Holt, Rinehart, and Winston, London, 1985, Chapter 6.

36. Pearce, T. R. A., Howes, T. D. and Stuart, T. V. 'The Application of Continuous Dressing in Creep Feed Grinding', *Proceedings of the Twentieth International Machine Tool Design and Research Conference*, 1979, p. 383.

37. Coes, L., Jr., *Abrasives*, Springer-Verlag, New York, 1971, Chapter 14.

38. Tsuwa, H. and Kawamura, S., 'On the Wear by Attrition of Abrasive Grains', *Bull. Japan Soc. Prec. Engg.*, 2, 1966, p. 40.

39. Goepfert, G. J. and Williams, J. L., *Mechanical Engineering*, April 1959, p. 69.

40. Kirk, J. A. and Syniuta, W. S., 'Scanning Electron Microscopy and Microprobe Wear Studies on Single Crystal Aluminum-Oxide Vitrified Wheels', *Proceedings Second North American Metalworking Research Conference*, 1974, p. 572.

41. Eiss, N. S. and Fabiniak, R. C., 'Chemical and Mechanical Mechanisms in Wear of Sapphire on Steel', *J. Amer. Ceram. Soc.*, 49, 1966, p. 221.

42. Komanduri, R., 'The Mechanism of Metal Build-up on Aluminum Oxide Abrasive', *Annals of the CIRP*, 25/1, 1976, p. 191.

43. Duwell, E. J., Hong, I. S. and McDonald, W. J., 'The Role of Chemical Reactions in the Preparation of Metal Surfaces by Abrasion', *Wear*, 9, 1966, p. 417.

44. Ikawa, N. and Tanaka, T., 'Thermal Aspects of Wear of Diamond Grain in Grinding', *Annals of the CIRP*, 19, 1971, p. 153.

45. Pepper, S. V., 'Shear Strength of Metal-Sapphire Contacts', *J. Appl. Phys.*, 47, 1967, p. 801.
46. Tanaka, Y. and Ueguchi, T., 'The Role of Oxygen in Grinding', *Annals of the CIRP*, 19, 1971, p. 449.
47. Outwater, J. O. and Shaw, M. C., 'Surface Temperatures in Grinding', *Trans. ASME*, 74, 1972, p. 73.
48. Duwell, E. J., Hong, I. S. and McDonald, W. J., 'The Effect of Oxygen and Water on the Dynamics of Chip Formation during Grinding', *ASLE Trans.*, 12, 1969, p. 86.
49. Yossifon, S. and Rubenstein, C., 'The Grinding of Workpieces Exhibiting High Adhesion. Part I: Mechanisms', *Trans. ASME, J. of Eng. for Ind.*, 103, 1981, p. 144.
50. Shaw, M. C. and Komanduri, R., 'Attritious Wear of Silicon Carbide', *Trans. ASME, J. of Eng. for Ind.*, 98, 1976, p. 1125.
51. Ahmed, O. I. and Dougdale, D. S., 'Performance of Silicon Carbide Wheels in Grinding Tool Steels', *Proceedings of the Seventeenth International Machine Tool Design and Research Conference*, 1976, p. 165.
52. König, W. and Messer, J., 'Influence of the Composition and Structure of Steels on Grinding Process', *Annals of the CIRP*, 30/2, 1981, p. 547.
53. Spath, H. P., 'Einfluss der Werkstockzusammensetzung und des Gefluges auf das Schleifverhalten von Stahlen', Dissertation, TU Berlin, 1969.
54. Komanduri, R. and Shaw, M. C., 'On the Grindability of AISI T-15 Tool Steel Produced by the Consolidation of Atomized Metal Powder', *Proceedings Third North American Metalworking Research Conference,* 1975, p. 481.
55. Badger, J., 'Grindability of Conventionally Produced and Powder-Metallurgy High-Speed Steels', *Annals of the CIRP*, 56/1, 2007, p. 353.
56. Osman, M. and Malkin, S., 'Lubrication by Grinding Fluids at Normal and High Wheel Speeds', *ASLE Trans.*, 15, 1972, p. 26.
57. Shaw, M. C., 'Grinding Fluids', SME Paper No. MR70-277, 1970.
58. Mittal, R. N., Porter, T. M. and Rowe, G. W., 'Lubrication with Cutting and Grinding Emulsions', *Lubrication Engineering*, 40, 1984, p. 160.
59. Tarasov, L. P., 'Grinding Fluids. Part 2 – How They Affect Grinding Action', *The Tool and Manufacturing Engineer*, July 1961, p. 60.
60. Opitz, H. and Guhring, K., 'High Speed Grinding', *Annals of the CIRP*, 16, 1968, p. 61.
61. Guhring, K., 'Hochleistungsschleifen – Eine Methode zur Leistungssteigerung der Schleifverfahren durch hohe Schnittgeschwindigkeiten', Dissertation, TH Aachen, 1967.
62. Komine, N., 'Effects of Surrounding Environment on Wear of Grinding Wheels', *Bull. Japan Soc. of Prec. Engg.*, 4, 1970, p. 115.
63. Imanaka, G., Fujino, S. and Shinohara, K., 'Effect of Environments on Fracture Strength of Aluminum Oxide Grain', *Bull. Japan Soc. of Prec. Engg.*, 2, 1966, p. 22.
64. Foerster, M. and Malkin, S., 'Wear Flats Generated during Grinding with Various Grinding Fluids', *Proceedings Second North American Metalworking Research Conference*, 1974, p. 601.
65. Tarasov, L. P., 'Factors Affecting the Grindability of Highly Alloyed Steels', *Proceedings International Conference on Manufacturing Technology*, ASTM, 1967, p. 689.
66. Shaw, M. C. and Yang, C. T., 'Inorganic Grinding Fluids for Titanium Alloys', *Trans. ASME*, 78, 1956, p. 861.

67. Yang, C. T. and Shaw, M. C., 'Grinding of Titanium Alloys', *Trans. ASME*, 77, 1955, p. 645.

68. Hong, I. S., Duwell, E. J., McDonald, W. J. and Mereness, C. E., 'Coated Abrasive Machining of Titanium Alloys with Inorganic Phosphate Solutions', *ASLE Trans.*, 14, 1971, p. 8.

69. Sayutin, G. I. and Nosenko, V. A., 'Micromechanical Changes on the Surface of Titanium Alloys in Grinding', *Trenie Iznos*, 4, 1983, p. 348.

70. Kaliszer, H. and Trmal, G., 'Mechanics of Grinding Fluid Delivery', SME Paper MR75-614, 1975.

71. Akiyama, T., Shibata, L. and Yonetzu, S., 'Behaviour of Grinding Fluid in the Gap of the Contact Area between a Grinding Wheel and a Workpiece', *Proceedings of the 5th International Conference on Production Engineering*, JSPE, Tokyo, 1984, p. 52.

72. Andrew, C., Howes, T. D. and Pearce, T. R. A., *Creep Feed Grinding*, Holt, Rinehart, and Winston, London, 1985, Chapter 5.

73. Hollands, S. J. and Schofield, R. E., 'Wheel Clogging When Grinding High Strength Heat and Corrosion Resistant Alloys', *Proceedings of the International Conference on Production Engineering*, Pt. 1, JSPE, Tokyo, 1974, p. 677.

74. Eda, H. and Kishi, K., 'Improvement of Grinding Process for Difficult-to-grind Materials by the Jet Infusion of Grinding Fluids', *Proceedings of the International Conference on Production Engineering*, Pt. 1, JSPE, Tokyo, 1974, p. 677.

75. Khudobin, L. V., 'Cutting Fluid and its Effect on Grinding Wheel Clogging', *Machines & Tooling*, 40, 1969, p. 54.

76. Satow, Y., 'Use of High Pressure Coolant Supply in Precision CBN Grinding', SME Paper MR86-643, 1986.

77. Howes, T. D., Tönshoff, H. K., and Heuer, W., 1991, 'Environmental Aspects of Grinding Fluids', *Annals of the CIRP*, 40/2, 1991, p. 623.

78. Oliveira, J. F. G., and Alves, S. M., 'Development of Environmentally Friendly Fluid for CBN Grinding', *Annals of the CIRP*, 55/1, 2006, p. 343.

79. Paul, S. and Chattopadhyay, A. B., 'A Study of Effect of Cryo-cooling in Grinding', *Int. J. Mach. Tools Manufact.*, 35, 1995, p. 109.

80. Paul, S. and Chattopadhyay, A. B., 'The Study of Cryogenic Cooling on Grinding Forces', *Int. J. Mach. Tools Manufact.*, 36, 1996, p. 63.

81. Baheti, U., Guo, C., and Malkin, S., 'Environmentally-Conscious Cooling and Lubrication for Grinding,' *Proceedings, International Seminar on Improving Machine Tool Performance,* San Sebastian, Spain, Vol. II, 1998, p. 643.

82. Brinksmeier, E., Brockhoff, T. and Walter, A., 'Minimum Quantity Lubrication in Grinding', *Proceedings of the 2nd International Machining and Grinding Conference*, SME, Dearborn, Michigan, 1997, p. 639.

83. Hafenbraedl, D. and Malkin, S., 'Environmentally-conscious Minimum Quantity Lubrication (MQL) for Internal Cylindrical Grinding', *Transactions of NAMRI/SME*, 28, 2000, p. 149.

84. Tang, J. S., Pu, X. F., Xu, H. J., and Zhang, Y. Z., 'Studies on Mechanisms and Improvement of Workpiece Burn during Grinding of Titanium Alloys', *Annals of the CIRP*, 39/1, 1990, p. 353.

85. Shaji, S. and Radhakrishnan, V., 'An Investigation on Solid Lubricant Moulded Grinding Wheels,' *Int. J. Mach. Tools Manufact.*, 43, 2003, p. 965.

86. Reichenbach, G. S., 'Wheel-Fluid-Work Interactions', ASTME Paper MR67-594, 1967.

87. Coes, L., Jr., *Abrasives*, Springer-Verlag, New York, 1971, Chapters 12 and 13.

Chapter 12

Grinding Deflections: Grinding Cycles, Inaccuracies, and Vibrations

12.1 INTRODUCTION

Forces generated during grinding cause elastic deformation and deflection of the machine, the grinding wheel, and the workpiece. Grinding deflections lead to geometrical inaccuracies in the components being ground and should be carefully considered in the design of grinding cycles. The deflection between the wheel and the workpiece may greatly exceed the depth of cut taken by the wheel. Although this is usually recovered to a greater or lesser degree by spark-out at the end of the grinding cycle, non-symmetric machine defections cause shape inaccuracies and taper, and periodic deflections associated with machine-tool vibrations cause chatter, which reduces surface and geometrical quality and limits the production rate.

The present chapter is concerned with the role of deflections in grinding. To provide a basis for characterizing and controlling the grinding cycle, simplified linear analyses are presented which account for the difference between the machine infeed and actual stock removal in both continuous and discrete infeed operations. Some examples are presented which illustrate how deflections can lead to geometric inaccuracies. The chapter concludes with a brief description of forced and self-excited vibrations and their causes, and of methods for vibration suppression to enhance part quality and productivity.

12.2 CONTINUOUS INFEED ANALYSIS

Elastic deflection of the grinding system causes the actual stock removal to be less than the controlled infeed input to the machine. For analyzing this phenomenon, we begin with an idealized model as illustrated in

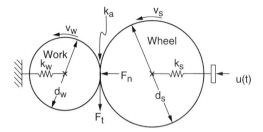

Figure 12-1 Idealized model of cylindrical plunge grinding.

Figure 12-1 for cylindrical plunge grinding. The machine structure supports the wheel with a linear spring of stiffness k_s and the workpiece with a linear spring of stiffness k_w. Together these two springs in series comprise the machine stiffness k_m in the infeed direction:

$$k_m^{-1} = k_s^{-1} + k_w^{-1} \tag{12-1}$$

Additional flexibility due to wheel and workpiece elasticity is considered to provide flexible contact of stiffness k_a at the grinding zone. Combining the contact stiffness with that of the machine, we obtain an overall effective stiffness k_e:

$$k_e^{-1} = k_m^{-1} + k_a^{-1} = k_s^{-1} + k_w^{-1} + k_a^{-1} \tag{12-2}$$

During grinding with controlled infeed, a time-dependent radial infeed velocity $u(t)$ is input to the machine (Figure 12-1), but the actual infeed velocity $v(t)$ corresponding to the radial size reduction rate of the workpiece is less[1]. Neglecting wheel wear for now, continuity requires that the difference between the controlled $u(t)$ and the actual $v(t)$ infeed velocities be equal to the time rate of change of the radial elastic deflection $\dot{\varepsilon}$ of the grinding system:

$$u(t) - v(t) = \dot{\varepsilon} \tag{12-3}$$

For the system in Figure 12-1, the deflection is given by

$$\varepsilon = \frac{F_n}{k_e} \tag{12-4}$$

---

[1] While the radial infeed velocity is referred to as v_f elsewhere in this book, here we use u and v to distinguish between the controlled and actual values, respectively.

where F_n is the normal force component and k_e is the effective stiffness (Eq. (12-2)).

To facilitate the analysis, we assume that the normal force component is proportional to the volumetric stock removal rate:

$$F_n = F_0 Q_w \tag{12-5}$$

or

$$F_n = F_0 b v_w a \tag{12-6}$$

For a given workpiece velocity v_w and grinding width b, the force is proportional to the wheel depth of cut a:

$$F_n = k_c a \tag{12-7}$$

where k_c is the cutting stiffness given by

$$k_c = F_0 b v_w \tag{12-8}$$

For cylindrical grinding, it may be more convenient to relate the normal force to the actual radial infeed velocity v (Figure 12-2). Writing the removal rate as

$$Q_w = \pi b d_w v \tag{12-9}$$

the normal force from Eq. (12-5) becomes

$$F_n = (\pi b d_w F_0) v \tag{12-10}$$

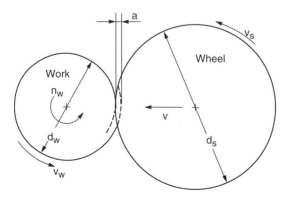

Figure 12-2 Illustration of cylindrical plunge grinding.

or

$$F_n = F_v v \qquad (12\text{-}11)$$

where F_v corresponds to the quantity within the parentheses in Eq. 12-10. Combining Eqs. (12-3), (12-4), and (12-11) leads to the differential equation

$$\dot{v} = \frac{1}{\tau}[u(t) - v(t)] \qquad (12\text{-}12)$$

where τ is a characteristic time constant:

$$\tau \equiv \frac{F_v}{k_e} \qquad (12\text{-}13)$$

The rate of workpiece radius reduction is equal to the actual infeed velocity

$$\dot{r} = v(t) \qquad (12\text{-}14)$$

where r is the accumulated reduction of the workpiece radius. For a controlled infeed velocity $u(t)$ input to the machine, the size-reduction process is described by Eqs. (12-12) and (12-14) together with the initial conditions

$$v(0) = v_0 \qquad (12\text{-}15)$$

and

$$r(0) = r_0 \qquad (12\text{-}16)$$

The same analysis can readily be modified to include the effect of wheel wear. For this purpose the radial wheel wear is assumed to be characterized by a grinding ratio G, which is defined as the ratio of volumetric stock removal to wheel consumption (see Chapter 11). For cylindrical plunge grinding

$$G = \frac{\pi d_w b v(t)}{\pi d_s b w(t)} = \frac{d_w v(t)}{d_s w(t)} \qquad (12\text{-}17)$$

where $w(t)$ is the radial wear rate of the wheel, and d_w and d_s are the workpiece and wheel diameters, respectively. Taking wheel wear into account, the continuity condition analogous to Eq. (12-3) is

$$u(t) - v(t) - w(t) = \dot{\varepsilon} \qquad (12\text{-}18)$$

Combining Eqs. (12-17) and (12-18) with Eqs. (12-3) and (12-4) yields

$$\dot{v} = \frac{1}{\tau'} [u'(t) - v(t)] \qquad (12\text{-}19)$$

where

$$\tau' \equiv \frac{\tau}{1 + \dfrac{d_w}{d_s G}} \qquad (12\text{-}20)$$

and

$$u'(t) \equiv \frac{u(t)}{1 + \dfrac{d_w}{d_s G}} \qquad (12\text{-}21)$$

Eq. (12-19) is the same as Eq. (12-12) but with τ replaced by τ' and $u(t)$ by $u'(t)$. The effect of wheel wear is insignificant when $G \gg d_w/d_s$.

12.3   GRINDING CYCLE BEHAVIOR

We will now use this analysis to predict the size-reduction behavior during infeed-controlled grinding cycles. We begin with the simple cycle illustrated in Figure 12-3, consisting of an initial roughing stage with a

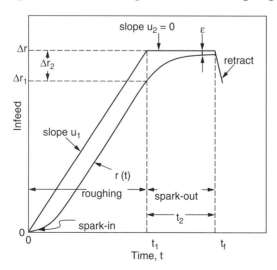

Figure 12-3 Simple grinding cycle consisting of roughing and spark-out stages.

constant controlled infeed velocity u_1 followed by spark-out with zero controlled infeed velocity ($u_2 = 0$). Solving Eqs. (12-12) and (12-14) for the first stage with $v(0) = 0$ and $r(0) = 0$ gives

$$v(t) = (1 - e^{-t/\tau})u_1 \qquad (12\text{-}22)$$

and

$$r(t) = (t + \tau e^{-t/\tau} - \tau)u_1 \qquad (12\text{-}23)$$

This behavior is shown in Figure 12-3. After an initial transient, the actual infeed velocity corresponding to the slope of the $r(t)$ curve in the first stage approaches the controlled infeed velocity, and the lag (deflection) of the actual infeed behind the accumulated controlled infeed approaches a steady-state value. How fast the transient occurs depends on the time constant τ. For $t \gg \tau$ in the steady state

$$v(t) = u_1 \qquad (12\text{-}24)$$

and

$$r(t) = u_1 t - u_1 \tau \qquad (12\text{-}25)$$

The quantity $u_1 \tau$ in Eq. (12-25) is the steady-state lag (deflection).

In theory, the time constant τ which characterizes the transient behavior may be estimated from Eq. (12-13) as the ratio of the force parameter F_v to the effective stiffness k_e, but accurate values for F_v and k_e are usually not available. One way to experimentally estimate the time constant is to measure the steady-state lag in the roughing stage by in-process gaging of the part diameter and dividing by u_1 [1]. Typical time constants might be $\tau \approx 0.5-1$ s for external grinding and $\tau \approx 1-10$ s for internal grinding, although these values vary widely. Longer time constants are usually obtained for internal grinding, owing mainly to a much lower wheel support stiffness k_s, since the wheel is mounted on the free end of a long and flexible spindle.

The initial roughing stage is followed by spark-out (Figure 12-3) with $u(t) = 0$, during which material removal continues at a decreasing rate until wheel retraction at $t = t_f$. Solving Eqs. (12-12) and (12-14) for spark-out ($t_1 < t < t_f$) with the initial conditions corresponding to the end of roughing (e.g. Eqs. (12-24) and (12-25) with $t = t_1$ for $t_1 \gg \tau$) leads to

$$v(t) = u_1 \exp\left(-\frac{t - t_1}{\tau}\right) \qquad (12\text{-}26)$$

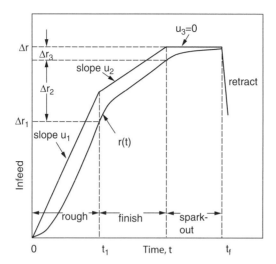

Figure 12-4 Grinding cycle with roughing, finishing, and spark-out stages.

and

$$r(t) = u_1 t_1 - u_1 \tau \exp\left(-\frac{t - t_1}{\tau}\right) \qquad (12\text{-}27)$$

The first term in Eq. (12-27) is the total controlled infeed Δr during the roughing stage, while the second term is the elastic deflection ε which exponentially decreases towards zero (Figure 12-3).

The same type of analysis can also be applied to more complex grinding cycles having one or more additional stages. Grinding cycles often have an intermediate finishing stage between roughing and spark-out, as illustrated in Figure 12-4. In each stage, the actual infeed velocity tends toward the controlled infeed velocity at a rate characterized by the time constant τ, and the steady-state lag (deflection) tends towards the product of the controlled infeed velocity and time constant. In practice, the time at which to switch from one stage to the next may be determined by the remaining stock to be removed, as monitored by a diametral size gage, or by other criteria.

12.4   DISCRETE INFEED ANALYSIS

A situation similar to that of cylindrical grinding applies to straight surface grinding with reciprocating table motion where the downfeed (depth of cut) is incremented in discrete steps between passes rather than continuously. Discrete infeed behavior can be analyzed by considering what happens during successive passes over the workpiece. For the first pass, the depth of cut is set at d, but the true depth removed a_1 is less owing to normal deflection of the grinding system. Neglecting wheel wear as before, continuity requires that

$$d - a_1 = \varepsilon_1 \tag{12-28}$$

where a_1 is the true depth of cut and ε_1 the deflection for the first pass. Assuming a proportional relationship between the normal force and true depth of cut (Eq. (12-7)) and a spring-like elastic system (Eq. (12-4)) we obtain

$$a_1 = \frac{d}{1 + \dfrac{k_c}{k_e}} \tag{12-29}$$

Prior to the second pass, the downfeed is incremented again by d, but there is also a residual thickness $(d - a_1)$ to be cut following the first pass. Analogous to Eq. (12-28), continuity requires that

$$d + (d - a_1) - a_2 = \varepsilon_2 \tag{12-30}$$

where a_2 is the true depth removed and ε_2 is the elastic deflection for the second pass. Combining Eqs. (12-30), (12-29), (12-7) and (12-4) gives

$$a_2 = \frac{2d - a_1}{1 + \dfrac{k_c}{k_e}} \tag{12-31}$$

Continuing the same procedure, the true depth of cut for the mth pass is obtained as

$$a_m = \frac{md - \displaystyle\sum_{n=1}^{m-1} a_n}{1 + \dfrac{k_c}{k_e}} \tag{12-32}$$

This can be written in a more convenient form as

$$a_m = d\left[1 - \left(\frac{1}{1 + \dfrac{k_e}{k_c}}\right)^m\right] \tag{12-33}$$

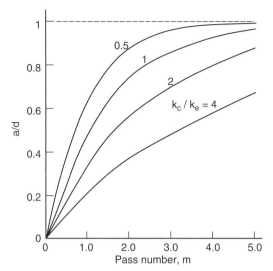

Figure 12-5 Ratio of true depth of cut to incremented downfeed versus number of grinding passes for different values of k_c/k_e.

This result is plotted in Figure 12-5 for different values of k_c/k_e. A steady-state condition is asymptotically approached where $a = d$ and the steady-state deflection is

$$\varepsilon_\infty = \frac{k_c d}{k_e} \qquad (12\text{-}34)$$

As with continuous infeed, elastic deflection can be recovered by taking additional spark-out passes over the workpiece without incrementing the downfeed. Assuming that the steady state was reached in prior rough grinding, the actual depth of cut a_1 for the first spark-out pass is less than ε_∞ owing to the elastic deflection ε_1 of the system:

$$a_1 = \varepsilon_\infty - \varepsilon_1 \qquad (12\text{-}35)$$

(Note that a_1 and ε_1 now refer to spark-out passes.) Substituting for ε_∞ from Eq. (12-33) and combining with Eqs. (12-4) and (12-7) leads to

$$a_1 = \frac{d}{1 + \dfrac{k_e}{k_c}} \qquad (12\text{-}36)$$

for the first spark-out pass. The allowance remaining for the second spark-out pass is reduced by the amount a_1 removed in the first spark-out pass, so the true depth of cut is

$$a_2 = (\varepsilon_\infty - a) - \varepsilon_2 \tag{12-37}$$

where ε_2 is now the elastic deflection. By analogy with the first spark-out pass:

$$a_2 = \frac{d - \dfrac{a_1 k_e}{k_c}}{1 + \dfrac{k_e}{k_c}} \tag{12-38}$$

Proceeding in this way, the true depth of cut for the pth spark-out pass is obtained as

$$a_p = \frac{d - \dfrac{k_e}{k_c}\displaystyle\sum_{n=1}^{p-1} a_n}{1 + \dfrac{k_e}{k_c}} \tag{12-39}$$

which is equivalent to

$$a_p = \frac{k_c d}{k_e}\left(\frac{1}{1 + \dfrac{k_e}{k_c}}\right)^p \tag{12-40}$$

What happens with discrete infeed grinding can also be analyzed in a different way using the continuous infeed results of the previous section. Since each discrete pass is analogous to one workpiece revolution in cylindrical grinding, the controlled downfeed for discrete grinding can be written as

$$d = \frac{u_1}{n_w} \tag{12-41}$$

and the true depth of cut as

$$a(t) = \frac{v(t)}{n_w} \tag{12-42}$$

where u_1 is the controlled infeed velocity and n_w is the rotational workspeed (see Figure 12-2). Combining with Eqs. (12-22) for $v(t)$ and noting that

$$t = mn_w^{-1} \tag{12-43}$$

leads to

$$a_m = (1 - e^{-m/m_0})d \qquad (12\text{-}44)$$

where m is the pass number as before and

$$m_0 \equiv \frac{k_c}{k_e} \qquad (12\text{-}45)$$

Likewise for spark-out

$$t - t_1 = pn_w^{-1} \qquad (12\text{-}46)$$

which combined with Eq. (12-26) yields

$$a_p = de^{-p/p_0} \qquad (12\text{-}47)$$

where p is spark-out pass number and

$$p_0 \equiv \frac{k_c}{k_e} \qquad (12\text{-}48)$$

These formulae for a_m and a_p (Eqs. (12-44) and (12-47)) can be shown to be very close to those obtained using the discrete infeed analysis (Eqs. (12-32) and (12-39)) provided that

$$p_0 = m_0 \approx \frac{k_c}{k_e} + 0.5 \qquad (12\text{-}49)$$

instead of Eqs. (12-45) and (12-48). This difference between the discrete and continuous infeed analyses arises because Eq. (12-42), relating the true depth of cut to true infeed velocity, is strictly valid only when $v(t)$ is constant. More precisely, the true depth of cut with continuous infeed is equal to the difference in accumulated infeed between successive workpiece revolutions, or

$$a(t) = r(t) - r(t - n_w^{-1}) \qquad (12\text{-}50)$$

For the continuous infeed analysis, the normal force was considered to be proportional to the true infeed velocity, which facilitates the analysis, but not to the true depth of cut or removal rate as in the discrete analysis. However, both analyses lead to the same type of transient behavior.

12.5   INACCURACIES AND ELASTIC DEFLECTIONS

The objective in cylindrical grinding operations is to produce axisymmetrical parts of revolution. Ideally a cross-section normal to the part axis should be round, but cylindrical grinding with radial infeed generates a spiral rather than a round shape, as seen in Figure 12-2. From simple geometrical considerations, it is apparent that the roundness and diametrical tolerances which can be held are inherently limited by the true depth of cut at the end of the grinding cycle. A more nearly round part and better size control should be possible with a smaller true infeed velocity at the end of the grinding cycle, and a faster rotational workspeed should also be beneficial for this purpose.

For a simple grinding cycle (Figure 12-3), one important role of spark-out is to reduce the true depth of cut as the elastic deflection is recovered in order to improve part geometry. While the degree of spark-out is related to the time constant τ, the true depth of cut after a given time under spark-out also depends upon the depth of cut in the prior roughing stage (Eqs. (12-26) and (12-42)).

Many practical grinding cycles have an intermediate finishing stage between roughing and spark-out (Figure 12-4) with a controlled infeed velocity which is about 20% of the value for roughing. Since the true depth of cut obtained is also correspondingly smaller, the roundness prior to final spark-out is also improved. With an intermediate finishing stage, acceptable roundness may be achieved in many cases with a very short spark-out, even less than τ. It should be realized, however, that reduction of the true depth of cut for roughing to a final required value may be accomplished more quickly by simple spark-out without any finishing stage. The necessity for a finishing stage may be for size control or to remove a thin layer of material thermally damaged during roughing.

There are numerous instances where elastic deflections of the grinding system cause part inaccuracies. Deflections in the infeed direction tend to cause oversized parts, although much of this error may be recovered during spark-out. Some of the more obvious inaccuracies associated with grinding deflections are illustrated in Figure 12-6. In internal grinding, bending of the wheel spindle under normal load causes taper (Figure 12-6(a)). A taper or profile error can also occur when grinding nonsymmetrical profiles or against the side of the wheel, which bends the wheel and possibly also the part being ground (Figure 12-6(b)) [2]. Another problem which arises particularly in creep-feed grinding with large depths of cut and long grinding zone contact lengths is illustrated in Figure 12-6(c) [3]. As the wheel passes off the end of the workpiece, the normal force F_n decreases which in turn reduces the deflection and increases the true depth of cut. A similar type of overcutting occurs in cylindrical grinding with traverse, where the force

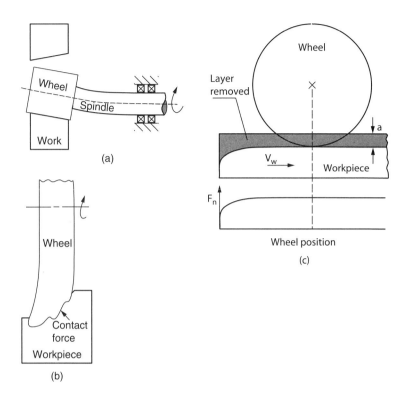

***Figure 12-6 Inaccuracies due to grinding deflections: (a) taper due to spindle deflection
in internal grinding; (b) shape and taper error due to non-symmetrical
force and wheel deflection; (c) overcut error due to decrease in force and
deflection as wheel passes off the end of the workpiece.***

also decreases as the wheel passes off the end of the workpiece. These geometrical inaccuracies are in addition to those caused by other factors such
as wheel wear and thermal distortion.

12.6 ACCELERATED SPARK-IN AND SPARK-OUT

Elastic deflection of the grinding system lengthens the total cycle
time. With the simple cycle in Figure 12-3, for example, time is lost both
during the spark-in transient at the start of the cycle as the wheel initially
engages the workpiece, and during the final spark-out transient at the end
of the cycle. The steady-state lag of $u_1\tau$ in the roughing stage (section 12.3)
indicates a time loss during spark-in equal to the time constant τ. For spark-
out, the time required depends on the degree of spark-out, as mentioned

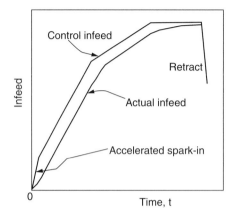

Figure 12-7 Grinding cycle with accelerated spark-in.

above. For example, recovering 90% of the elastic deflection by spark-out would theoretically add about 2.3τ to the cycle time.

One method to accelerate the initial spark-in transient is by the use of controlled force grinding (section 5.7), especially for internal grinding operations which have relatively long time constants. With a controlled normal force input to the system, the true infeed velocity adjusts itself to the preset normal force according to the characteristic grinding behavior (e.g. Eq. (12-24), Figures 5-20 and 5-21), whereas with infeed-controlled grinding the normal force gradually builds up during spark-in as the true infeed velocity increases toward its 'steady state' value. However the response of controlled-force grinding machines may be slowed down somewhat by damping which is necessary for removing the initial out-of-roundness of the workpiece. Some infeed-controlled grinders are also equipped for accelerated spark-in with either pre-set controlled force or spindle power during the initial portion of the grinding cycle.

Another method to accelerate spark-in at the start of grinding is to add an initial short stage with a high controlled infeed rate at the start of grinding. This is illustrated in Figure 12-7 which shows a three-stage cycle which has been modified by the addition of a fourth stage at the start of grinding to quickly accelerate the actual infeed velocity up to the desired value for roughing. The time to initially accelerate the actual infeed up to the desired value can be estimated from the continuous infeed analysis (Eq. (12-22)) as:

$$t = \tau \ln \left(1 - \frac{v}{u_1} \right) \qquad (12\text{-}51)$$

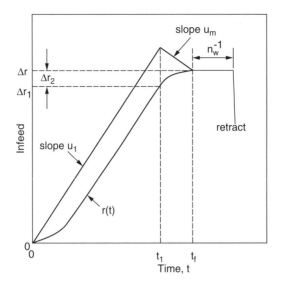

Figure 12-8 Grinding cycle with accelerated spark-out.

For example, this analysis predicts that it will take only about 30% of the time constant $(t = 0.3\tau)$ to reach the desired infeed velocity if the control infeed velocity during accelerated spark-in is four times the desired value $(u_1/v = 4)$

One concept for accelerating the spark-out is illustrated in Figure 12-8 [1, 4] for cylindrical grinding. The idea is to overshoot the controlled infeed and then back off at a controlled rate to the final dimension and complete spark-out such that the true infeed velocity is reduced to zero. A dwell of duration n_w^{-1} corresponding to one workpiece revolution may also be added prior to retraction at the end of the cycle in order to ensure complete spark-out around the entire workpiece periphery especially with slowly rotating (e.g. large diameter) parts.

Now let us proceed to analyze what happens during accelerated spark-out using the continuous infeed results (sections 12.2 and 12.3). At the end of the initial roughing stage $(t = t_1)$, the true infeed velocity and accumulated removal are (Eqs. (12-22) and (12-23))

$$v(t_1) = (1 - e^{t_1/\tau})u_1 \tag{12-52}$$

and

$$r(t_1) = (t_1 + \tau e^{-t_1/\tau} - \tau)u_1 \tag{12-53}$$

Taking this as the initial condition, the behavior during accelerated spark-out obtained by solving Eqs. (12-12) and (12-14) is:

$$v(t') = e^{-t'/t}(1 - e^{-t_1/\tau}) + (1 - e^{-t'/\tau})u_m \qquad (12\text{-}54)$$

and

$$r(t') = r(t_1) + \tau(1 - e^{-t'/t})(1 - e^{-t_1/\tau}) + (t' + \tau e^{-t'/\tau} - \tau)u_m \qquad (12\text{-}55)$$

where u_m is the controlled infeed velocity (negative for backing off) and the time t' is measured from t_1 (i.e. $t' = t - t_1$). Assuming that a steady-state condition was reached at the end of roughing, these relationships simplify to

$$v(t') = u_1 e^{-t_1/\tau} + (1 - e^{-t'/t})u_m \qquad (12\text{-}56)$$

and

$$r(t') = r(t_1) + \tau(1 - e^{-t'/t})u_1 + (t' + \tau e^{-t'/\tau} - \tau)u_m \qquad (12\text{-}57)$$

The final condition after backing off is

$$v(t_2) = v(t_f - t_1) = 0 \qquad (12\text{-}58)$$

and

$$r(t_2) = r(t_f - t_1) = \Delta r - r(t_1) \qquad (12\text{-}59)$$

The total time required for each stage is obtained from Eqs. (12-56)–(12-59) as

$$t_1 = \frac{\Delta r}{u_1} - \frac{\tau}{u_1}\left(u_m \ln \frac{u_1 - u_m}{-u_m}\right) \qquad (12\text{-}60)$$

for roughing and

$$t_2 = t_f - t_1 = \tau \ln\left(\frac{u_1 - u_m}{-u_m}\right) \qquad (12\text{-}61)$$

for accelerated spark-out. Adding the contribution from each stage, the total cycle time (neglecting any final dwell) can be written as

$$t_f = \frac{\Delta r}{u_1} + \tau\left[\left(1 - \frac{u_m}{u_1}\right)\ln\left(\frac{u_1 - u_m}{-u_m}\right)\right] \qquad (12\text{-}62)$$

The first term would be the time to remove the total allowance Δr with an infinitely stiff system, and the second term the additional time for the transients.

In order to implement the accelerated spark-out strategy, we need to specify the time t_1 at which to switch from the roughing infeed to the back-off. As a practical matter, the time t_1 cannot be directly measured with sufficient accuracy owing to initial part-to-part size variation. It has been proposed to switch when the remaining allowance Δr_2, measured with an in-process diameter gage, reaches a specified value. Combining Eqs. (12-25) and (12-60):

$$\Delta r_2 = \Delta rt - r(t_1) = u_1\tau + u_m\tau \, \ell n\left(\frac{u_1 - u_m}{-u_m}\right) \quad (12\text{-}63)$$

In general, the system performance may be limited by how accurately Δr_2 can be measured in-process, insofar as a smaller value of Δr_2 reduces the total cycle time. Once the value of Δr_2 is specified, the required back-off velocity u_m may be obtained by solving Eq. (12-63). This calculation is very tedious, but it can be shown for $0.2 < \Delta r_2/u_1\tau < 0.8$ to a very good approximation that [1]

$$u_m = -u_1 \exp\left(1.65 - 5.15\frac{\Delta r_2}{u_1\tau}\right) \quad (12\text{-}64)$$

The time constant τ which is needed to calculate u_m may be estimated in-process by measuring the steady-state lag in the roughing stage and dividing by u_1 as previously mentioned.

For this analysis, we have assumed linear behavior, but non-linearities particularly in the force versus infeed behavior would cause the system to behave more sluggishly than theoretically expected. Therefore, accelerated spark-out can be expected to result in slightly oversize parts, which may necessitate an additional size correction at the end of the cycle.

The potential benefits of accelerated spark-in and spark-out strategies are greatest for systems having long time constants, especially internal grinding. Accelerated spark-in using a high initial control velocity as in Figure 12-7 has been successfully applied in industry. Accelerated spark-out as described above has apparently not been applied in production, although some advanced grinding systems successfully use a simpler approach of infeed overshoot followed by rapid back-off to quickly recover the system deflection. The accelerated spark-out strategy as described here should achieve this more efficiently.

12.7   GRINDING VIBRATIONS

Up to this point we have considered how deflections of the grinding system affect the grinding cycle and workpiece accuracy. Other important aspects of machine flexibility are related to grinding vibrations. Periodic deflections associated with vibrations can adversely affect the part quality and limit the production rate. Such vibrations are commonly referred to as chatter. In this section, we present a brief introduction to grinding chatter vibrations and their suppression. A comprehensive review of this topic can be found in a CIRP keynote paper [5].

As with other machine tools, grinding machine vibrations are usually classified into two types: forced vibrations and self-excited vibrations. Forced vibrations are caused by periodic disturbances external to the cutting process such as from an unbalanced wheel or spindle, electric motors, bearings, hydraulic systems, or even other nearby machines [6-14]. The resulting associated chatter frequency corresponds to that of the vibration source or some harmonic thereof, and the amplitude depends on the strength of the vibration source and the compliance of the machine tool at the particular chatter frequency. The compliance of the machine, which is the inverse of its stiffness, is frequency dependent. In the foregoing discussion, the machine stiffness was defined as k_m (Eq. (12-1)), but henceforth we will refer to this as the 'static' stiffness (zero excitation frequency). The dynamic stiffness may be significantly less than k_m particularly at excitation frequencies close to natural frequencies of the machine structure. Aside from static stiffness, other important factors are the prevailing vibration modes and the degree of structural damping.

The causes of self-excited vibrations are much more complicated than forced vibrations, and a great deal of research has been devoted to understanding them [5-8, 15-27]. Self-excited vibrations are generally associated with natural vibration modes of the machine-tool structure. The grinding instability is attributed to regenerative feedback effects on the workpiece and the wheel. Any irregularities in the cutting process cause variations in the cutting force which can dynamically excite the machine tool structure. This leads to variations in the local depth of cut during successive passes of the wheel, thereby regenerating undulations or lobes on the workpiece. Wheel regeneration can occur in a similar manner with periodic wear rate variations and lobes developing around the wheel periphery.

With other types of machining, such as turning and milling, self-excited vibrations usually occur near the natural frequencies of the machine structure. In grinding, the local deformation between the wheel and the workpiece must also be taken into account. Flexible contact between the wheel and the workpiece (see Figure 12-1) has the effect of raising the chatter frequencies above the resonant frequencies of the machine structure [7, 26-28]. In most cases, one particular chatter frequency predominates. The shift to a higher resonant frequency can make it difficult to relate the chatter to a particular vibration mode.

Self-excited vibrations are stable if the amplitude of regenerative waves becomes progressively smaller, and unstable if the wave amplitude grows. The threshold condition between stable and unstable behavior can be derived by applying classical feedback techniques to a mathematical model of the grinding system. For plunge grinding, a structural model like the one in Figure 12-1 has been used [8,15]. Also included in the model are a proportional relationship between the normal force and actual depth of cut analogous to Eq. (12-7), and a continuity condition analogous to Eq. (12-18). The dynamic response of the machine structure is introduced in terms of its directional frequency response in the complex plane, $G_m(j\omega)/k_m$, which can be considered as the dynamic deflection due to a frequency-dependent unit excitation force between the wheel and the workpiece. The limiting stability condition which is obtained can be written as

$$\frac{|Re_m|}{k_m} < \frac{1}{2k_c}\left(1 + \frac{v_w}{v_sG}\right) + \frac{1}{k_a} \qquad (12\text{-}65)$$

where Re_m is the negative real part of $G_m(j\omega)$, k_m is the static machine stiffness (Eq. (12-1)), G is the grinding ratio (Chapter 11), and v_w and v_s are the workpiece and wheel velocities, respectively. The left side of Eq. (12-65) may be considered to represent the dynamic compliance of the machine in the direction of the normal force. On the right-hand side, the first term equals half the combined cutting and wheel-wear compliances, and the second is the contact compliance. In many practical cases, the wheel-wear compliance is much less than the cutting compliance ($v_w/v_sG \ll 1$) in which case the stability condition simplifies to

$$\frac{|Re_m|}{K_m} < \frac{1}{2k_c} + \frac{1}{k_a} \qquad (12\text{-}66)$$

In order to apply the stability condition, we need appropriate values for the parameters in Eq. (12-66). The range of parameters normally encountered for external cylindrical grinding of hardened steels is as follows [15]:

static machine stiffness, k_m = 10-100 kN/mm
cutting stiffness per unit width, k_c/b = 2-10 kN/mm^2
contact stiffness per unit width, k_a/b = 1-10 kN/mm^2
dynamic machine characteristic, $|Re_m|$ = 1-10

The cutting and contact stiffnesses are proportional to the grinding width, so the values quoted are per unit width b. The cutting stiffness is approximately proportional to the ratio of workspeed to wheelspeed, and Eq. (12-8) also indicates that increasing the workspeed can be expected to raise the cutting stiffness proportionally. The lower limit of $|Re_m| = 1$ would be for a highly damped machine whereas the upper limit of $|Re_m| = 10$ would apply to a machine with very little damping.

Now substituting some typical values into Eq. (12-66), i.e. $k_m = 30$ kN/mm, $k_c/b = 5$ kN/mm^2, $k_c/b = 5$ kN/mm^2, and $|Re_m| = 2.5$, the stability condition becomes $b < 3.6$ mm. In this case, stability requires a grinding width less than 3.6 mm. A wider cut would make k_c and k_a proportionally bigger and the process would be unstable. In many practical cases, the grinding width is likely to be bigger than the critical value and the process is prone to regenerative chatter. Whether or not the regenerative chatter actually becomes a problem depends on how quickly the chatter amplitude grows to an unacceptable level.

Compared with external cylindrical grinding, the stability condition (Eq. (12-66)) would seem to suggest a worse situation for internal and straight surface plunge grinding. The machine stiffness k_m with internal grinding is usually much smaller than with external grinding, as previously mentioned, and both the cutting and the contact stiffnesses tend to be bigger owing to the higher degree of wheel-workpiece conformity. Furthermore, the necessity for a small wheel will result in a more rapid rate of wheel regeneration. With straight surface grinding, static machine stiffnesses are only about half as big as for external cylindrical grinders, but this may be offset by lower cutting stiffnesses due to the use of slower workspeeds. An additional factor which lessens the growth of regenerative chatter with straight surface grinding is the lack of a fixed phase relationship between the wheel and workpiece. With cylindrical grinding, the wheel and workpiece rotate continually with a fixed speed ratio and phase, but re-engagement of the wheel with the workpiece for each pass with straight surface grinding causes a random phase shift between the vibration motion and regenerative wave generation which interrupts regenerative feedback.

Regenerative chatter waves can develop both on the wheel and on the workpiece. However, the vibration frequency causing regeneration on either body is limited by wheel-workpiece contact. Vibration frequencies with half wavelengths shorter than the contact length should be strongly attenuated by a mechanical filtering effect as illustrated in Figure 12-9 for workpiece wave filtering, and a similar effect applies to the wheel [8, 15]. The break frequencies above which filtering should occur in this way are readily obtained as

$$f_w = \frac{v_w}{2l_c} \tag{12-67}$$

for the workpiece and

$$f_s = \frac{v_s}{2l_c} \tag{12-68}$$

for the wheel, where l_c is the contact length (e.g. Eq. (3-8)). For typical external grinding conditions, f_w might be about 200 Hz and f_s about 50-100

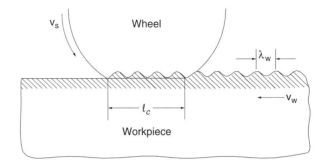

***Figure 12-9** Mechanical filtering of regenerative workpiece vibrations of wavelength λ_w by a grinding zone having a contact length l_c.*

times higher, which corresponds to typical ratios of v_s to v_w. The break frequencies for internal grinding tend to be somewhat lower owing to longer contact lengths. Many practical cases arise where the first (lowest) natural frequency of the grinding system is higher than f_w, thereby attenuating workpiece regeneration and favoring stability. Grinding chatter at frequencies between f_w and f_s can only occur by wheel regeneration, which may grow quite slowly owing to the inherent wheel-wear resistance. During each wheel revolution, the wheel wear is usually much less than the stock removal, and the variable part of the wear responsible for wheel regeneration comprises only a very small fraction of the total. Despite the instability, it is often possible to remove a significant amount of material without objectionable chatter.

12.8   VIBRATION SUPPRESSION

The obvious way to suppress forced vibration is to eliminate or isolate the vibration source. Other possibilities for reducing the amplitude of forced vibrations include increasing the static and dynamic stiffnesses of weak structural elements, adding damping, and shifting the excitation frequency.

Some effects of various factors on self-excited vibrations can be inferred from the stability criterion of Eq. (12-65) or (12-66). Increasing the right-hand side of the equation or decreasing the left will tend to favor stability and lessen regenerative chatter. Parameters which may be varied are related to the machine structure, the wheel-workpiece combination, and the operating conditions.

For stability considerations, the machine structure is characterized by k_m and $|Re_m|$. The static stiffness k_m depends upon the configuration of the machine and the materials from which it is built. On internal grinders, the wheel spindle is usually the weakest element which limits the stiffness, as mentioned previously. The flexibility of the component being ground may also be an important factor, in which case supporting the workpiece more rigidly lengthens the chatter-free grinding time while also reducing wheel consumption and spark-out time [28]. The dynamic stiffness can also be increased by reducing $|Re_m|$ which generally requires more damping. Structural damping depends on the materials from which the machine is built and also localized damping at interfaces between machine elements. Cast iron provides better damping than structural steel, although machine-tool builders often find it more convenient to work with welded-steel structures. Composite materials provide much better damping and have become quite popular [29]. However, one disadvantage with composites is the need to cast metallic interface elements into the structure, which limits possibilities for subsequent modifications. Dynamic weaknesses associated with particular vibration modes may also be corrected with auxiliary dampers.

The particular wheel-workpiece combination mainly affects the cutting stiffness k_c and contact stiffness k_a. More difficult-to-grind materials generally require larger forces, which means a larger cutting stiffness and more chatter problems. Decreasing the wheel hardness lowers the contact stiffness which makes the system more stable, and this has the additional benefit of reducing the cutting stiffness. Of course, a softer wheel will usually reduce the G-ratio, although this is likely to have only a secondary influence on regenerative chatter. Improved chatter performance with superabrasive wheels has been obtained by adding a flexible coupling between the abrasive rim and the hub and by the use of a more flexible hub material [30]. This not only decreases the contact stiffness, but also improves the dynamic behavior by adding more damping to the system.

Insofar as the frequency of self-excited vibrations is nearly constant, the number of waves regenerated on a grinding wheel rotating at a constant speed remains the same. Changing to a different wheelspeed should cause the number of waves to vary such that the previous wave configuration becomes obliterated. It has been demonstrated that regenerative chatter can be effectively suppressed by continuously varying the rotational wheelspeed so as to cause a more random regeneration process [31-33]. Similar benefits are also obtained with workspeed variations [24, 34]. Although this method of vibration suppression should be rather easy to implement, it has not been widely adopted in production.

REFERENCES

1. Malkin, S. and Koren, Y., 'Optimal Infeed Control for Accelerated Spark-out in Grinding', *Trans. ASME, J. of Eng. for Ind.*, 106, 1984, p. 70.
2. Saljé, E., Damlos, H. H. and Teiwes, H., 'Problems in Profile Grinding, Angular Plunge Grinding and Surface Grinding', *Annals of the CIRP*, 30/1, 1981, p. 219.

3. Andrew, C., Howes, T. D. and Pearce, T. R. A., *Creep Feed Grinding*, Holt, Rinehart, and Winston, London, 1985, p. 172.

4. Levin, A. I. and Mashnistov, V. M., 'Optimization of a Plunge Grinding Cycle', *Machines and Tooling*, 48, No. 12, 1977, p. 36.

5. Inasaki, I., Karpuschewski, B. and Lee, H. S., 'Grinding Chatter–Origin and Suppression', *Annals of the CIRP*, 50/2, 2001, p. 515.

6. Hahn, R. S., 'Vibration Problems and Solutions in Grinding', ASTME Paper No. MR69-246, 1969.

7. Leist, T. H. and Lemon, J. R., 'Practical Approaches to Obtain Improved Machine Stability', *Industrial Diamond Review*, 30, No. 353, 1970, p. 129.

8. Snoeys, R. 'Dominating Parameters of Grinding Machine Stability', *Het lngenieurs-blad*, 40, No. 4, 1971, p. 87.

9. Kaliszer, H. and Trmal, G., 'Forced Vibration During Plunge Grinding and its Effect on Surface Topography', *Proceedings of the International Grinding Conference*, Pittsburgh, 1972, p. 708.

10. Kaliszer, H. and Sindwani, A. D., 'The Effect of Grinding Wheel Unbalance on Workpiece Waviness', *Proceedings of the Tenth International Machine Tool Design and Research Conference*, 1969, p. 395.

11. Peters, J. and Vanherck, P., 'Unbalance of Electromotors and their Influence on the Surface Geometry in Surface Grinding', *Annals of the CIRP*, 19, 1971, p. 585.

12. Nikulin, B. I., 'Effect of Wheel Imbalance on Diamond Grinding', *Machines and Tooling*, 4, No. 12, 1970, p. 49.

13. Arantes, A. A. and Da Cunha, P. M. R., 'Two Cases of Forced Vibrations in Grinding Machines', *Annals of the CIRP*, 18, 1970, p. 425.

14. Rubinchik, S. I. and Soloveichik, Y. S., 'Effect of Spindle Unbalance on Surface Waviness in Internal Grinding', *Machines and Tooling*, 41, No. 2, 1970, p. 2.

15. Snoeys, R., and Brown, D., 'Dominating Parameters in Grinding Wheel and Workpiece Regenerative Chatter', *Proceedings of the Tenth International Machine Tool Design and Research Conference*, 1969, p. 325.

16. Hahn, R. S., 'On the Theory of Regenerative Chatter in Precision Grinding Operations', *Trans. ASME*, 76, 1954, p. 563.

17. Gurney, J. P, 'An Analysis of Surface Wave Instability in Grinding', *Journal of Mechanical Engineering Science*, 7, 1965, p. 8.

18. Bartalucci, B. and Lisini, G. G., 'Grinding Process Instability', *Trans. ASME, J. of Eng. for Ind.*, 91,1969, p. 597.

19. Shiozaki, S., Miyashita, M. and Furukawa, Y., 'Generation and Growing Up Process of Self Excited Vibrations', *Bulletin of the JSME*, 13, 1970, p. 1139.

20. Ohno, S., 'Self-excited Vibration in Cylindrical Grinding, Parts 1 and 2', *Bulletin of the JSME*, 13, 1970, p. 616 and p. 623.

21. Thompson, R. A., 'The Dynamic Behavior of Surface Grinding, Parts 1 and 2', *Trans. ASME, J. of Eng. for Ind.*, 93, 1971, p. 485 and p. 492.

22. Thompson, R. A., 'The Character of Regenerative Chatter in Cylindrical Grinding', *Trans. ASME, J. of Eng. for Ind.*, 95, 1973, p. 858.

23. Thompson, R. A., 'On the Doubly Regenerative Stability of a Grinder', *Trans. ASME, J. of Eng. for Ind.*, 96, 1974, p. 275.

24. Takayanagi, K., Inasaki, I. and Yonetsu, S., Regenerative Chatter Behavior During One Cycle of Cylindrical Plunge Grinding', *Bull. Japan Soc. of Prec. Eng.*, 12, 1978, p. 121.

25. Saljé, E. and Dietrich, W., 'Analysis of Self-excited Vibrations in External Cylindrical Plunge Grinding', *Annals of the CIRP*, 31/1, 1982, p. 255.
26. Inoue, H., 'Chattering Phenomena in Grinding', *Bull. of the Japan Soc. of Prec. Engg.*, 3, 1969, p. 67.
27. Hoshi, T. and Koumoto, Y., 'Mechanism of Vibration in Plunge-cut Cylindrical Grinding', *Proceedings of the 5th International Conference on Production Engineering*, Tokyo, 1984, p. 314.
28. Colding, B., 'How Stiffness Affecting Grinding Performance', *Machinery* (NY), March 1970, p. 57.
29. Renker, H. J., 'The Future of the Grinding Process', *Second International Grinding Conference*, Philadelphia, SME, 1986.
30. Sexton, J. S., Howes, T. D. and Stone, B. J., 'The Use of Increased Wheel Flexibility to Improve Chatter Performance in Grinding', *Proc. Instn. Mech. Engrs*, 196, 1982, p. 291.
31. Bartalucci, B., Lisini, G. G. and Pinotti, P. C. 'Grinding at Variable Speed', *Proceedings of the Eleventh International Machine Tool Design and Research Conference*, 1970, p. 169.
32. Inasaki, I. and Yonetsu, S., 'Surface Waves Generated on the Grinding Wheel', *Bulletin of the JSME*, 12, 1969, p. 918.
33. Cegrell, G., 'Variable Wheel Speed – a Way to Increase the Metal Removal Rate', *Proceedings of the Fourteenth International Machine Tool Design and Research Conference*, 1973, p. 653.
34. Inasaki, I., Cheng, C. and Yonetsu, S., 'Suppression of Chatter in Grinding', *Bull. Japan Soc. of Prec. Engg.*, 9, 1976, p. 133.

BIBLIOGRAPHY

Hahn, R. S., 'Vibration Problems and Solutions in Grinding', ASTME Paper No. MR69-246, 1969.

Inasaki, I., Karpuschewski, B. and Lee, H. S., 'Grinding Chatter – Origin and Suppression', *Annals of the CIRP*, 50/2, 2001, p. 515.

Leist, T. H. and Lemon, J. R., 'Practical Methods to Obtain Improved Grinding Machine Stability', *Industrial Diamond Review*, 30, No. 353, 1970, p. 129.

Malkin, S. and Koren, Y., 'Optimal Infeed Control for Accelerated Spark-out in Grinding', *Trans. ASME, J. of Eng. for Ind.*, 106, 1984, p. 70.

Okamura, K., 'Analytical Description of Accumulation Phenomenon in Grinding', *Proceedings of the Tenth International Conference on Production Engineering, Part 2*, Tokyo, 1974, p. 58.

Snoeys, R., 'Dominating Parameters of Grinding Machine Stability', *Het Ingenieursblad*, 40, No. 4, 1971, p. 87.

Snoeys, R. and Brown, 'Dominating Parameters in Grinding Wheel and Workpiece Regenerative Chatter', *Proceedings of the Tenth International Machine Tool Design and Research Conference*, 1969, p. 325.

Tlusty, J., 'Grinding Machine Ability to Reduce Workpiece Form Error', *Proceedings of the International Grinding Conference*, Pittsburgh, 1972, p. 366.

Chapter 13

Simulation, Optimization, and Intelligent Control

13.1 INTRODUCTION

In the previous chapters, a critical evaluation has been presented of what we know about the grinding process. Quantitative models have been shown to provide a relatively clear understanding of the many diverse aspects of the process. But utilization of these models to predict or control what happens during a grinding process presents a daunting challenge because most of the models are interrelated and the process changes as grinding proceeds. This can be appreciated in a somewhat oversimplified way by considering the sequence of events in a typical grinding operation. For example, using the grinding models to understand what happens during a cylindrical grinding process might begin with selection of the grinding wheel, and then proceed to dressing of the wheel, implementing the cycle by infeeding the wheel into the workpiece, calculating the actual removal rate and machine deflection during the cycle, finding the corresponding forces and energy, considering wheel wear and dulling, calculating the temperatures and predicting possible thermal damage, estimating the roughness and roundness, proceeding to the next part with the used wheel, etc. Any attempt to apply grinding theory may seem like a hopeless task.

In order to address this challenge, simulation software has been developed which integrates the various models to predict what happens during a grinding operation. While some 'simulations' may deal with only a particular aspect of the grinding process [1], the present simulation considered here is intended to work as a virtual grinder to quantitatively replicate the many diverse and interrelated aspects of what occurs throughout a grinding cycle, and also from cycle to cycle. This simulation software can also utilize actual grinding data, if available, to calibrate the models and thereby improve the accuracy of the simulation. The updated knowledge base provided by the calibrated simulation can then be used to identify optimal grinding conditions and for computer control of the process.

Simulation software packages have been developed for cylindrical grinding, creep-feed form grinding, double disk grinding, and helical groove grinding [2, 9]. In this chapter, we will focus mainly on cylindrical and creep-feed form grinding simulations. It should be noted that these Windows©-based software packages are not just academic curiosities, but are actually used in industry. The cylindrical grinding simulation (GrindSim©) has been acquired by bearing, automotive, and machine tool companies. The creep-feed form grinding simulation software has been applied, up to now, mainly in the aerospace industry for continuous-dress creep-feed (CDCF) grinding of turbine blades and vanes.

Simulation software is normally applied off-line for prediction and optimization of grinding processes. This chapter concludes with a consideration of direct on-line computerized control of grinding. As with off-line simulation, the ultimate objective is to utilize our knowledge of the grinding process (grinding models) to optimize the grinding process. A model based intelligent system for optimal computer control of grinding is described which has been implemented in industry. With the advent of PC-based open-architecture machine tool control, a more advanced on-line system has also been developed which utilizes the simulation software as the knowledge base for intelligent optimal control of cylindrical grinding.

13.2   ORIGINAL SIMULATION SOFTWARE FOR CYLINDRICAL GRINDING

Simulation software was originally developed for cylindrical grinding (Figure 13-1) [1]. The concept for this model-based simulation is shown in Figure 13-2. The simulation is analogous to that of an actual grinding operation. It begins with **Input** information related to the job requirements, grinding system, and operating parameters. The workpiece is then 'ground' by execution of the **Grinding Model** to predict what happens. The **Grinding Model** includes a collection of interrelated models in the form of mathematical equations which describe the various aspects of the process. Finally the predicted grinding behavior is presented in the **Output,** which includes the actual infeed, forces, power, deflection, and temperatures versus time during the cycle, and the final roughness and out-of-roundness.

The input parameters for the simulation software are sequentially introduced according to the flow chart shown in Figure 13-3 [1]. Specification of the type of cylindrical grinding - external or internal - is first requested. The fixed parameters are then provided: wheel diameter,

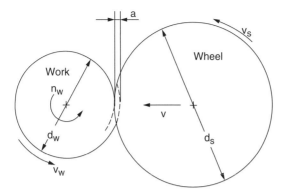

Figure 13-1 Illustration of cylindrical plunge grinding.

initial and final workpiece diameters, grinding width, wheel velocity, workpiece velocity, and system stiffness. This is followed by information about the workpiece and grinding wheel.

The next input concerns the lubricating and cooling characteristics of the grinding fluid. The role of the fluid as a lubricant is mainly to keep the wheel sharp (Chapter 11). This is reflected in the simulation by a relative ranking of lubricant effectiveness from 0 to 10: dry grinding is 0, a water based soluble oil is about 5, and a heavy duty straight oil 9 or 10. The main role of the fluid as a coolant is to reduce thermal expansion by lowering the bulk temperature of the workpiece.

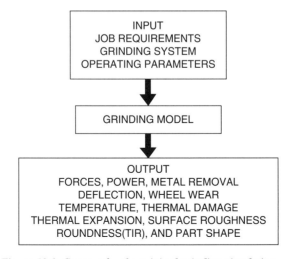

Figure 13-2 Concept for the original grinding simulation.

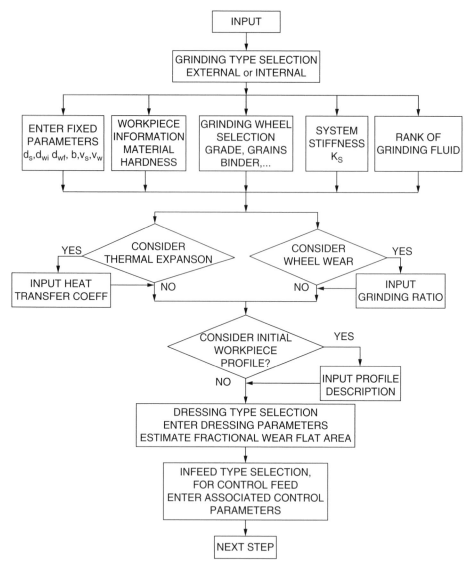

Figure 13-3 **Input** *flow chart for the original simulation.*

The user is asked whether to consider thermal expansion of the work-piece in the simulation. Thermal expansion is usually of concern with long grinding cycles, when using Minimum Quantity Lubrication (MQL) (Chapter 11), or for dry grinding. If thermal expansion is to be considered, then the heat transfer coefficient for cooling by the grinding fluid is requested. Water-based fluids have much higher convection coefficients and are more effective coolants than straight oils due to their much higher thermal conductivities. The user then decides whether to consider wheel wear. The effect of wheel wear on grinding performance is especially significant for internal grinding.

Next comes the controlled infeed cycle, which is usually divided into a number of discrete stages (Chapter 12), although the prospects for continuous variation of infeed control has also been investigated [10,11]. Input for each stage of the grinding cycle includes the controlled infeed rate and the point at which to switch to the next stage. Switching points can utilize either infeed control or size control criteria. For infeed control, switching from one stage to the next is activated upon reaching a specified infeed (cross-slide) position, which is equivalent to specifying the time duration of each stage. With size control, the switching point at the end of each stage is specified by the remaining radial stock allowance, which is analogous to using an in-process size gage. Final wheel retraction to terminate the cycle is set either by fixing the final spark-out duration or by reaching the required size tolerance.

Before grinding, the initial workpiece will generally have significant out-of-roundness. This will cause the actual infeed to fluctuate, especially during initial grinding. If this effect is to be considered by the simulation, the user can quantitatively specify the initial non-round workpiece shape as either wavy with lobes, eccentric due to inaccurate mounting, a superposition of lobes and eccentricity, or by the radius variation around the workpiece.

The last step in the **Input** is to dress the wheel. The user selects the type of dressing (single point or rotary diamond) and enters the associated dressing parameters. The computer then estimates the effective dullness (wear flat area) of the dressed wheel. Finer dressing, a harder wheel grade, and finer grit size generally result in a duller wheel (Chapter 4). Dressing conditions also affect the surface roughness (Chapter 10).

With the wheel now dressed, the software proceeds to the **Grinding Model** illustrated by the flow chart in Figure 13-4 [2]. The computer program generates discrete grinding data at N points equally spaced around the workpiece periphery. The number N is specified by the user. If initial out-of-roundness is considered, the workpiece radii at the N points around the workpiece are obtained from the input, and grinding is assumed to begin at the maximum radius.

The **Grinding Model** in Figure 13-4 includes four interrelated aspects: forces, material removal, wheel wear, and thermal. As the wheel interacts with the workpiece, grinding forces are generated and power is dissipated. The normal force causes a deflection, so that the actual radial infeed of the wheel into the workpiece lags behind the controlled infeed to

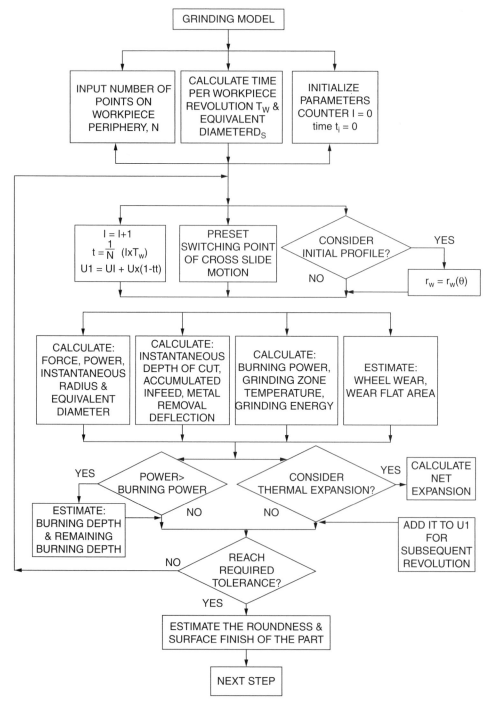

Figure 13-4 **Grinding Model** *flow chart for the original simulation.*

the machine (Chapter 12). Bulk wheel wear adds to the lag as part of the infeed is taken up by the receding wheel surface. Localized attritious wear leads to dulling, thereby increasing the forces and power, which may be off-set to some degree by self-sharpening. Virtually all the energy input is converted to heat, which causes elevated temperatures at the grinding zone and possible thermal damage to the workpiece (Chapter 6). The bulk temperature rise of the workpiece causes thermal expansion, which also affects the actual material removal and the final size.

Calculating the forces, material removal, wheel wear, and temperatures begins with the continuity condition (Chapter 12). Neglecting thermal expansion of the workpiece, the continuity condition for cylindrical grinding (Eq. (12-18)) is given by:

$$u(t) - v(t) - w(t) = \dot{\varepsilon} \qquad (13\text{-}1)$$

where $u(t)$ is the controlled infeed rate, $v(t)$ is the actual infeed rate, $w(t)$ is the radial wear rate, and ε is the system deflection. The actual infeed rate $v(t)$ can be expressed as the product of the wheel depth of cut a and the work rotational speed n_w:

$$v(t) = an_w \qquad (13\text{-}2)$$

and the deflection is equal to the normal force F_n divided by the system stiffness k_e (Eq. (12-4):

$$\varepsilon = \frac{F_n}{k_e} \qquad (13\text{-}3)$$

Satisfying the continuity condition to obtain the actual infeed rate or depth of cut also requires solving for the normal force and wheel wear, which are in turn dependent on the actual infeed rate or depth of cut. This is a further indication of how the various aspects of the process are interrelated. Iterative procedures are incorporated into the simulation software in order to quickly obtain the solution to this problem for each workpiece rotation at all N points around the periphery during the grinding cycle, and these results are then used to calculate grinding temperatures.

13.3   GRINDSIM©: *SIMULATION, CALIBRATION, AND OPTIMIZATION OF CYLINDRICAL GRINDING*

The simulation software described above was developed before the widespread availability of personal computers (PCs) with significant computing power. An interactive updated Windows©-based PC version of the

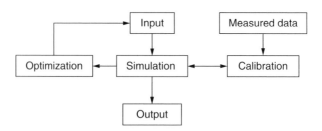

Figure 13-5 GrindSim© flow chart for simulation, calibration, and optimization.

software (**GrindSim©**) was subsequently developed as illustrated in Figure 13-5 [7-9]. The capability of the original software was greatly enhanced mainly by the addition of **Calibration** and **Optimization** modules, and also by being able to consider successive grinding of multiple parts between dressings. **Simulation** predicts the grinding behavior based upon physical and empirical models which describe various aspects of the process. **Calibration** enhances predictions by improving the accuracy of the models using actual grinding data. As such, the software 'learns' from the calibration. **Optimization** can then be applied to identify the best grinding conditions for maximum productivity. Further details and examples are presented to illustrate each of these modules.

Simulation

The **Simulation** module for GrindSim© works very much like the original simulation software described in the previous section. The software takes input data related to the wheel, workpiece, grinding process, and dressing. Grinding models are then executed to predict forces, power, deflection, temperature, specific energy, surface roughness, out-of-roundness, etc. These results are stored in a number of data files, thereby allowing the user to obtain various plots in the output module.

To illustrate the simulation, consider a 3-stage internal plunge grinding cycle consisting of roughing, finishing, and spark-out stages (Chapter 12). The input parameters for this example are summarized in Table 13-1. Figure 13-6 shows simulated results for the controlled and actual infeed, power (P), normal and tangential force components (F_n and F_t), deflection, and temperature versus grinding time during a single cycle. Also included is an estimation of the power to cause workpiece burn (P_b) (Chapter 6). Initially during the roughing stage, the grinding power, forces, deflection and temperatures increase rapidly as the infeed rate increases towards a nearly constant value. After this initial transient, the power and force components continue to increase but at a much slower rate due to dulling of the wheel. This causes a corresponding increase in the deflection and the

Table 13-1 Grinding conditions for simulation

	Rough	Finish	Spark-out
Infeed velocity	0.012 mm/s	0.004 mm/s	0
Time	17.4 s	10.2 s	4.3 s

Wheel:

32A80M6, d_s = 50 mm

Workpiece:

AISI52100 (HRC55), d_w = 70 mm, b_w = 9.0 mm

Single Point Dressing:

5 passes, lead s_d = 0.137 mm, depth a_d = 0.02 mm.

Grinding parameters:

v_s = 30 m/s, v_w = 150 mm/s, radial stock allowance 0.24 mm, system stiffness k_e = 6500 N/mm.

temperature. Workpiece burn is most likely to occur during the roughing stage, but should not happen in the present case since the grinding power P is less than the corresponding burning power P_b throughout the entire cycle. The power, forces, deflection and temperature decrease during the finishing stage and the final spark-out. For this operation, the simulation predicts a surface roughness of R_a = 0.71 μm, out of roundness of 2.2 μm, radial wheel wear of 0.01 mm, and a system time constant (Chapter 12) of $\tau \approx 1.8$ s.

This simulation example is for grinding only one part after dressing the wheel. In production, multiple parts are often ground after each dress, so the wheel condition after grinding the first part becomes the initial condition for the second part, etc. The grinding behavior and part quality can vary significantly from part to part between wheel dressings due to changes in the wheel topography. The simulation can also take this situation into account as seen in Figure 13-7 for grinding of 21 parts using the same grinding conditions as in Table 13-1 but with intermittent dressing every 7 parts. The maximum grinding power, temperature, roughness, out-of roundness, and wheel wear all increase from part to part until the grinding behavior is restored by dressing. Note that the indicated wheel wear results also include wheel consumption by dressing.

Calibration

Simulation predicts the grinding behavior based upon physical and empirical models which describe the various aspects of the process.

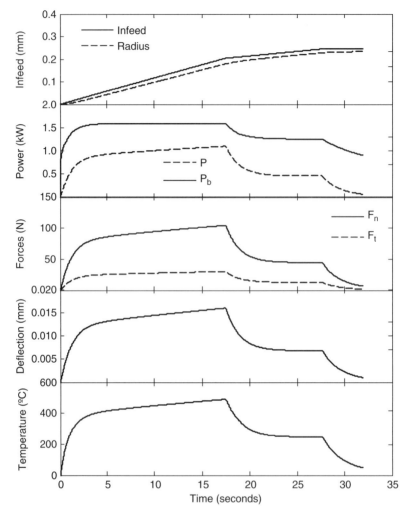

Figure 13-6 Simulation results showing infeed, power, forces, deflection, and temperature versus time.

However variations in grinding behavior arise due to uncontrolled variations related to the wheel properties, the dressing tools, changes in the chemistry of the grinding fluid, etc. In order to deal with this situation, predictions from the simulation can be enhanced if the models are calibrated against actual grinding data. As such, the software can 'learn' from the calibration, while also providing a data base which can be saved for future use.

For cylindrical grinding operations, the most useful parameter for calibration is the grinding power. Power is relatively easy to measure even

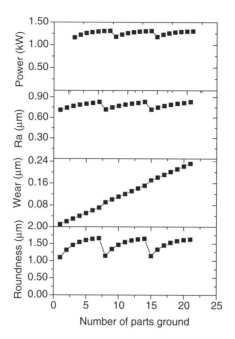

Figure 13-7 Maximum power, roughness, wheel wear, and out-of-roundness for multi-part grinding simulation.

in a production environment. During the initial transient at the start of the grinding cycle, the rate at which the power increases depends mainly on the system stiffness and also the grinding characteristics. Good agreement between predicted and measured power during the transient period of the roughing stage indicates that the system stiffness specified in the simulation is close to reality. For calibration, the system first compares the measured and predicted power during the initial transient period of the roughing stage to estimate the system stiffness. The average power during the steady period of the roughing stage is then used to calibrate the initial wear flat area model. Other parameters in the dulling and self-sharpening models can also be calibrated by examining how the power varies during this period.

Figure 13-8(a) shows both simulated and measured power for the grinding cycle in Table 13-1. The simulation in this case utilized default parameters in the models. The measured power is somewhat smaller than the simulated power. However after calibration, the simulated and measured grinding power in Figure 13-8(b) match very well. Furthermore the calibrated models have been found to result in accurate simulations even when the grinding and dressing conditions are varied over a very wide range of conditions. The surface roughness and out-of-roundness models can also be readily calibrated.

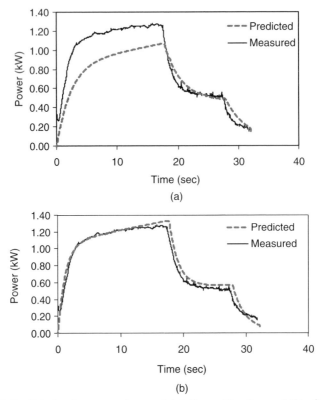

Figure 13-8 Predicted and measured power (a) before calibration and (b) after calibration for the grinding conditions in Table 1.

Optimization

The simulation software provides a comprehensive quantitative description of the grinding process which can be used not only to predict grinding behavior but also for optimization. Optimization of the grinding process is intended to identify the 'best' grinding and dressing conditions while satisfying constraints associated with part quality and machine limitations. [12-14]. For cylindrical plunge grinding, the usual optimization objective is to minimize the cycle time while satisfying constraints associated with workpiece burn, surface roughness, and out-of-roundness. Two other objective functions, targeting either the average cycle time or the combined grinding plus dressing time to achieve a balanced process flow, are also built into the simulation software.

Optimization requires consideration of all the stages in the cycle and their interactions. For example, specification of a grinding operation with a three-stage cycle requires selection of ten independent variables: infeed rates, u_1 and u_2, for the roughing and finishing stages; time durations for the roughing and finishing stages, t_1 and t_2, which are related to the radial stock allowance for each of the stages; spark-out time t_3; dressing depth a_d; dressing lead s_d; dressing interval N_d (parts/dress); wheel velocity v_s; and workpiece velocity v_w [12]. Some of these parameters are usually not varied, which greatly simplifies the optimization. The wheel velocity v_s is usually limited by safety considerations and is kept constant. Simulation results indicate that the workpiece velocity v_w has a secondary influence on cycle time, so it may also be kept constant. In practice, the dressing lead s_d is usually varied to alter the dressing severity while the dressing depth a_d is unchanged.

To optimize the grinding cycle, it is necessary to minimize the cycle time per part with respect to the independent operating parameters. Let us first consider a situation where the dressing interval (number of parts per dress) N_d is fixed. In this case, the objective function Φ is the sum of the grinding time t_g and dressing time t_d per part [12] and the optimization requires:

$$Min\{\Phi = t_g + t_d\}$$

with respect to: $u_1, u_2, t_1, t_2, t_3, s_d$

$$(13\text{-}4)$$

subject to constraints: no workpiece burn, surface roughness, out-of-roundness, wheel wear, and size. Times for part loading/unloading and setup are omitted from the objective function because they are independent of the grinding and dressing conditions.

Some constraints are tight for all grinding conditions at the optimal point and some are tight only under certain conditions. Monotonicity analyses can be used to identify the tight constraints and turn them into equalities to obtain a closed-form solution of the optimization problem [12]. The out-of-roundness usually puts an upper limit on the finishing infeed velocity u_2. The duration of t_3 is only sufficient to reach the final size requirement as measured in process. The roughing time t_1 is related to the radial stock allowance and the roughing infeed rate u_1. The wheel wear constraint typically requires that the radial wear be less than about half the dressing depth a_d in order to locate the wheel surface for dressing, although a tighter constraint may be needed to maintain a required form. Roughness, thermal damage, and/or wheel wear constraints are usually the active constraints for minimization of the objective function to find u_1, t_2, and s_d.

While the optimization minimizes cycle time for a fixed dressing interval, a further reduction in cycle time may be obtained by also optimizing the dressing interval. For this purpose, the dressing time per part t_d includes not only the time for actual dressing, but also the time to change the wheel [12]:

$$t_d = \frac{T_d}{N_d} + \frac{T_a}{n} \qquad (13\text{-}5)$$

where T_a is the time per wheel change, n is the number of the workpieces ground per wheel life, T_d is the time per wheel dressing, and N_d is the number of workpieces ground per wheel dressing. The wheel change time may become significant for internal grinding. The modified objective function now becomes:

$$\Phi = t_g + t_d = t_g + \frac{mT_d + T_a}{mN_d} \tag{13-6}$$

where m is the number of times a wheel can be dressed during its life. Taking the derivative with respect to N_d leads to

$$[t_g(N_d^* + 1) - t_g(N_d^*)] - \frac{mT_d + T_a}{mN_d^{*2}} = 0 \tag{13-7}$$

where $t_g(N_d^*)$ is the optimal grinding time per part for a fixed dressing interval N_d, which can be obtained with the optimum strategy described above. The optimal dressing interval N_d^* can be found as the integer number which most closely satisfies Eq. (13-7).

　　To illustrate the optimization, the software was applied to the same process considered above in Table 13-1. The constraints for the optimization required no workpiece burn throughout the cycle, maximum allowable surface roughness of 0.8 μm, size tolerance of 0.01 mm, and maximum out-of-roundness of 4 μm. The results for this optimization are summarized in Table 13-2. It can be seen that the cycle time (rough, finish, and spark-out) originally 31.9 seconds is now reduced to 23.1 seconds. For this optimization, both the burning and surface finish constraints were tight, which means that the grinding power reached the allowable burning limit at the end of the roughing stage and the final surface roughness was at its limit of 0.8 μm. The predicted out-of-roundness of 2.6 μm was less than the 4 μm limit.

Table 13-2 Summary of optimization results

	Rough	**Finish**	**Spark-out**
u(t) (mm/s)	0.018	0.002	0
t (s)	12.6	5.2	5.3

Wheel: 32A80M6. Workpiece: 52100 (HRC55). Dressing: lead $s_d = 0.10$ mm/rev, depth $a_d = 0.02$ mm. $v_s = 30$ m/s, $v_w = 150$ mm/s, stock allowance 0.24 mm, grinding width = 9.0 mm, system stiffness = 6500 N/mm.

The optimal grinding conditions are usually very sensitive to workpiece burn and surface roughness constraints. The workpiece burn constraint requires that the grinding power should not exceed the corresponding burning power throughout the grinding cycle. Normally this thermal limit would be reached at the end of the roughing stage. This constraint can be relaxed and the cycle time reduced by allowing some workpiece burn to occur towards the end of the roughing stage with subsequent removal of the thermally damaged layer of workpiece material during the subsequent finishing stage. To illustrate this behavior, consider optimization of the same grinding cycle as above but with the allowable grinding power at the end of the roughing stage equal to 1.3 times the corresponding burning power. This further reduces the optimal cycle time from 23.1 seconds to 15.5 seconds. The simulated results for this optimum cycle are shown in Figure 13-9. Note that the grinding power P now exceeds the estimated burning power P_b at the end of the roughing stage, thereby causing a thermally damaged depth of 7 microns. The thermally damaged material is subsequently removed during the beginning of the finishing stage.

The simulation examples presented above utilized a 50 mm diameter wheel to internally grind a 70 mm diameter workpiece. This particular size combination was selected because it matched the instrumented grinder used for testing and calibration of the software. It should be noted however that internal cylindrical grinding operations are usually performed in production with the largest possible wheel diameter and least possible clearance between the wheel and workpiece, which might be only about 5 mm, just sufficient to allow for fluid access and a size gage. For internal grinding, the wheel diameter has a big effect on the equivalent wheel diameter (Chapter 3) and grinding behavior, so it also affects the optimal removal rate. For internal grinding a 70 mm diameter workpiece, the equivalent diameter (Eq. 3-9) increases from 175 mm with a 50 mm diameter wheel to 420 mm with a 60 mm wheel and to 910 mm with a 65 mm wheel. According to the software, the minimum optimal cycle time of 15.5 seconds for the above operation with the 50 mm wheel decreases to about 12.1 seconds with the 60 mm wheel and to only about 8.1 seconds with the 65 mm wheel. It is especially important to use the maximum possible wheel diameter for internal grinding, and also to frequently replace the worn wheel as it becomes smaller.

13.4 SIMULATION OF CREEP-FEED FORM GRINDING

Simulation software has been developed not only for cylindrical grinding, but also for creep-feed, double-disk, and helical grinding operations [2-9]. In this section we focus on simulation software for creep-feed form grinding and how it can be used to enhance productivity [2, 8, 9].

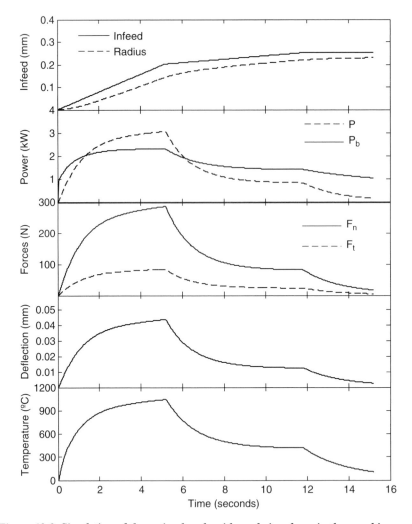

Figure 13-9 Simulation of the optimal cycle with workpiece burn in the roughing stage. The burned material is removed during the subsequent finishing stage.

Creep-feed grinding with large depths of cut and slow (creep) workpiece velocities is often applied to machining of components with complex profiles across the grinding direction. The process is widely used for aerospace engine components, such as turbine blades and vanes, which are usually made of nickel-based alloys [15]. Grinding of these difficult-to-machine materials is facilitated by Continuous Dress Creep Feed (CDCF) form

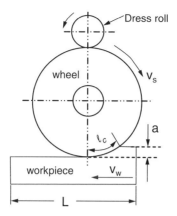

Figure 13-10 Illustration of CDCF grinding with over the wheel rotary dresser.

grinding, whereby the vitrified aluminum oxide wheel is dressed while grinding using a diamond impregnated dressing roll, as illustrated in Figure 13-10, in order to maintain the required wheel profile and sharpness (Chapter 7). The dressing roll velocity is usually 70 to 90% of the wheel surface speed with both the dresser and wheel velocities in the same direction at their contact (down mode), and the feed of the dressing roll into the wheel is typically 0.4 to 1.0 μm per wheel revolution.

Figure 13-11 shows the root serration profile of a IN100 nickel alloy turbine blade to be ground to the required form from a casting with two

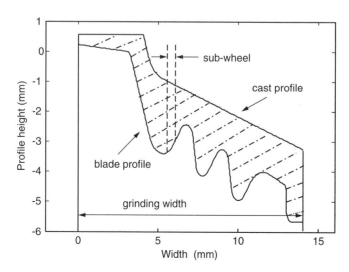

Figure 13-11 Blade and casting cross-sectional profiles.

grinding passes: rough and finish [16, 17]. The hatched area in Figure 13-11 designates the total amount of material to be removed. Most of this material is usually removed in the roughing pass by CDCF form grinding, leaving only 20 to 50 microns for the finishing pass at a faster workpiece velocity without any dressing.

Creep-feed grinding of profiles is extremely complex because the grinding conditions vary across the wheel width. A further complication arises because the process may not reach the quasi-steady state condition with full wheel engagement for much or even all of the grinding pass, so that the actual depth of cut varies not only across the wheel width but also at each instant during the grinding pass. This transient situation is common for creep-feed grinding of relatively short workpieces with large depths of cut (Chapter 7) such that the instantaneous wheel depth of cut never reaches the overall depth removed and the actual wheel-workpiece contact length is less than the theoretical length l_c (Figure 7-9). Model based simulation is very helpful in such cases due to the complex time-dependent interaction between the wheel and the workpiece.

13.4.1   Simulation

Similar to cylindrical grinding simulation, Windows©-based software was developed to simulate the form (profile) grinding process [3]. The software package consists of **Input**, **Simulate**, and **Output** modules. Besides the grinding conditions, including workpiece velocity, wheel parameters, and workpiece properties, the input module also deals with both the profile to be ground and the initial workpiece profile before grinding in order to determine the amount of material to be removed at each point across the grinding width. Both profiles can be downloaded from a data file (CAD or ASCII). Profiles consisting of multiple segments of straight lines and arcs can also be created using a built-in CAD tool in the software.

For analyzing the grinding process, the grinding wheel is divided into a number of sub-wheels as shown in Figure 13-11. Grinding by each of these sub-wheels is then modeled as grinding of an inclined surface (see Figure 3-8). Depth of cut, wheel speed, and dressing conditions can be significantly different for each sub-wheel, depending on the profile being ground. By using the concept of effective wheel diameter and effective depth of cut, the inclined grinding can be modeled as straight surface grinding (see Chapter 3), so that the various grinding models presented in the previous chapters can then be applied to form grinding. Much more data are needed for simulating form grinding than for cylindrical grinding since predictions are needed for each sub-wheel at each instant during each grinding pass.

To illustrate the software, consider form grinding of the blade serration profile shown in Figure 13-11. The grinding consists of a CDCF roughing pass to remove most of the material with a maximum depth of cut in the vertical direction (downfeed) of 2.8 mm followed by a finishing pass without dressing to remove the remaining 0.05 mm. The grinding conditions are: maximum wheel velocity v_s = 30 m/s with a maximum wheel diameter of d_s = 500 mm, and workpiece velocity v_w = 2.54 mm/s for the roughing pass and v_w = 8.5 mm/s for the finishing pass. For the CDCF roughing pass, the dress ratio (ratio of maximum dresser velocity to maximum wheel velocity) is 0.80 (down mode) and the dressing infeed is 0.4 μm per wheel revolution. The total wheel width of 14 mm is divided into 100 sub wheels.

As stated above, most of the grinding parameters will be different for each sub-wheel and should also vary with time as the engagement between the wheel and the workpiece changes. Figure 13-12 shows the instantaneous depth of cut distribution across the wheel width at 1.5 seconds and 8.3 seconds of the CDCF roughing pass after the wheel starts to actually grind. After 1.5 seconds, only part the wheel is actually grinding and that the maximum depth of cut at any point across the profile is less than 0.3 mm. Later on after 8.3 seconds, the maximum depth of cut increases to about 2.0 mm. This particular time of 8.3 seconds was chosen for illustration because it is close to when the maximum depth of cut is biggest. This can be seen in Figure 13-13 where the maximum depth of cut is plotted versus the grinding time during the cycle. The maximum depth of 2.2 mm, which is reached at about 10 seconds into the roughing pass, is less than the maximum depth removed (downfeed) of 2.8 mm for this pass. For this pass, the workpiece is geometrically short (Chapter 7), so the quasi-steady state condition is not reached.

Simulated results are shown in Figure 13-14 for the distribution of the power, forces, depth of cut, and maximum temperature at 8.3 seconds into the cut. Simulation predictions for these same parameters can also be

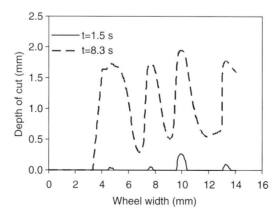

Figure 13-12 Distribution of depth of cut across the wheel width after 1.5 and 8.3 seconds.

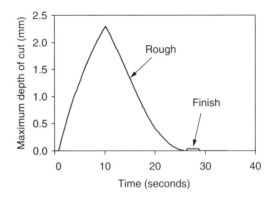

Figure 13-13 Maximum depth of cut during the grinding cycle for roughing and finishing passes.

obtained at any other time during the grinding cycle. Note that the maximum grinding temperature is estimated at about 135 °C, which should be close to the fluid burn-out limit for grinding with the soluble oil. Also included in the figure are results for the final surface roughness. Smaller surface roughness is predicted at locations having a bigger profile angle. A bigger profile angle generally results in less severe dressing, since the infeed rate of the dresser normal to the wheel surface is smaller at those locations.

The total power and forces during the cycle are obtained by integrating the corresponding distributions across the grinding width at each instant. Results obtained in this way for the total grinding power throughout the grinding cycle for both the roughing and finishing passes are shown in Figure 13-15.

13.4.2   Calibration

As with cylindrical grinding, the accuracy of the simulation software for form grinding can be improved by calibration. For this purpose, it is most convenient to use the grinding power, which is relatively easy to measure in a production environment, although grinding forces can also be used. The power measured during CDCF form grinding includes not only the actual grinding power, but also additional power for idling, dressing, and fluid interaction with the wheel. These additional power components can be found by taking a 'spark-out pass' which removes virtually no material. The actual net power for grinding can then be obtained by subtracting the power trace for spark out from the total measured power. Figure 13-16 shows a comparison after calibration of the predicted and measured grinding power for CDCF rough grinding of the serrated wheel profile in

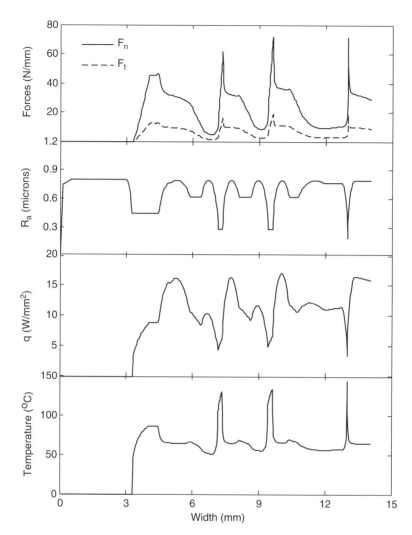

Figure 13-14 Simulation results for forces, surface roughness, heat flux, and tempera-
ture across the grinding profile at 8.3 seconds.

Figure 13-11. The calibrated results obtained in this case were used for the simulation example presented above.

13.4.3 Optimization and Process Monitoring

The simulation software for CDCF form grinding can be readily used to optimize and also to monitor the process. The usual optimization objective is to minimize the cycle time, which corresponds to maximizing the

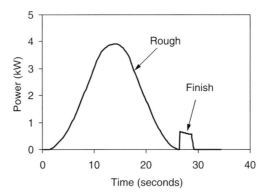

Figure 13-15 Power during the grinding cycle for roughing and finishing passes.

wokpiece velocity. Other objectives may include minimization of process cost or wheel consumption, but minimizing the cycle time is likely to come close to also satisfying these other objectives. The extent to which the cycle time can be minimized is limited by various constraints. For CDCF form grinding, the most common constraint is thermal damage to the workpiece, which is usually associated with fluid burn-out (Chapter 7). Other constraints might be related to the machine capability, surface roughness, and geometrical tolerances.

For CDCF form grinding, there are only a few process parameters which can be manipulated. The wheel depth of cut is usually pre-determined so that most of the stock is removed during the roughing pass leaving only 20-50 microns for finishing. The wheel velocity v_s may be limited by the rated speed of the grinding wheel. The remaining parameters include work

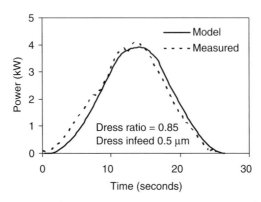

Figure 13-16 Comparison of measured and predicted power traces after calibration for CDCF grinding of the serration in Figure 13-11.

velocity v_w and the dressing condition (dress ratio and infeed). Therefore satisfying the optimization objective of minimum time becomes equivalent to maximizing the workpiece velocity. Normally it is desirable to use the maximum possible dress ratio (down mode), so that a smaller dress infeed can be used to minimize wheel consumption which is usually an important cost factor in CDCF form grinding. This scenario suggests that the grinding time can be minimized and the process optimized by progressively increasing the workpiece velocity until the thermal damage limit is reached at the severest point during the grinding cycle. If the surface roughness is better than required at this point, then coarser dressing can be used so as to obtain a sharper wheel and reduce the grinding power, thereby allowing for a further increase in the workpiece velocity and reduction in grinding time. The calibrated simulation can be used in this way to identify the optimal workpiece velocity and dressing infeed.

A further time reduction for CDCF form grinding may be achieved by the use of variable workpiece velocity and dressing infeed during the grinding pass [16]. For the example presented above, simulation predicted that the thermal damage limit would be reached at about 8.3 seconds into the grinding pass. During most of the pass, the temperature is below the thermal limit. Therefore it should be possible to reduce the grinding time by using a strategy of variable workpiece velocity and dressing so that the critical thermal condition is reached at each point along the grinding pass. Implementing this strategy reduced the grinding time by about 40%. The required variation in the workpiece velocity and dress infeed to bring the process near to its thermal limit at each point during CDCF form grinding was estimated using the simulation software.

Process variations associated with initial workpiece size and property variations can significantly affect the grinding power and temperatures generated during CDCF form grinding. Power monitoring can provide a relatively simple and robust solution to this problem [17]. For this purpose, the power measured during grinding is compared with the power predicted according to the CDCF grinding simulation in order to identify any abnormalities. This approach can also take into account the progressively decreasing wheel size due to continuous dressing.

13.5   MACHINE TOOL CONTROL

The simulation software packages for cylindrical and creep-feed form grinding integrate various grinding models to predict what happens during the process. Calibration of the simulations with actual grinding data enhances the accuracy of the predictions. This detailed knowledge of the process can be used to improve productivity in an off-line manner without any direct link between the computer and the machine tool. The present section is concerned with direct on-line computer control of grinding.

Most modern machine tools utilize Computer Numerical Control (CNC). With CNC machine tools, the cutting path, feed rates, and speeds are usually programmed so that the operation is carried out in a predetermined way. However the computational capability with CNC also allows for possible 'adaptation' or Adaptive Control of machine settings so as to enhance productivity. Adaptive Control (AC) as applied to machine-tool systems refers to control of the operating parameters in response to the actual machining behavior in order to enhance productivity [18]. Some or all of the operating parameters are not preset, but rather 'adapt' themselves automatically to the actual behavior of the process. Adaptive control is particularly suitable for those applications, such as grinding, where the process behavior may be subject to uncontrolled variations.

Adaptive Control for machine tools may be further classified as either Adaptive Controlled Optimization (ACO) or Adaptive Controlled Constraint (ACC). With ACO, the operating condition is selected by extremizing a performance index (e.g. maximize production rate) subject to process and system constraints. With ACC, the operating condition is usually determined according to a single process constraint (e.g. normal force) without using a performance index. In principle, ACO should provide better performance than ACC.

There has been considerable research on in-process ACO systems for grinding machines [19-27], but few are found in production. Successful implementation of an ACO system requires a realistic optimization strategy, which can respond to inputs from sensors capable of reliably characterizing the grinding process in a production environment. Practically all AC machining systems in use are of the simpler ACC type having only a single constraint and control of only one parameter. For cylindrical grinding, the most common example of AC is the Controlled-Force system, which uses a preset normal force rather than a prescribed infeed cycle [28]. With application of a controlled normal force, the infeed velocity quickly responds to the applied normal force, thereby saving the transient time associated with an infeed controlled process (see Chapter 12). However such systems need some damping and additional stages in order to reduce the out-of-roundness to an acceptable level.

One of the early model-based ACO systems for cylindrical grinding consisted of a grinder interfaced to a computer. The optimization strategy was designed to maximize the removal rate subject to constraints on workpiece burn and surface roughness [19]. This system utilized feedback measurements of both the grinding power and surface roughness. The measured grinding power was fed directly to the computer, but the surface roughness was measured off-line and input by the operator to the computer owing to the lack of a suitable in-process roughness transducer. Therefore the system actually operated in a mixture of on-line and off-line modes. As compared with conventional control, a unique feature of this ACO system was the internal estimation of the burning power limit (Eq. (6-14)). A system which is smart enough to generate its own control references is sometimes referred to as an 'intelligent' system, although the term is often applied to many other types of control systems [20,21].

A more advanced ACO intelligent grinding system was subsequently developed as illustrated in Figure 13-17 [29-32]. In addition to having a power sensor as in the previous ACO system, a workpiece diameter gage system was also included. A model-based control methodology was developed for this system according to the overall structure illustrated in Figure 13-18. With this arrangement, the independent grinding parameters regulated from the **Axis Control** block are the radial infeed velocity $u(t)$ and the workpiece velocity v_w. The reference objectives for controlling these parameters are selected by **Meta-Control** which constitutes the 'brain' of the system. The exogenous inputs to **Meta-Control** are the constraint requirements for size tolerance, roughness, out-of-roundness, and the feedback inputs including on-line measurements of power and size, post process measurements of roughness and out-of-roundness, and additional parameters derived from sensory data in **System Identification** including the time constant, wheel dullness, wheel wear, and depth of thermal damage.

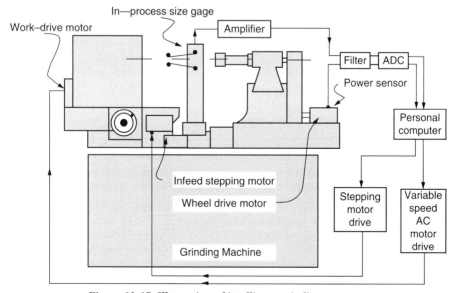

Figure 13-17 Illustration of intelligent grinding system.

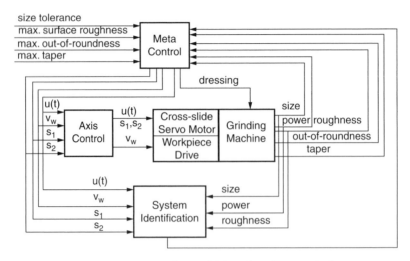

Figure 13-18 Control system for model-based intelligent grinding system.

The rationale for **Meta-Control** to select and adjust the operating parameters to reduce and minimize the cycle time is illustrated in Figure 13-19. The left side in this figure shows the sequential stages in the cycle. At each stage the operating parameters to be specified are constrained either by part quality requirements or machine limitations. The right side shows the role of sensory data and **System Identification**. The machine control begins with gap elimination, which should be performed at the maximum available infeed velocity. The end of gap elimination and beginning of roughing can be identified from the spindle power, although an acoustic emission sensor can be used for this purpose in order to obtain a faster response. The amount of material to be removed during the roughing stage should be maximized so that the amount of material removed during the slower finishing and spark-out stages to follow is minimized. Some additional time may be saved by using accelerated spark-in (see Chapter 12). Other than possible machine limitations, the infeed rate for the roughing stage is constrained by workpiece quality requirements including thermal damage and surface roughness. The thermal damage constraint may require no thermal damage (workpiece burn) during roughing, or it may be less restrictive and allow for a shallow depth of thermal damage to be removed from the workpiece during the subsequent finishing stage.

The roughing stage is followed by an intermediate finishing stage which may be required to remove any thermally damaged material from the previous roughing stage, to control size, to improve roughness, and to control roundness. After finishing, the process then proceeds to spark-out with

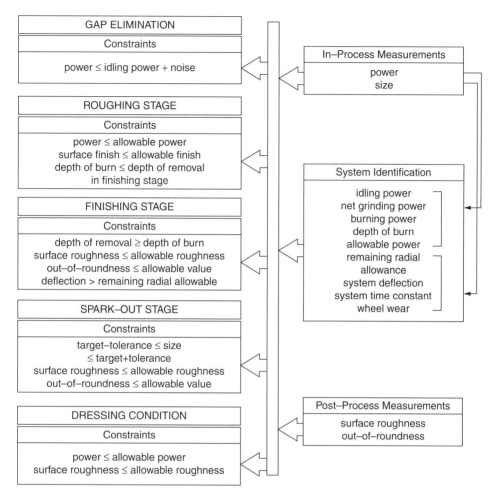

Figure 13-19 Rationale for selecting and adjusting the operating parameters to mini-mize cycle time. Sequential stages and constraints are shown on the left side, and the role of sensory measurements and System Identification on the right.

zero programmed infeed rate, which further improves roundness and roughness at a diminishing rate as the elastic deflection of the machine is recovered. The cycle is terminated and the wheel is retracted when the final size is reached as indicated by the size gage measurement.

This intelligent grinding system was implemented in production on a grinding machine equipped with conventional ladder logic computer control. However the control methodology for this system was developed in

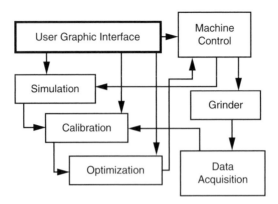

Figure 13-20 Intelligent Grinding System with simulation knowledge base.

the laboratory using a grinding machine interfaced to a personal computer (PC) controller (Figure 13-18). PC-based control has emerged in recent years as a widely used technology for machine tools. A distinctive characteristic of PC based control is that the control algorithm is implemented by software rather than hardware as with conventional (proprietary) controllers. This type of software-driven PC control, which allows access to its internal variables, is referred to as Open Architecture Control (OAC).

More recently, a new approach to PC based intelligent control for cylindrical grinding was implemented which utilizes the simulation software (GrindSim©) as the knowledge base. The concept for this system, which is ideally suited to OAC, is illustrated in Figure 13-20 [33]. This integrated sensor-based intelligent grinding system can perform grinding process simulation, model calibration, process optimization, machine control, and data acquisition. The machine control module directly loads process parameters from any specified or simulated cycle to execute the grinding cycle. During actual grinding, power and workpiece size data are collected on-line. After grinding is finished, the calibration module is invoked to calibrate the model coefficients based on measured data, and the calibrated system is then optimized. Process parameters obtained from the optimization can be saved and/or fed into the simulation or machine control for grinding the next part.

REFERENCES

1. Brinksmeier, E., Aurich, J. C., Govekar, E., Hoffmeister, H. W., Klocke, F., Peters, J., Rentsch, R., Stephenson, D. J., Ullman, E., Weinert, K. and Wittman, M., 'Advances in Modeling and Simulation of Grinding Processes', *Annals of the CIRP*, 55/2, 2006, p. 667.

2. Chiu, N. and Malkin, S., 'Computer Simulation for Cylindrical Plunge Grinding', *Annals of the CIRP*, 42/1, 1993, p. 383.

3. Chiu, N. and Malkin, S., 'Computer Simulation for Creep-Feed Form Grinding', *Transactions of NAMRI/SME*, 22, 1994, p. 119.

4. Sheth, D. S. and Malkin, S., 'CAD/CAM for Geometry and Process Analysis of Helical Groove Machining', *Annals of the CIRP*, 39/1, 1990, p. 129.

5. Shi, Z. and Malkin, S., 'Valid Machine Tool Setup for Helical Groove Machining', *Trans. of NAMRI/SME*, 31, 2003, p. 193; and *Journal of Manufacturing Processes*, 6, 2004, p. 148.

6. Shanbhag, N., Rajan, M., Manjunathaiah, J., Krishnamurty, S. and Malkin, S., 'Analysis and Simulation of Double Disc Grinding', *Trans. of NAMRI/SME*, 26, 1998, p. 111.

7. Xiao, G. and Malkin, S, 'Grinding Software for Simulation, Calibration, and Optimization', SME Paper No. MR97-215, 2nd International Machining and Grinding Conference, SME, 1997.

8. Guo, C. and Malkin, S., 'Cylindrical Grinding Simulation, Optimization, and Control', SME Paper No. MR01-334, 4th International Machining and Grinding Conference, SME, 2001.

9. Guo, C. and Malkin, S., 'Model Based Simulation of Grinding', European Conference on Grinding, Bremen, Germany, 2006.

10. Dong, S., Danai, K., Malkin, S. and Deshmukh, A., 'Continuous Optimal Infeed Control for Cylindrical Plunge Grinding – Part I: Methodology', *Trans. ASME, Journal of Manufacturing Science and Engineering*, 126, 2004, p. 327.

11. Dong, S., Danai, K. and Malkin, S., 'Continuous Optimal Infeed Control for Cylindrical Plunge Grinding – Part II: Controller Design and Implementation', *Trans. ASME, Journal of Manufacturing Science and Engineering*, 126, 2004, p. 334.

12. Xiao, G. and Malkin, S., 'On-Line Optimization of Internal Cylindrical Grinding Cycles', *Annals of the CIRP*, 45/1, 1996, p. 287.

13. Peters, J. and Aerens, R., 'Optimization Procedure of Three Phase Grinding Cycles of a Series Without Intermediate Dressing', *Annals of the CIRP*, 29/1, 1980, p. 195.

14. Iwata K., Takazawa K., Yamamoto Y. and Horike, M., 'Determination of Optimum Machining Conditions in Cylindrical Plunge Grinding', *Proc. of the 5th North American Metalworking Research Conference*, 1977, p. 284.

15. Andrew, C. A., Howes, T. D. and Pearce, T. R. A., *Creep Feed Grinding*, Holt, Rinehart and Winston, London, 1985.

16. Guo, C., Campomanese, M., McIntosh, G. D., Becze, C. and Green, T., 'Optimization of Continuous Dress Creep-Feed Form Grinding Process', *Annals of the CIRP*, 53/1, 2003, p. 259.

17. Guo, C., Campomanese, M, McIntosh, G. D. and Becze, E. C., 'Model–Based Monitoring and Control of Continuous Dress Creep-Feed Form Grinding', *Annals of the CIRP*, 53/1, 2004, p. 263.

18. Koren, Y., *Computer Control of Manufacturing Systems*, McGraw-Hill, New York, 1983, Chapter 8, p. 193.

19. Amitay, G., Malkin, S. and Koren, Y., 'Adaptive Control Optimization of Grinding', *Trans. ASME, J. of Eng. for Ind.*, 103, 1981, p. 103.

20. Tönshoff, H. K., Friemuth, T. and Becker, J. C., 'Process Monitoring in Grinding', *Annals of the CIRP*, 51/2, 2002, p. 551.

21. Rowe, W. B., Yan, L., Inasaki, I. and Malkin, S, 'Applications of Artificial Intelligence in Grinding', *Annals of the CIRP*, 43/2, 1994, p. 521.
22. Wada, R. and Kodama, H., 'Adaptive Control in Grinding', *Bull. Japan Society of Prec. Eng.*, 11, 1977, p. 1.
23. König, W. and Werner, G., 'Adaptive Control Optimization of High Efficiency External Grinding-concept: Technological Basis and Application', *Annals of the CIRP*, 23/1, 1974, p. 101.
24. Inoue, H., Tamakohri, K., Sutoh, T., Noguchi, H., Waida, T. and Sata, T., 'An Adaptive Control System of Grinding Process', *Proc. International Conference on Production Engineering, Pt. I*, Tokyo, 1974, p. 671.
25. Smith, R. L., 'Energy Adaptive Grinding', *American Machinist*, July 1977, p. 115.
26. Kaliszer, H., 'Adaptive Control in Grinding Processes', *Proceedings of the 4th International Conference on Production Engineering*, Tokyo, 1980, p. 579.
27. Kim, S. W. and Shunsheruddin, A. A., 'Adaptive Computerized Numerical Control of the Grinding Process for Industrial Applications', *Proceedings of the Twenty Fourth International Machine Tool Design and Research Conference*, 1983, p. 239.
28. Hahn, R. S., 'Controlled Force Grinding – a New Technique for Precision Internal Grinding', *Trans. ASME, J. of Eng. for Ind.*, 86, 1964, p. 287.
29. Xiao, G. and Malkin, S., 'On-Line Optimization for Internal Plunge Grinding,' *Annals of the CIRP*, Vol. 45/1, 1996, p. 287.
30. Xiao, G., Malkin, S. and Danai, K., 'Intelligent System for Cylindrical Plunge Grinding', *Fifth International Grinding Conference*; 1993.
31. Xiao, G., Malkin, S. and Danai, K., 'Autonomous System for Multistage Cylindrical Grinding', *Trans. ASME, J. of Dynamic Systems, Measurement and Control*, 115, 1993, p. 667.
32. Ivester, R., Danai, K and Malkin, S., 'Cycle Time Reduction in Machining by Recursive Constraint Bounding', *Trans. ASME, Journal of Manufacturing Science and Engineering*, 119, 1997, p. 201.
33. Zhou, X., 'Simulation-Based Control of Cylindrical Grinding', MS Thesis, University of Massachusetts, 2000.

Index